Internet of Things

Edge AI in Future Computing

Series Editors: Arun Kumar Sangaiah

SCOPE, VIT University, Tamil Nadu Mamta Mittal, G. B. Pant Government Engineering College, Okhla, New Delhi

Soft Computing Techniques in Engineering, Health, Mathematical and Social Sciences

Pradip Debnath and S. A. Mohiuddine

Machine Learning for Edge Computing: Frameworks, Patterns and Best Practices

Amitoj Singh, Vinay Kukreja, Taghi Javdani Gandomani

Internet of Things: Frameworks for Enabling and Emerging Technologies

Bharat Bhushan, Sudhir Kumar Sharma, Bhuvan Unhelkar, Muhammad Fazal Ijaz, Lamia Karim

For more information about this series, please visit: https://www.routledge.com/Edge-AI-in-Future-Computing/book-series/EAIFC

Internet of Things

Frameworks for Enabling and Emerging Technologies

Edited by

Bharat Bhushan, Sudhir Kumar Sharma, Bhuvan Unhelkar, Muhammad Fazal Ijaz, and Lamia Karim

CRC Press is an imprint of the
Taylor & Francis Group, an **informa** business

First edition published 2022
by CRC Press
6000 Broken Sound Parkway NW, Suite 300, Boca Raton, FL 33487-2742

and by CRC Press
4 Park Square, Milton Park, Abingdon, Oxon, OX14 4RN

CRC Press is an imprint of Taylor & Francis Group, LLC

ISBN: 978-1-032-10431-7 (hbk)
ISBN: 978-1-032-11382-1 (pbk)
ISBN: 978-1-003-21962-0 (ebk)

DOI: 10.1201/9781003219620

Typeset in Times LT Std
by KnowledgeWorks Global Ltd.

Contents

Preface

Internet of Things (IoT) is a fast-emerging research paradigm with significant practical applications that connects anything, anytime and anywhere. Apart from the connected sensors and devices in a wireless or wired network, IoT is an integration of real and virtual world where people and devices can communicate. IoT devices have been interconnected for consumer applications (e.g., smart city, smart home, smart grid, smart transportation, mobile devices) dedicated to provide intelligence, efficiency and convenience to consumers by efficiently handling their personal resources and time. Furthermore, Industry 4.0 envisions the adoption of IoT as it has the potential to improve intelligence, safety, efficiency and productivity of industrial factories to manifolds. However, the existing IoT systems are vulnerable to malicious attacks and single point of failure and thereby cannot provide the desired services. Owing to the massive numbers of deployed IoT devices and inadequate data security, the impact of security breaches (such as denial of service attack, Sybil attack, selective forwarding attack, network availability attack, data interception) turns out to be humongous leading to severe negative impacts. The existing security solutions are insufficient and therefore, it is necessary to enable the IoT devices to dynamically counter the threats and save the system. Owing to the security benefits provided by the blockchain technology, the idea of combining IoT and blockchain has gained momentum in the recent past. This book aims to discuss the role of IoT as applied in practice to real-life organizations. This book considers IoT as a revolutionary (disruptive) technology that requires study and sharing of use case around integration of multiple domains like Cyber-Physical System (CPS), networking, ubiquitous distributed computing, sensor technologies, wireless sensor networks, data security and privacy, big data, machine learning, blockchain. These are the topics envisaged by us in this book so that the reader discovers new ideas and sidesteps risks associated with the use of IoT in practice.

Editor Biographies

Dr. Bharat Bhushan is an Assistant Professor of Department of Computer Science and Engineering (CSE) at School of Engineering and Technology, Sharda University, Greater Noida, India. He is an Alumnus of Birla Institute of Technology, Mesra, Ranchi, India. He received his Undergraduate Degree (BTech in Computer Science and Engineering) with Distinction in 2012, received his Postgraduate Degree (MTech in Information Security) with Distinction in 2015 and Doctorate Degree (PhD Computer Science and Engineering) in 2021 from Birla Institute of Technology, Mesra, India. He earned numerous international certifications such as CCNA, MCTS, MCITP, RHCE and CCNP. In the last three years, he has published more than 80 research papers in various renowned international conferences and SCI-indexed journals. He has contributed with more than 25 book chapters in various books and has edited 11 books from the most famed publishers like Elsevier, IGI Global and CRC Press. In the past, he worked as an Assistant Professor at HMR Institute of Technology and Management, New Delhi and a Network Engineer in HCL Infosystems Ltd., Noida. He is also a member of numerous renowned bodies, including IEEE, IAENG, CSTA, SCIEI, IAE and UACEE.

Prof. (Dr.) Sudhir Kumar Sharma is currently a Professor and Head of the Department of Computer Science, Institute of Information Technology & Management affiliated to GGSIPU, New Delhi, India. He has extensive experience for over 21 years in the field of Computer Science and Engineering. He obtained his PhD degree in Information Technology in 2013 from USICT, Guru Gobind Singh Indraprastha University, New Delhi, India. Dr. Sharma obtained his MTech degree in Computer Science & Engineering in 1999 from the Guru Jambheshwar University, Hisar, India and MSc degree in Physics from the University of Roorkee (now IIT Roorkee), Roorkee, in 1997. His research interests include Machine Learning, Data Mining and Security. He has published more than 60 research papers in various prestigious International Journals and International Conferences. He is a life member of CSI and IETE. Dr. Sharma is a lead guest editor of special issue in Multimedia Tools & Applications, Springer. He was a convener and volume Editor of two international conferences, namely ICETIT-2019 and ICRIHE-2020. He authored and edited seven Computer Science books in the field of Internet of Things, WSN, Blockchain, Cyber-Physical Systems of Elsevier, Springer, CRC Press, USA.

Prof. (Dr.) Bhuvan Unhelkar (BE, MDBA, MSc, PhD; FACS; PSM-I, CBAP®) is an accomplished IT professional and Professor of IT at the University of South Florida, Sarasota-Manatee (Lead Faculty). He is also Founding Consultant at MethodScience and a Co-Founder/Director at PlatiFi. He has mastery in Business Analysis & Requirements Modeling, Software Engineering, Big Data Strategies, Agile Processes, Mobile Business and Green IT. His domain experience is banking, finance, insurance, government and telecommunications. Bhuvan is a thought-leader and a prolific author of 20 books – including Big Data Strategies for Agile Business and The Art of Agile Practice (Taylor and Francis/CRC Press, USA). He is a winner of the Computerworld Object Developer Award (1995), Consensus IT Professional Award (2006) and IT Writer Award (2010). He has a Doctorate in the area of "Object Orientation" from the University of Technology, Sydney, in 1997. Bhuvan is Fellow of the Australian Computer Society, IEEE Senior Member, Professional Scrum Master, Life Member of Computer Society of India and Baroda Management Association, Member of SDPS, Past President of Rotary Sarasota Sunrise (Florida) & St. Ives (Sydney), Paul Harris Fellow (+6), Discovery Volunteer at NSW parks and wildlife and a previous TiE Mentor. Dr. Unhelkar is the winner of the Computerworld Object Developer Award (1995), Consensus IT Professional Award (2006) and IT Writer Award (2010). He also chaired the Business Analysis Specialism Group of the Australian Computer Society.

Dr. Muhammad Fazal Ijaz received his BEng degree in Industrial Engineering and Management from University of the Punjab, Lahore, Pakistan, in 2011, and Dr Eng degree in Industrial and Systems Engineering from Dongguk University, Seoul, South Korea, in 2019. From 2019 to 2020, he worked as an Assistant Professor in Department of Industrial and Systems Engineering, Dongguk University, Seoul, South Korea. Currently, he is working as an Assistant Professor in Department of Intelligent Mechatronics Engineering, Sejong University, Seoul, South Korea. He has published numerous research articles in several international peer-reviewed journals, including IEEE Access, Sensors, Journal of Food Engineering, Applied Sciences, Asia Pacific Journal of Marketing and Logistics, and Sustainability. His research interests include Machine learning, Blockchain, Healthcare Engineering, Internet of Things, Supply Chain Management, Big Data and Data mining.

Prof. (Dr.) Lamia Karim is a Professor of Computer Science at the National School of Applied Sciences Berrechid (ENSAB) Hassan I University. In 2020, she obtained the HDR diploma at ENSAB Hassan I University. In 2015, she obtained the PhD degree in Computer Science at the Faculty of Science and Technology of Mohammedia, Hassan II University of Casablanca, Morocco. She obtained Masters of Science and Technical in Computer Sciences in 2006 from Faculty of Sciences and Technologies (FSTM), Morocco, and Computer Engineering degree in Software Engineering from National School of Mineral Industry (ENIM), Rabat, Morocco in 2008. Her research interest includes spatial data engineering, mobile and web geo-computing, real-time intelligent transportation systems, complexes systems computing, soft computing, reactive intelligent systems.

Contributors

M. Appadurai
Department of Mechanical Engineering
Dr. Sivanthi Aditanar College of Engineering
Tamil Nadu, India

M. Agus Syamsul A.
University of Sriwijaya
Palembang, Indonesia
and
University of Bina Insan
Lubuklinggau, Indonesia

J. A. Benson
NHS England
Keynes, England

A. Boulmakoul
LIM Lab, FSTM
Hassan II University
Casablanca, Morocco

Rahmat Budiarto
University of AlBaha
Al Bahah, Saudi Arabia

Brijesh Kumar Chaurasia
IIIT
Lucknow, India

Murthy Cherukuri
Department of Electrical and Electronics Engineering
NIST (Autonomous)
Berhampur, India

Ghyzlane Cherradi
LIM Lab, FSTM
Hassan II University
Casablanca, Morocco

Linda Chibuzor
NHS Milton Keynes CCG
Keynes, United Kingdom

S. Darwin
Department of Electronics and Communication Engineering
Dr. Sivanthi Aditanar College of Engineering
Tamil Nadu, India

Nirmala Devi
Amrita Vishwa Vidyapeetham
Ettimadai, India

Hemlata Goyal
Manipal University Jaipur
Jaipur, India

Mohd. Yazid Idris
Universiti Teknologi Malaysia
Johor, Malaysia

Kusumlata Jain
Manipal University Jaipur
Jaipur, India

M. Jayakumar
Amrita Vishwa Vidyapeetham
Ettimadai, India

A. Kałowski
University of Social Sciences
Warsaw, Poland

Lamia Karim
LISA Lab, ENSAB
Hassan First University
Settat, Morocco

Ritu Khandelwal
Manipal University Jaipur
Jaipur, India

Anju V. Kulkarni
Dr. D. Y. Patil Institute of Technology
Pune, India

Sachin Kumar
Department of Computer Science & IT
University of Jammu
Jammu, India

Yashwardhan Kumar
Department of Electronics and Communication Engineering
NIST (Autonomous)
Berhampur, India

Sandipan Mallik
Department of Electronics and Communication Engineering
NIST (Autonomous)
Berhampur, India

Meriem Mandar
LISA Lab, ENSAB
Hassan First University
Settat, Morocco

Vibhakar Mansotra
Department of Computer Science & IT
University of Jammu
Jammu, India

Radhika Menon
Dr. D. Y. Patil Institute of Technology
Pune, India

Dimitrios V. Moysidis
Hippokration University Hospital
Aristotle University of Thessaloniki
Thessaloniki, Greece

Neelu
IIIT
Lucknow, India

J. O. Nehinbe
ICT Security Solutions
West Africa

Utpal Pandey
IIIT
Lucknow, India

Prakash Panigrahi
Department of Electronics and Communication Engineering
NIST (Autonomous)
Berhampur, India

Andreas S. Papazoglou
Athens Naval Hospital,
Athens, Greece

W. Pizło
Warsaw University of Life Sciences
Warsaw, Poland

E. Fantin Irudaya Raj
Department of Electrical and Electronics Engineering
Dr. Sivanthi Aditanar College of Engineering
Tamil Nadu, India

Sree Ranjani Rajendran
University of Florida
Gainesville, Florida, USA

E. Francy Irudaya Rani
Department of Electronics and Communication Engineering
Francis Xavier Engineering College
Tamil Nadu, India

Rehab A. Rayan
Department of Epidemiology
High Institute of Public Health
Alexandria University
Alexandria, Egypt

Hemant Kumar Saini
Government Engineering College
Banswara, India

Sourabh Shastri
Department of Computer Science & IT
University of Jammu
Jammu, India

Rajveer Singh Shekhawat
Manipal University Jaipur
Jaipur, India

Kuljeet Singh
Department of Computer Science & IT
University of Jammu
Jammu, India

Shivendra Pratap Singh
Department of Electronics and Communication Engineering
NIST (Autonomous)
Berhampur, India

Vivek Kumar Srivastav
IIIT
Lucknow, India

Deris Stiawan
University of Sriwijaya
Palembang, Indonesia

Susanto
University of Sriwijaya
Palembang, Indonesia
and
University of Bina Insan
Lubuklinggau, Indonesia

Kunjabihari Swain
NIST (Autonomous)
Berhampur, India

Himanshu Swarnakar
Government Engineering College
Banswara, India

Edyta Karolina Szczepaniuk
Military University of Aviation
Dęblin, Poland

Hubert Szczepaniuk
Warsaw University of Life Sciences
Warsaw, Poland

Christos Tsagkaris
Novel Global Community Educational Foundation
Hebersham, NSW, Australia

Mithra Venkatesan
Dr. D. Y. Patil Institute of Technology
Pune, India

A. Zarzycka
Warsaw School of Economics
Warsaw, Poland

1 IoT Conceptual Model and Application

Utpal Pandey, Vivek Kumar Srivastav, Brijesh Kumar Chaurasia, and Neelu
IIIT, Lucknow, India

CONTENTS

DOI: 10.1201/9781003219620-1

1.1 INTRODUCTION

Smart technologies, such as smartphones, home, city, business and entertainment applications, now have over two billion users [1]. The capabilities of these smart things allow machines to connect with or without the use of a user intermediary, giving rise to the name "Internet of Things" (IoT). The IoT is a network of interconnected computing systems, animals, objects or people with unique identifiers, digital and mechanical devices and the capacity to share data without having human-to-computer or human-to-human interaction [2–4]. When household appliances are linked to a network, they will collaborate to deliver the best service possible, rather than as a series of individually operating machines. This function is helpful for a variety of real-world technologies and utilities, such as building a smart home; for example, windows can be automatically opened to maintain oxygen saturation when the gas oven is turned on or closed when the air conditioner is switched on.

IoT provides a lucid and broad platform for actuation, device and process integration, turning them into intelligent systems that can perform and learn independently. A massive array of unstructured, structured and semi-structured data is generated, which requires improved data space and a more comprehensive range of processing and storage systems. These IoT systems have surfaced as dynamic and global network transport with self-configuring ability, including things that can work in coherence and communicate with them and the environment through sensor swinging and the Internet. But heterogeneity in terms of sensing, actuation and variety in applications poses state-of-the-art problems in IoT architecture definitions, communication and routing protocol designing and security. IoT has transformed the world into an ecosystem of intelligent devices with HetNets, and its penultimate goal is to provide ease to the end user with simple plug and play options.

1.1.1 History of IoT

IoT is not a modern concept in computer science. Still, it has evolved into a new standard that incorporates many smart objects – which are rapidly increasing, and their capability to be remotely linked – with data sharing from multiple sources.

In 1990, Simon Hackett and John Romkey created the Internet Toaster, the first connected toaster appliance powered by the Internet [5]. Interop also added a small robotic arm to pick up a slice of bread into the toaster in 1991, making it a fully automated unit.

The Internet Toaster was linked to the Internet ten years later, in 1999. As Kevin Ashton invented the word "Internet of things," the name "IoT" became well-known. In the same year, Arlen Nipper of Arcom (now Eurotech) and Dr Andy Stanford-Clark of IBM presented Telemetry Transport, the first machine-to-machine protocol for connected devices [6].

LG Company unveiled plans for the first refrigerator linked to the Internet, called the LG Internet Refrigerator [7], a year later. After more than 20 years of birth of the Internet Toaster, more than 40 years of Internet growth, and more than ten years since the word "Internet of Things" was coined, there are currently 13 billion devices linked worldwide (2 devices per person), with 50 billion devices. These devices are predicted to be connected to the Internet (six devices per person) by 2020 [8].

1.2 CONCEPTUAL MODEL

The IoT is a network of physical objects and sensors that collect data at distant locations and interact with units that manage, acquire, organise and analyse the data in services and processes [8]. The defined Equation 1.1 describes the IoT's fundamental conceptual framework.

$$f\,(\text{Physical object, sensor, controller, actuators, Internet}) = \text{Internet of Thing} \quad (1.1)$$

The Internet, which consists of actuators, sensors, controllers and the Internet for connectivity by any web services and mobile service provider, is conceptually defined by the abovementioned equation.

The computers that communicate data to the cloud server or data centre make up the overall framework. The communication and action of data at successive stages in IoT consisting of interconnected objects and devices are represented by another computational architectural framework for services and enterprise processes in Equation 1.2.

$$\begin{array}{c}\text{Gather + enrich + stream + manage + acquire + organise and analyse =}\\ \text{Internet of Things enterprise and business applications, integration and SoA}\end{array} \quad (1.2)$$

Since Asthon has coined the term Internet of Things, research communities and organisations like IBM, Microsoft etc. are trying to conceptualise generic and application-specific architecture. Platforms like Azure, AWS cloud and SmartCloud framework etc. offer generic cloud computing services to fulfil intensive computing requirements for versatile IoT applications. Problem-centric architectures for management of smart cities, Internet of medical things (IoMT), transportation, smart agriculture etc. are also proposed. For example, Ayaz M. *et al.* have discussed state-of-the-art architecture for smart agriculture for services like pest control, irrigation management, fertilisation and post-harvest management [12]. These integrated systems rely on deep learning, machine learning and Blockchain technologies to process and protect data integrity [9–15]. A three-layer architecture is being considered by IEEE P2413 which has application, networking and data communication and sensing layers from top to bottom [13]. Cisco has proposed an innovative IoT-based seven-layer business model, which attempts to provide a generic solution for versatile applications [14].

1.3 APPLICATIONS OF IoT

1.3.1 IoT in Smart Homes

With the help of sensors, smart gadgets, IoT establishes an intelligent communication network among domestic products used in daily life, and they can serve us smartly without any human input. Context-awareness w.r.t. family house, usability, heterogeneity in appliances and services, security, privacy, intelligence are few aspects of a smart house [9].

1.3.2 IIoT (Industrial Internet of Things)

Industrial environment is a composite of active and passive sensors, mechanical compressors and pumps, electrical drives, motors etc. It relies on sensors for data gathering, open and closed loops for actuation. When the Internet assists this process of actuation and data gathering, then the scenario transforms to Industrial Internet of Things (IIoT). Wireless sensor networks, wireless sensor actuation networks, virtual sensing, cyber-physical systems and big data analytics are key components [10]. It can monitor and maintain in-house air quality within chemical plants, level of oxygen, level of toxic gases, and observe the silo temperatures for product protection. During the drying process of meat in food factories, ozone levels are monitored. CAN bus data is collected to transmit real-time alarms in the event of an emergency or to give directions to drivers [11].

1.3.3 IoT in Agriculture

Implementation of IoT in the agricultural sector is the need of the day. It collects and analyses the information gathered from different environment sensors, e.g. temperature sensor, humidity sensor, proximity sensor, gas sensor, smoke sensor etc. After this, it suggests the methods for soil preparation, fertilisation and harvesting timing, which improves crop efficiency, health and production [12]. It also safeguards them against disease, insect attack, use of excessive pesticides and weather change etc. IoT-enabled agricultural system not only suggests the best method of harvesting but also enhances efficiency by investigating the harvest statistics.

1.3.4 IoT in Health Management

IoMT platform is assisted by smart body area network sensors, lightweight communication protocols and smart devices, which acquire biomedical signals, process and transmit them for analysis and report generation [15]. It employs artificial intelligence-based schemes to distinguish critical and subcritical conditions of the patient, detect diseases like cancer and accordingly find an appointment with a concerned specialist [16]. It is not less than a blessing for the patients who reside at remote locations. It facilitates the remote patients to get a consultation from different health experts around the globe. It can get the best healthcare solution for various diseases, e.g. Alzheimer, COVID, diabetes, dementia etc. It saves time, cost and increases the quality of treatment.

1.3.5 IoT in Transport Sector

IoT can also be a great contributor to minimising traffic congestion in the city. IoT can establish a smart transport network in a city. By processing the data collected by GPS and different sensors of the smart network in a city, it can suggest various solutions for traffic management. For example, creating a digital map of the city displays the citywide view of various routes with real-time traffic, probable bus arrival-departure time at different stops, alternate routes. It regulates the traffic lights depending upon traffic flow. It helps reduce traffic congestion and is a smart and smooth transport solution [17].

1.3.6 IoT in Handy Devices

Handy devices that are wearable can be a smart solution for better attribute human life, which could not be done by smartphone alone. Nowadays, people also have a great desire to wear these gadgets. IoT can upgrade our regular wearables like phones, jewellery, glasses, skin patches, clothing and so on into smart ones. For example, smart electronic clothing, glasses or other smart gadgets can find, accumulate and record physiological data that can be used for our betterment. All these are not possible to be done by smartphones alone. These handy wearable gadgets save our effort and time to perform micro-tasks, e.g. sending and receiving important messages and identifying the urgent data in no time. They are crafted to supply purposive information to the owner. Figure 1.1 depicts few wearable smart gadgets. IoT-enabled handy wearable devices are for everyone in almost every field [18, 19].

1.3.6.1 Head

It is the most important and uppermost part of the human body. It carries all-human sensors, e.g. ear, nose, tongue, skin and eye. The IoT-enabled devices used here are goggles, headbands, hearing aid, glasses, earrings, hats, skin patches and lenses etc.

1.3.6.2 Torso

It is the middle part of the human body. It can carry IoT-enabled devices, e.g. clothing, waist belts, gloves and undergarments. These smart gloves, smart jackets, smart uniforms etc. can find, accumulate and record the body's temperature, electrocardiogram, oxygen level, pulse rate and other physiological data. Even tattoos can be a wireless sensor, smart uniforms for defence personals can communicate with computer or mobile, other suits with UV sensors also fall into this category. For example, to regulate the glucose level in the blood, a wearable gadget is available named Genesis.

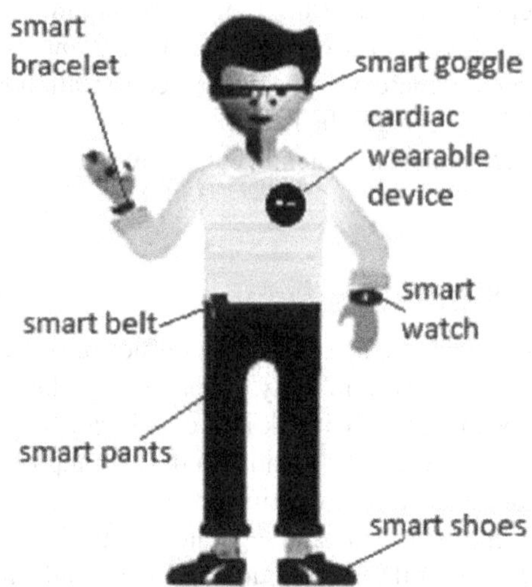

FIGURE 1.1 IoT wearable devices.

1.3.6.3 Foot, Leg and Arm

Pulse rate, sugar level, body temperature, UV exposure level, heart rate, blood pressure and various daily activities can be gathered by smart jewellery, smart wristband, smartwatch and smart armband. Smart socks, shoes and sleeves etc. not only process the physiological data but also safeguard against injury.

1.3.6.4 Cardiac Wearable Devices

It is a handy biomechanical device that maintains blood circulation throughout the blood vessels of the circulatory system of the human body by efficient pumping. It removes unwanted materials exerted from other organs and also supplies necessary nutrients. The heart rate comprises different mechanical, optical, electrical and acoustic signals that form pulsatile hemodynamic flow, which is detected by proper conversion.

1.4 SIMULATION MODEL AND FRAMEWORK

The expansion of the digital universe is twofold every two years from 2012 to 2020, which is a clear indication of the "Big Data" era assisted by the IoT [20]. These "Data Explosions" have spawned the "Big Data" issue [21, 22], which is described as the practice of gathering and analysing unstructured data flowing at a considerable volume and velocity, managing storage, analysing manually or through conventional management applications with rational applicability and reasonable timing constraints.

Many simulation frameworks have been built over the last few decades to help researchers better understand the behaviour of large-scale distributed systems that host various application services, viz. web hosting, social networking, content delivery and scientific applications [23]. Simulators, it is well known, provide an atmosphere in which performance assessment studies can be carried out in an iterative and controlled manner.

1.4.1 Design of IoT Simulator

An extensible modelling java-based toolkit CloudSim [20] facilitates design and replication of cloud computing environments and application purveying. It allows for the modelling and development of multiple virtual machines (VMs) on a simulated node of a data centre with various hardware configurations, as well as the mapping of cloud-based tasks to appropriate VMs. It also provides flexibility for the simulation of several data centres, allowing for a study on federated and related policies for VM migration and automated scaling of applications. The IoTSim simulator is built on a layered architecture and includes a big data processing platform.

1.4.2 General Layered Architecture of IoTSim [20]

The general layered architecture of IoTSim is illustrated in Figure 1.2 constitutes of core simulation engine, simulation,storage,big data processing and user code layers.

1.4.2.1 CloudSim Core Simulation Engine Layer

The lowest in this architecture is thc simulation engine layer that performs many core functions like queuing, processing of events, development of cloud system entities, creation of VMs, hosting, data centre management, broker and connection establishment between blocks and components, and simulation and administration of the clock etc. [20, 23].

1.4.2.2 CloudSim Simulation Layer

Designing and simulation of virtualised cloud-based data centre environments comprising dedicated interfaces for VMs, memory allocation, storage distribution strategies, bandwidth management are offered by this layer. The fundamental problems like host provisioning, controlling program execution and supervision of dynamic states are also handled within this layer. The data centre, cloud coordinator and network topology are all represented by the bottom sublayers. These blocks help in developing Infrastructure-as-a-Service (IaaS) environments. The VM Services and Cloud Services include the ability to design and manage VMs and an application scheduling algorithm [20, 22, 24].

1.4.2.3 Storage Layer

This layer can be used to model various types of storage; few such models are Amazon S3, Azure Blob Storage and HDFS, which are used to store large datasets created by devices. These data files are then utilised by IoT-based applications to copy during execution and save generated data as per the requirement with unavoidable storage delay [20, 24].

1.4.2.4 Big Data Processing Layer

It is divided into stream computing and MapReduce sublayers. The streaming computing sublayer addresses the real-time processing requirements and helps in

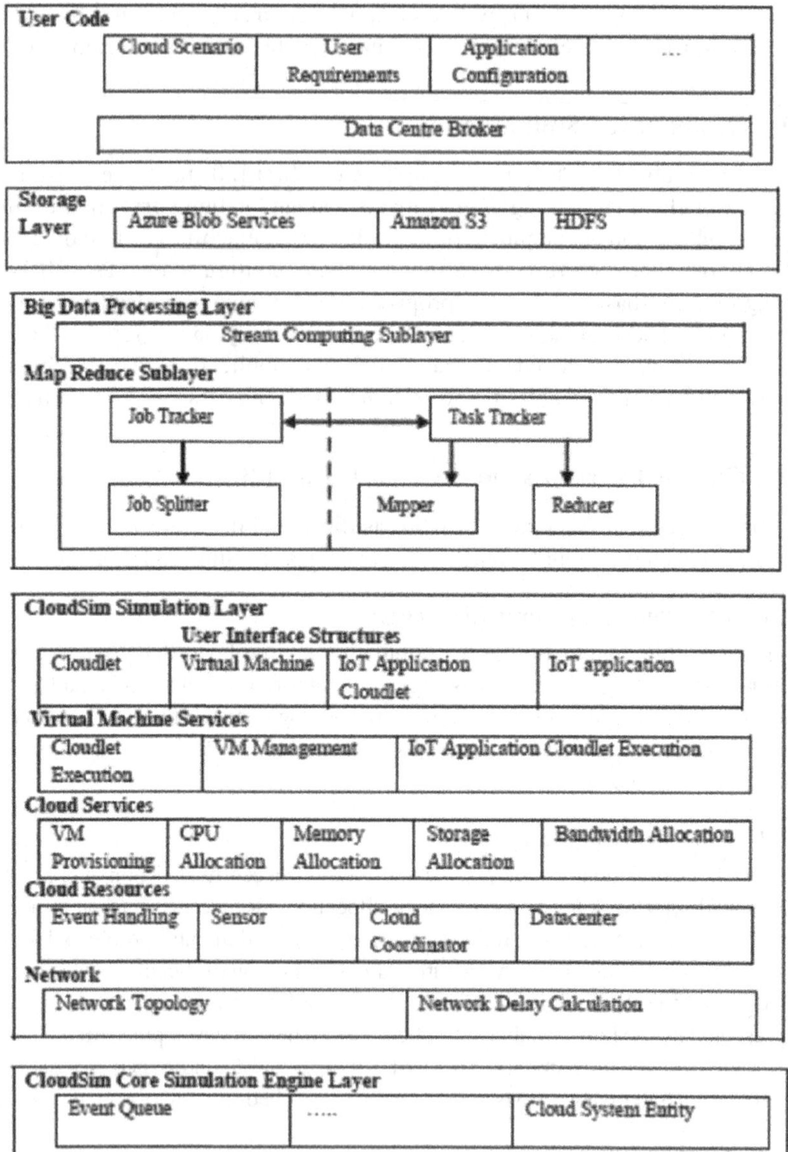

FIGURE 1.2 Layered architecture of IoTSim [20].

processing enormous data produced by actuators and sensors using streaming computing models. Hence this layer is critical and has been in high demand for simulating and analysing IoT-based live streaming applications.

The MapReduce sublayer is for applications that involve batch-oriented data processing, and these types of compatible models are completely implemented here [20, 22]. When a model is submitted through brokerage, it is broken into smaller pieces requiring tracking, reduction and appropriate mapping; this is accomplished using trackers and mappers. These entities work in the same way as real Hadoop does, and finally, the MapReduction operation is performed over events occurring in a particular order.

1.4.2.5 User Code Layer

The top-level layer consists of two sublayers, the topmost simulation specification layer deals with several devices, application requirements, user specifications etc., and the second one is related to scheduling and brokerage policies. It assesses job length and requirements, creates VMs according to users' number, provides flexibility to acquire resources [22] and implements broker scheduling policies. Here users can build their simulation setups for testing their algorithms.

1.5 IoT PROTOCOLS

Protocols define the format for sending data between two different devices on the same or other networks. IoT necessitates different protocols for various activities, such as protocols for collecting data sensors and issuing control commands to actuators, protocols for Machine to Machine (M2M), Device to People (D2P) and Device to Device (D2D) interactions and protocols for sending data to servers. Some important IoT protocols are described in detail.

1.5.1 Constrained Application Protocol (COAP)

It's a web transfer protocol based on REpresentational State Transfer (REST) designed for low-power and constrained nodes. For IoT-related applications, it's also known as an application layer protocol [25]. Constrained Application Protocol (COAP) uses an interaction model that is similar to HTTP's client/server model. This model was created specifically for M2M applications that use the REST architecture, such as smart building and farm applications. COAP is divided into two layers: the top layer and the bottom layer. The request/response sublayer handles REST-based communication. The messaging sublayer manages single messages sent between end users and delivers them reliably over the user datagram protocol (UDP) transport layer. Resource scrutiny, block by block transport and discovery, HTTP interaction, security are some of the main attributes [26].

1.5.2 Message Queue Telemetry (MQTT)

Message Queue Telemetry (MQTT) is TCP/IP-based transport protocol. Subscriber, publisher and broker are the three elements of MQTT's hub and spoke architecture. With the quality of service (QoS), it utilises published and subscribed messaging patterns. Much of this takes place in the broker's office, where they track all publications and subscriptions. So, if a publisher detects a new data change, he broadcasts a post, and the broker is in charge of delivering the new data to all subscribers. Because of the inherent support for QoS, the broker will ensure that the message is delivered.

1.5.3 Advanced Message Queuing Protocol (AMQP)

It's an IoT messaging application layer protocol that focuses on message-oriented environments. This protocol was created to allow clients and message middleware servers to communicate with one another. Advanced Message Queuing Protocol (AMQP) is a binary protocol that enables secure and efficient communication. It has routing, confidentiality and reliability capabilities. It also supports P2P routing and the publish/subscribe model [27] and can connect with different organisational set-ups using different technologies in distant space and time.

1.5.4 Extensible Messaging and Presence Protocol (XMPP)

It's an Internet Engineering Task Force (IETF) instant messaging (IM) standardisation for sending and receiving messages and chatting. It is a well-established protocol that allows users to interact with one another over the Internet. It is compatible with publish/subscribe communication protocols as well as quest/response message systems. It has transport layer security/successor security protocol (TLS/SSL) protection and supports low latency message exchange [28]. It is an impractical M2M communication that enables access control, end-to-end encryption and authentication for IM applications. It uses message stanzas [29] to bind client and server. Message stanzas use the push method to get data like source and destination addresses, as well as Extensible Messaging and Presence Protocol (XMPP) entity IDs. This protocol is better suited for IoT since it is widely supported around the Internet.

1.5.5 Data Distribution Service (DDS)

Data Distribution Service (DDS) is a public subscribe protocol for M2M communication developed by the Object Management Group (OMG), allowing interoperability in data exchange among publishers and subscribers. It's a data centric protocol that uses a broken-less architecture to offer its application in better QoS (supporting 23 services) and high consistency. Data-Centric Publish-Subscribe (DCPS) and Data Local Reconstruction Layer (DLR) are the two layers defined by DDS (DLRL). DCPS helps ensure that useful data is sent to the desired node at accurate timing, and DLRL is an optional layer that allows easy integration with the application layer. It can send the majority of data and information to different receivers at the same time.

1.5.6 Zigbee

It is a low-cost solution with low-speed, low-power, high-level communication designed specifically for Personal Area Networks (PANs) with low-rate sensors and 128-bit advance encryption standard (AES) encryption. It was created by the Zigbee Alliance using IEEE802.15.4 and includes new functionality to satisfy the requirements. This protocol lacks security because of its simplicity and low cost; its maximum range is 10–100 m with a maximum throughput of 250 kbps and a network capacity of 65,000 nodes.

<table>
<tr><td colspan="2">Session</td><td>MQTT, XMPP, SMQTT, CoRE, DDS, AMQP, CoAP etc</td></tr>
<tr><td rowspan="2">Network</td><td>Encapsulation</td><td>6LowPAN,6TiSCH,6Lo etc</td></tr>
<tr><td>Routing</td><td>RPL, COPRL, CARP etc</td></tr>
<tr><td colspan="2">Perception Layer</td><td>WiFi, Bluetooth, Z-Wave, ZigBee Smart, NFC etc</td></tr>
</table>

FIGURE 1.3 Layered structure of IoT protocols.

1.5.7 Z-WAVE

It is a low-power wireless networking solution to remote control applications and small-scale commercial domains [30]. It supports 232 nodes in a network, and coverage is about 32. Its operational frequency is about 900 MHz and has a transmission rate of 40 kbps. Because of its low capacity, it is not useful in streaming applications as well as transferring massive data since it does not support full-duplex communication.

1.5.8 LOW-POWER WIRELESS PERSONAL AREA NETWORK (LOWPAN)

Low-Power Wireless Personal Area Networks (LoWPANs) are wireless networks with unique features such as low latency, packet size limitations and a wide range of address lengths. It's an adaptation layer that enables IPv6 packets to travel over 802.15.4 networks. IEEE 802.15.4-2003 is the foundation for this protocol. To use the 6LoWPAN protocol in a network, each node must identify the same protocol. 6LoWPAN also provides better compatibility with pre-existing architecture.

Apart from the above-described protocols, Figure 1.3 presents a glimpse of IoT protocols viz. their applicability layers in totality [31].

While developing an IoT-based system, suitable technology, operating frequency, range and data rate are critical parameters and need optimisation. These requirements directly affect the cost of the system. Table 1.1 provides a comparative view for the above metrices.

1.6 SOFTWARE ARCHITECTURE OF IoT

The seven-layer architecture of an IoT system is depicted in Figure 1.4. IoT hardware devices are arranged from bottom to top. Lowest layer that deals with physically realisable nodes are divided into two layers: bottommost takes care of sensors and sensor network, actuators sensing termed as act device system layer. Next, the one above is dedicated to LAN; IoT setups are called intelligent device layers [29–31]. The physical information layer and the logical information layer are connected to the information of objects. The service layer performs planning and

TABLE 1.1
Technology, Frequency, Data Rate, Range, Power Usage and Cost Comparison

Technology	Frequency	Data Rate	Range	Power Usage	Cost
2G/3G	Cellular bands	10 Mbps	Several miles	Low	High
Bluetooth	2.4 GHz	1,2,3 Mbps	~300 ft	Low	Low
802.15.4	Sub GHz	40,250 kbps	>100 mi^2	Low	Low
LoRa	Sub GHz	Less than 50 kbps	1–3 mi	Medium	Medium
LTE Cat 0/1	Cellular bands	1–10 Mbps	Several miles	Medium	High
NB-IoT	Cellular bands	0.1–1 Mbps	Several miles	Medium	High
SigFox	Sub GHz	<1 Mbps	Several miles	Low	Medium
Wi-Fi	Sub GHz 2.4 GHz, 5 GHz	0.1–54 Mbps	<300 ft	Low	Low
Wireless HART	2.4 GHz	250 kbps	~300 ft	Medium	Medium

Source: While developing an IoT-based system, suitable technology, operating frequency, range and data rate are critical parameters and need optimization. These requirements directly affect the cost of the system.

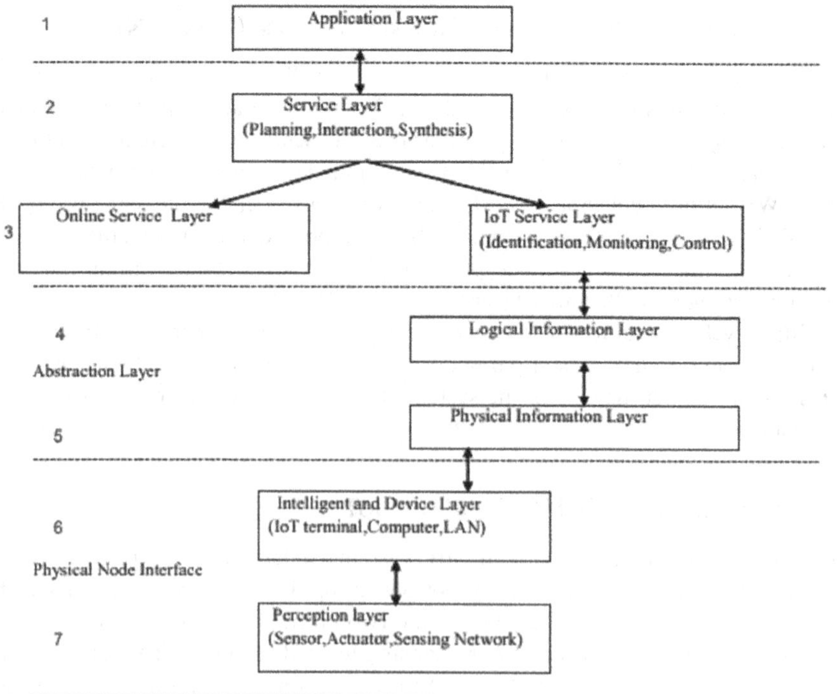

FIGURE 1.4 Architecture of IoT.

synthesis. It is assisted by the IoT basic service layer (BSL), which is responsible for monitoring and control; the topmost in this structure is the application layer.

The sensing and executing devices at the system layer collect data through the sensing devices and issue commands to the executing devices. The sensing device is a sensor directly or indirectly connected to an intelligent device through a bus. It converts the signals of the external physical world into a specific format that the intelligent device can detect. It can also be a network of sensors, which can collect data and take it to different paths before sending it to intelligent devices. The execution system will receive control commands from intelligent devices and use them to control the actions of objects.

The intelligent system layer includes intelligent devices that can collect details of objects and access the Internet directly. IoT terminal, computer and LAN are three types of intelligent machines. The IoT terminal is an integrated machine with an intelligent chip. The computer may be a general-purpose computer (GPU), an industrial computer or a personal mobile device. To provide information about activities, the LAN must be linked to the monitoring and control equipment.

The physical information layer of things is created by the device layer of the IoT, which provides physical information to the upper layer. It is preferable to represent data that is easier to comprehend, which programmers do while creating application databases in general. As a result, the logical information layer of things is a composite of the device information. In other words, logical information of things describes attributes of virtualised things. The most critical layer for linking items and the Internet is the BSL. There are three types of basic services offered by things with Internet connectivity.

1.6.1 Thing's Identification

It provides the most basic service, i.e. identification of IoT objects. To identify objects, first and foremost, they must be given unique identifiers. At the moment, there is no universal standard for identifying IoT objects.

1.6.2 Data Collection of Things

Industrial control equipment, environmental monitoring systems, emergency shutdown systems etc. are among the first to use this method. Due to data specificity and protection, expanding all of these data to the Internet is not possible. But nowadays, data is readily available, and its integration can be easily accomplished by BSL.

1.6.3 Thing's Behaviour Control

One of the earliest behaviour control methods is the use of items. This type of control is an active activity used mostly in industrial, water management, crop failure predictions etc. If they are applied to the Internet, they can be effectively monitored using IoT frameworks linked with higher security concerns.

Basic services based on things can be organically combined with Internet services, followed by a middle layer. The middle-layer services shape the service hierarchy that the application layer needs. A middle-layer service can be decomposed into sub-services and merge sub-services to complete the service. It exhibits the preparation and synthesis capabilities of the middle-layer service. Also, planning capability determines how services communicate, and this type of interaction can be limited, with bottlenecks in service.

1.7 MIDDLEWARE IN IoT

A middleware conceals the complexities of the system and its hardware, allowing the developer to work seamlessly without any orthogonal concerns at any level either system or hardware, which increases efficiency and productivity [32, 33]. It provides a soft interface for coordination and cooperative processing among operating system (OS), network communication layers and applications. Middleware is an essential support for the technologies used in the system and operational communication technologies since both are likely to be quite diverse [34].

There are two types of middleware service specifications for IoT: one is functional and the other is non-functional. Non-functional requirements encompass the functions or services like resource management, abstractions, while functional requirements capture the functions or services like QoS support, security, performance issues, reliability and availability [35].

1.8 IoT SIMULATORS AND EMULATORS

1.8.1 IoT Simulator

The term "simulation" refers to examining and exploring available tools and their methodologies. A dedicated programming model or device capable of real-time simulation and producing outcomes for various considerations is known as a simulator. It concludes the outcome without exposing the experimenter to any ambiguity.

A base engine and GUI (Graphical User Interface) are the essential parts of many simulators, which allow the users to get going with the simulation process expeditiously and also aid in the assessment of IoT simulators and their offerings to the operational devices [6–8]. Simulators have many benefits, e.g. integration, manageability, scalability, configurability, repeatability and accessibility. The simulators are flexible enough to simulate a few hundred devices simultaneously that can communicate continuously with each other without any problem [11, 17, 19, 20, 24]. A parameter-based comparative table of simulators based on Java, C++, C#, Python and Matlab is given in Table 1.2.

1.8.2 IoT Emulator

Testing requires us to replicate the system that mimics the original one. To achieve this, we use an IoT emulator [1–3] (which may be a hardware or software or combo). Emulator oversees the actual simulation process sitting in the middle and allows the replication (host) device to operate the hardware and/or the software program of the main (visitor) device. It also observes the outcome of wireless test beds [29–31]. Nowadays, the availability of improved and authentic test beds in the market is in abundance. Tables 1.3 and 1.4 compare IoT simulators

TABLE 1.2
Comparison of Some IoT Simulators

Simulators	Scope	Type	Programming Language	IoT Architecture Layers	Built-in IoT Standards	API Integration	Cyber Resilience Simulation	Service Domain	Security Measures
DPWSim	IoT	Open source	Java	Application	Secure Web Services Messaging	SOAP	No	Generic	Medium
iFogSim	Fog	Discrete-event	Java	Perceptual Network application	No	SOAP	No	Generic	Medium
IoTSim	Data analysis	MapReduce model	Java	Application	No	REST	No	Generic	Medium
Cup Carbon	Network	Agent-based discrete-event	Java custom scripting	Perceptual network	802.15.4 LoRaWAN	UDX	No	Smart city	High
SimIoT	Data analysis	Discrete-event	Java	Application	No	REST	No	Generic	High
OMNeT++	Network	Discrete-event	C++	Perceptual network	Manual extension	SOAP	Custom extensions	Generic	Medium
NS-Series	Network	Discrete-event	C++	Perceptual network	802.15.4 LoRaWAN	REST	No	Generic	High
QualNet	Network	Discrete-event	C++	Perceptual network	802.15.4 (Zigbee only)	REST	Yes	Generic	Medium
Atomiton IoT	IoT IIoT	Edge	Go Java	Communication network	Socialise	REST	No	MQIdentity	High
SWE-IoT	WSN	Sensor Observation Service	C C++	Communication network	Collision detection	SOAP	No	Human interface	High

(Continued)

TABLE 1.2 *(Continued)*
Comparison of Some IoT Simulators

Simulators	Scope	Type	Programming Language	IoT Architecture Layers	Built-in IoT Standards	API Integration	Cyber Resilience Simulation	Service Domain	Security Measures
TOSSIM	TinyOS	Sensor Observation Service	C Python	Communication network	Injecting packets	REST	Yes	Generic	High
Bevywise IoT	IoT device	Broker	Python Java	Network	Real time	REST	No	Smart city	Medium
EdgeCloudSim	Edge WLAN	Realistic	Matlab	Network	Mist Computing	SOAP	No	Edge Orchestrator	High
IoTIFY	Hardware connection	Mobile app	Python Java	Application network	Real time	REST	Yes	Smart City	High
MobIoTSim	IoT Networks	Research based	C++ C Sharp	Application network	Devices Profile for Web Services (DPWS)	REST	No	Generic	Medium
Ansys-IoT	IoT industry	Autonomous	Python Java	Network	Real time	REST	Yes	Industry	High

TABLE 1.3
Comparison of Selected IoT Emulators

Emulators	Scope	Type	Programming Language	IoT Architecture Layers	Built-in IoT Standards	API Integration	Cyber Resilience Simulation	Service Domain	Security Measures
Cooja	Network	Discrete-event	C/Java	Perceptual network	Protocols supported by Contiki OS	REST	Custom infrastructure	Enables real world	High
MAMMotH	IoT device	M2M	Python Java	Application network	Cost-efficient	REST	No	Generic	Medium
NetSim	IoT networks	Research based	C Code	Perceptual network	802.15.4 LTE MANETs	SOAP	Yes	Military utilities	High
NCTUns 6.0	Sensor networks	Discrete-event	C++	Network data-link	802.11p WiMAX MANETS *Optical network*	SOAP	Yes	Open source	High

TABLE 1.4
Comparison of Selected IoT Test Beds

Test Beds	Scope	Type	Programming Language	IoT Architecture Layers	Built-in IoT Standards	API Integration	Cyber Resilience Simulation	Service Domain	Support for Visualisations
MBTAAS	IoT platform	Service oriented	OCL	All	Model based	REST	No	Smart city	IoT dashboard
WHYNET	Wireless network	Network protocol	Java	Network	Application based	SOAP	No	Energy efficient	Web portal
FIT IoT-LAB	IoT network	IoT spectrum	nesC Java	Perceptual network	802.15.4 LoRaWAN	REST	No	Heterogeneous platform	FIT cloud
FIESTA-IoT	Energy	Sensor observation service	C, Java Python	Communication network	Energy consumption	REST	Yes	Ambient environment	Meta-cloud
SmartSantander	IoT mobile sensing	Map data map data image	Java JavaScript	Application network	IEEE 802.15.4 GPRS RFID tags	REST	Yes	Smart city	Management console
JOSE	IoT WSN SDN	Smart ICT service platform	C, Java JavaScript	Virtualised network services	Sensor networks	SOAP	Yes	Real time	Distributed cloud

and test beds to develop comprehension for their architecture standards and API integration.

1.9 NEED OF FIRMWARE AND IoT PROGRAMMING LANGUAGES

The IoT design architecture is divided by Skerrett. Critical three sections of IoT architectural setting are the data produced by sensors, local gateways and hubs and the geographical distance from where the data are collected and centralised servers. There are many facets in determining the language of choice; some important considerations are hardware, speed, development tools and supports in terms of library sets, schemes and community groups. Sensors and actuators are portable connected end devices, the "fieldwork" of which necessitates all firmware to gather metrics, perform some simple actions to turn on/off, and so on. They have finite memory capacity and low-power computation. The language selection to program devices and applications is critical in terms of computation load, modularity, support and licensing. Few open-source languages are C#, Go, Swift, Ruby, Rust, LUA etc.

1.10 SECURITY IN IoT

For information security in any communication system, we have to fulfil the following three requirements: data availability, data integrity and confidentiality. These are abbreviated as CIA triad [35]. These fundamentals find their applicability for IoT but not in entirety due to heterogeneity and distributed nature. Moreover, devices and actuators are nearer to users and a potential source of information than servers; if they are compromised, they can lead to convulsion. In addition, IoT communication parties communicate over an open-access public network, and entities like devices, fog and edge are not trusted. An adversary can modify, delete and can even capture nodes physically to extract information from memory. Each application of IoT has a unique set of characteristics in terms of sensor usage, deployment and level of criticality and thus attack environment has also evolved. Different types of attacks mapped w.r.t. three-layer architecture of IoT in Figure 1.5.

Application Layer		Phishing, Worms, Virus etc
Network Layer	Encapsulation	Denial of Service, Man in Middle, Sybil, Hello Flooding, Buffer Overflow etc.
	Routing	
Perception Layer		Malicious code injection, Sybil, Node capturing, Interference and Jamming, Interception.

FIGURE 1.5 Attacks in IoT.

In this lyceum of security, Blockchain, machine learning, deep learning and cryptography can provide potential solutions. Fog- and edge-dependent lightweight detection addresses attacks of distributed nature and computation extensive cloud-based solutions for centralised security management.

1.11 CONCLUSION

IoT spans almost every field and successfully amalgamates with existing technologies. Nowadays, IoT-centric technologies are trending and adding millions and billions of devices to our planet. This poses a serious threat to security at each level of architecture. Though many solutions have been proposed as a remedy, it is still a futuristic field of research. Hardware- and software-based intrusion detection systems are being developed to address this problem, but suitable placement for these centralised and distributed detection systems inside IoT architecture is yet a research problem.

REFERENCES

1. Miorandi, D., Sicari, S., Pellegrini, F. D., & Chlamtac, I. (2012). Internet of things: Vision, applications and research challenges. *Ad Hoc Networks, 10*, 1497–1516.
2. Sharma, V., & Tiwari, R. (2016). "IOT" & it's smart application. *International Journal of Science, Engineering and Technology Research (IJSETR), 5*, 472–776.
3. Silva, B., Khan, M., & Han, K. (2018). Internet of Things: A comprehensive review of enabling technologies, architecture, and challenges. IETE Technical Review, 35(2), 205–220, doi: 10.1080/02564602.2016.1276416.
4. Taneja, D. (2019). The Internet of Things: Overview & analysis. *International Journal of Electronics Engineering, 11*, 407–413.
5. Sachs, K., Petrov, I., & Guerrero, P. (2010). *In From Active Data Management to Event-Based Systems and More: Papers in Honor of Alejandro Buchmann on the Occasion of His 60th Birthday* (Vol. 6462). Springer.
6. Collina, M., Corazza, G., & Vanelli-Coralli, A. (2012). Introducing the QEST broker: Scaling the IoT by bridging MQTT and REST. In *2012 IEEE 23rd International Symposium on Personal, Indoor and Mobile Radio Communications-(PIMRC)* (pp. 36–41). IEEE.
7. Raduescu, C. (2004). Towards an ethnographic study of information appliances: The case of LG internet refrigerator. *Annals Computer Science Series Journal*, 2(1). http://anale-informatica.tibiscus.ro/download/lucrari/2-1-21-Raduescu.pdf [17 March 2016]
8. Evans, D. (2011). The Internet of Things: How the next evolution of the Internet is changing everything. In *The Internet of Things: How the Next Evolution of the Internet is Changing Everything* (pp. 1–11). http://www.cisco.com/web/about/ac79/docs/innov/IoT_IBSG_0411FINAL.pdf
9. Zielonka, A., Woźniak, M., Garg, S., Kaddoum, G., Piran, M., & Muhammad, G. (2021). Smart homes: How much will they support us? A research on recent trends and advances. *IEEE Access, 9*, 26388–26419.
10. Aazam, M., Zeadally, S., & Harras, K. (2018). Deploying fog computing in Industrial Internet of Things and Industry 4.0. *IEEE Transactions on Industrial Informatics, 14*, 4674–4682.
11. Dudhe, P., Kadam, N., Hushangabade, R., & Deshmukh, M. (2017). Internet of Things (IOT): An overview and its applications. In 2017 International Conference on Energy, Communication, Data Analytics and Soft Computing (ICECDS) (pp. 2650–2653). IEEE.

12. Ayaz, M., Ammad-Uddin, M., Sharif, Z., Mansour, A., & Aggoune, E. H. (2019). Internet-of-Things (IoT)-based smart agriculture: Toward making the fields talk. *IEEE Access*, *7*, 129551–129583.
13. Meddeb, A. (July 2016). Internet of Things standards: Who stands out from the crowd? *IEEE Communications Magazine*, 54.
14. Z. Bakhshi, A. B. (2018). Industrial IoT security threats and concerns by considering Cisco and Microsoft IoT reference models. In *IEEE Wireless Communications and Networking Conference Workshops (WCNCW)*.
15. Vishnu, S., Ramson, S., & Jegan, R. (2020). Internet of Medical Things (IoMT) – An overview. In *2020 5th International Conference on Devices, Circuits and Systems (ICDCS)* (pp. 101–104).
16. Khamparia, A., Singh, P., Rani, P., Samanta, D., Khanna, A., & Bhushan, B. (2021). An Internet of Health Things-driven deep learning framework for detection and classification of skin cancer using transfer learning. In *Transactions on Emerging Telecommunications Technologies* (Vol. 32).
17. Yogitha, K., & Alamelumangai, V. (2016). Recent trends and issues in IoT. *International Journal of Advances in Engineering Research (IJAER)*, *11*(1), January e-ISSN: 2231-5152/ p-ISSN: 2454-1796.
18. Poongodi, T., Krishnamurthi, R., Indrakumari, R., Suresh, P., & Balusamy, B. (2020). Wearable devices and IoT. In V. E. Balas, V. K. Solanki, R. Kumar, & M. A. Ahad (Eds.), *A Handbook of Internet of Things in Biomedical and Cyber Physical System* (pp. 245–273). Cham: Springer International Publishing.
19. Calo, S. B., Touna, M., Verma, D. C., & Cullen, A. (2017). Edge computing architecture for applying AI to IoT. In *2017 IEEE International Conference on Big Data (Big Data)* (pp. 3012–3016).
20. Deng, Z., Wu, X., Wang, L., Chen, X., Ranjan, R., Zomaya, A., et al. (2015). Parallel processing of dynamic continuous queries over streaming data flows. *IEEE Transactions on Parallel and Distributed Systems*, *26*, 834–846.
21. Fan, W., & Bifet, A. (2013). Mining big data: Current status, and forecast to the future. *ACM SIGKDD Explorations Newsletter*, *14*, 1–5.
22. Perera, C., Zaslavsky, A., Christen, P., & Georgakopoulos, D. (2014). Sensing as a service model for smart cities supported by Internet of Things. *ArXiv*, abs/1307.8198.
23. Calheiros, R., Ranjan, R., Beloglazov, A., De Rose, C., & Buyya, R. (2011). CloudSim: A toolkit for modeling and simulation of cloud computing environments and evaluation of resource provisioning algorithms. *Software*, *41*, 23–50.
24. Shelby, Z., Hartke, K., Bormann, C., & Frank, B. (2013). *Constrained Application Protocol (CoAP), Draft-Ietf-Corecoap-18, Work in Progress.*
25. Al-Fuqaha, A., Guizani, M., Mohammadi, M., Aledhari, M., & Ayyash, M. (2015). Internet of Things: A survey on enabling technologies, protocols, and applications. *IEEE Communications Surveys & Tutorials*, *17*, 2347–2376.
26. Schneider, S. (2009). Understanding the protocol behind the Internet of Things. *Understanding the Protocol behind the Internet of Things*. https://www.electronicdesign.com/technologies/iot/article/21798493/understanding-the-protocols-behind-the-internet-of-things.
27. Bendel, S., Springer, T., Schuster, D., Schill, A., Ackermann, R., & Ameling, M. (2013). A service infrastructure for the Internet of Things based on XMPP. In *2013 IEEE International Conference on Pervasive Computing and Communications Workshops (PERCOM Workshops)* (pp. 385–388).
28. Saint-Andre, P. (2004). *Extensible Messaging and Presence Protocol (XMPP): Core. Internet Engineering Task Force*. RFC, 3920.
29. Gomez, C., & Josepparadells. (2010). Wireless home automation networks: A survey of architectures and technologies. *IEEE Communications Magazine*, *48*, 92–101.

30. Bhushan, B., & Sahoo, G. (2019). Routing protocols in wireless sensor networks. In B. B. Mishra, S. Dehuri, B. K. Panigrahi, A. K. Nayak, B. S. Mishra, & H. Das (Eds.), *Computational Intelligence in Sensor Networks* (pp. 215–248). Berlin, Heidelberg: Springer Berlin Heidelberg.
31. Lv, W., Meng, F., Zhang, C., Lv, Y., Cao, N., & Jiang, J. (2017). A general architecture of IoT system. In *2017 IEEE International Conference on Computational Science and Engineering (CSE) and IEEE International Conference on Embedded and Ubiquitous Computing (EUC)* (Vol. 1, pp. 659–664).
32. Neely, S., Dobson, S., & Nixon, P. (2006). Adaptive middleware for autonomic systems. *Annales des télécommunications*, *61*, 1099–1118.
33. Bhushan, B., Sahoo, C., & Sinha, P. (2021). Unification of blockchain and Internet of Things (BIoT): Requirements, working model, challenges and future directions. *Wireless Network*, *27*, 55–90.
34. Razzaque, M. A., Milojevic, M., Palade, A., & Clarke, S. (2015). Middleware for Internet of Things: A survey. *IEEE Internet of Things Journal*, 3(1), 1. 10.1109/JIOT.2015.2498900.
35. Roy, M., Chowdhury, C., & Aslam, N. (2020). Security and privacy issues in wireless sensor and body area networks. In B. B. Gupta, G. M. Perez, D. P. Agrawal, & D. Gupta (Eds.), *Handbook of Computer Networks and Cyber Security: Principles and Paradigms* (pp. 173–200). Cham: Springer International Publishing.

2 Standardization of IoT Ecosystems

Open Challenges, Current Solutions, and Future Directions

Hubert Szczepaniuk
Warsaw University of Life Sciences,
Warsaw, Poland

Edyta Karolina Szczepaniuk
Military University of Aviation, Dęblin, Poland

CONTENTS

2.1 INTRODUCTION

The term "standardization" is defined in the Cambridge Dictionary as "the process of making things of the same type all have the same basic features" [1]. In the area of computer science, there is standardization in formal terms. As a result of the standardization process, formal standards are created for various areas of IT systems.

DOI: 10.1201/9781003219620-2

Standardization processes can take many years. Standards are developed by international organizations, including the International Standard Organization (ISO), the American National Standards Institute (ANSI), the National Institute of Standards and Technology (NIST), the Institute of Electrical and Electronics Engineers (IEEE), and the European Telecommunications Standards Institute (ETSI). The purpose of standardization is to normalize selected system parameters so as to provide users with the highest quality products.

The scientific community is also intensifying research on the multidimensional meaning of standardization, reflected in the growing number of available publications in this area. Standardization is an essential determinant of technological changes (see e.g. [2]). Correctly carried out standardization can bring significant benefits in key social, economic, economic, and technical areas (see e.g. [2–5]). Examples of specific technical and formal solutions with a global scope resulting from standardization in computer science can be TCP/IP network protocols, a set of computer network standards of the IEEE 802 series, or the ISO/IEC 25010 software quality standard. In computer science, standardization ensures the compatibility of hardware and software coming from various suppliers.

The Internet of Things (IoT) market is clearly in opposition to the considerations related to standardization in IT. In quantitative terms, it should be noted that in 2021 there are over 10 billion active IoT devices [6]. It is estimated that the number of active IoT devices will exceed 25.4 billion in 2030 [6]. In turn, studies by Manyika et al. suggest that the IoT market will account for around 4–11% of total global GDP in 2025 [7]. Long-term projections based on the growth account framework for the contribution of IoT to economic growth suggest a global average annual contribution of 0.99% per annum by 2030, which is approximately $849 billion per year in world GDP (2018 prices) [8]. The development of IoT ecosystems also directly affects the issues related to Big Data. Available estimates suggest that the amount of data generated by IoT devices will reach 73.1 ZB (zettabytes) by 2025 [7]. The presented statistics illustrate the scale of various IoT solutions and technologies available on the market. In many cases, the available solutions are based on a variety of technologies and architectural assumptions. Consequently, it leads to difficulties in the free implementation of IoT ecosystems using components from different vendors. In 2016, an article appeared in the IEEE newsletter on the need to take action based on standardization in IoT solutions [9].

Given the above, there is a justified need for research into the holistic standardization of IoT ecosystems to improve the compatibility, scalability, interoperability, and security of IoT systems. Implementing the IoT standardization process must be preceded by identifying the requirements and possible solutions in this area. Therefore, the subject matter of this chapter is in line with the objectives of international organizations. The importance of the research topic undertaken is also enhanced by the fact that the development of the IoT concept is one of the foundations of sustainable development and is supported by many countries and international organizations, including the European Union and the World Economic Forum (see e.g. [10, 11]).

The main finding of the chapter is to develop a framework for systemic standardization of IoT ecosystems to improve compatibility, scalability, interoperability, and cybersecurity. The model covers the requirements of logical and technical architecture, programming solutions, network protocols, and cybersecurity. The proposed framework of IoT systemic standardization may constitute a general reference model for the design, implementation, and improvement of individual components of the IoT architecture.

The structure of the chapter corresponds to the adopted specific objectives of the research. The chapter consists of an introduction, methodological basics of research, six substantive sections, conclusions, and references. The first substantive section defines the prerequisites for the IoT system standardization metamodel to ensure compatibility, scalability, interoperability, and cybersecurity. The second substantive section deals with programming standardization. The section contains analyses of programming languages, programming paradigms, and Integrated Development Environments (IDEs). In the third substantive section, communication standards and protocols are analyzed, emphasizing new, developing wireless communication technologies offering low power consumption. The next subchapter concerns the requirements of IoT cybersecurity standardization. The section is based on the requirements defined by the theoretical foundations of safety sciences. The final subchapter analyzes the open challenges and future directions of IoT standardization in the areas of architecture, programming, communication, and cybersecurity. Based on research findings, a framework for IoT systemic standardization was developed. The chapter ends with conclusions, which contain a synthetic summary of the conducted research.

2.2 RESEARCH METHODOLOGY

The research methodology adopted in the chapter is based on system analysis, which is a set of methods, techniques, and tools used to solve complex decision-making situations [12]. The research procedure consistent with the system analysis requires the definition of the context, goals, and requirements, thanks to which they will enable a holistic approach to the reality under study. Furthermore, the obtained research findings constitute the basis for developing a model that enables the explanation and improvement of the tested system. The final effect of the system analysis is the recommendation of specific methods of action [12].

The main research aim of the chapter is to identify and analyze critical areas of IoT ecosystems in terms of their standardization and also identify the requirements and possible solutions in this area. Therefore, the following specific objectives are defined:

- Define the requirements for the IoT standardization metamodel.
- Analysis of IoT architectural standardization.
- Analysis of programming languages, paradigms, and frameworks used in IoT programming.
- Analysis of network communication standards.
- Distinguishing IoT cybersecurity requirements.
- Development of a framework for systemic IoT standardization.

The structure of the research procedure is based on the system analysis method. The applied research methods were based on the analysis of real IoT projects, observation of phenomena occurring on the IoT market, analysis of programming concepts, and study of technical documentation and normative acts.

The research used an interdisciplinary approach taking into account the perspective of computer science (programming standardization, logical and technical architectures IoT), security sciences (IoT cybersecurity requirements), and management sciences (IoT quality requirements).

2.3 REQUIREMENTS FOR THE IoT ECOSYSTEM STANDARDIZATION METAMODEL

The research approach based on system analysis requires that a set of system requirements and standardization objectives be defined holistically. Defining the standardization objectives will allow for assessing the adopted solutions and their continuous improvement in terms of the quality of products and services provided to users.

The chapter defines four key objectives of standardization, which are also specific systemic features of the studied IoT ecosystems:

- compatibility – the system's ability to cooperate and exchange data with other systems;
- scalability – the ability of the system to operate in conditions of dynamically changing number of users and processed data;
- interoperability – the ability of systems to cooperate in order to ensure access to services provided by these systems;
- cybersecurity – the ability of the system to ensure the protection of the processed information in terms of the attribute of confidentiality, availability, and integrity.

The next research step was to define the requirements for the IoT standardization process. In the IEEE newsletter, it was noted that in the IoT standardization process, particular stages of IoT implementation should be taken into account, such as sensors, communication networks, technological and regulatory standards, intelligent analysis, and intelligent actions [9]. This chapter adopts similar assumptions, extending them with additional elements related to the implementation of IoT ecosystems. The implementation of IoT ecosystems considered in the perspective of IT projects is aimed at ensuring the quality of the products and services offered. In particular, the analysis of requirements covered:

- architectural standardization – logical models, hardware architectures, service orientation;
- programming standardization – methodologies and programming languages for IoT ecosystems;
- communication standardization – transmission protocols;
- standardization of cybersecurity – ensuring confidentiality, integrity, and availability of information.

A metamodel for the standardization of IoT ecosystems was developed by synthesizing the defined goals and requirements (Figure 2.1).

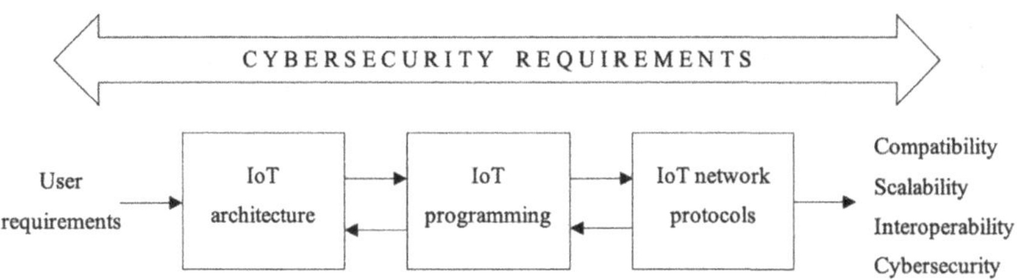

FIGURE 2.1 IoT ecosystem standardization metamodel.

Source: Own work.

2.4 ARCHITECTURAL STANDARDIZATION

The architecture of IoT ecosystems can be analyzed from two perspectives: logical and technical. Logical architecture concerns the model of business logic implementation by IoT ecosystems. Technical architecture refers to the hardware architecture of IoT devices.

2.4.1 IoT Logical Models

IoT ecosystems are embedded in the architecture of next-generation computer networks. Standardization of technologies related to computer networks is often embedded in a layered model. An example would be the ISO/OSI (Open System Interconnection) model or the TCP/IP layered reference model. In the initial period of establishing the scientific theory for IoT, a three-layer model of the IoT architecture was developed (see e.g. [13–16]). The model includes the application layer, the network layer, and the perception layer. These layers cooperate and cover the key areas of the functioning of IoT ecosystems. The application layer is responsible for the implementation of functional and business logic implemented by IoT ecosystems. The network layer carries out the data transmission between the perception and the application layers. In the network layer, classic network communication protocols are used and technologies dedicated to applications in the IoT environment, such as 6LoWPAN, ZigBee, and M2M. The lowest perception layer is responsible for handling IoT end devices, which recognize and record signals from the surrounding environment.

The global dissemination and development of technology have led to new IoT conceptual models (see e.g. [17]). For example, ITU-U Y.4000 describes the IoT four-tier reference model [18]. The model includes the application, support, network, and device layers [18]. As in the architecture discussed earlier, the application layer deals with utility applications that implement IoT ecosystems' business logic and functional requirements. The support layer includes support for data processing and storage as well as dedicated functionalities required by individual IoT applications. The network layer provides data and information transmission, mobility management as well as authentication and authorization mechanisms. In turn, the lowest layer includes support for end devices and IoT gateways.

The presented layered models have standard features in terms of the logical IoT architecture. It is possible to distinguish in the analyzed IoT models key areas important from the standardization perspective. These are in particular:

- software;
- computer networks;
- technical device architecture.

Software is a key component of IoT ecosystems. Properly defined functional and nonfunctional requirements should ensure the implementation of business logic and maximize the added value of IoT. Software in IoT ecosystems is often based on the client-server model. In real IoT projects, individual programming of server logic and client applications is often required. The software is responsible for device control and data processing operations obtained in IoT end devices. The software used by IoT often includes Database Management Systems (DBMS), Data Warehousing (DW), Artificial Intelligence (AI) with Machine Learning (ML) tools, and cloud-based services.

2.4.2 Technical Architecture

In terms of IoT technical architecture, two key elements essential from the standardization perspective can be distinguished:

- hardware architecture of the microcontroller installed in the end devices;
- operating system (OS) installed on end devices.

There is a wide range of devices on the market based on various microprocessor systems and architectures. The microcontroller is an integrated circuit consisting of a microprocessor, RAM, ROM, and I/O circuits. The most important element of microcontrollers is a central processing unit (CPU), which is responsible for the system's performance. There is a wide range of CPUs on the market. IoT creates a demand for new types of CPUs that can function efficiently with minimal demand for electricity. CPU manufacturers quickly noticed the market demand for a new segment of processors dedicated to IoT. Currently, the largest CPU manufacturers offer and constantly improve this type of microprocessor. The selection of the correct CPU architecture determines the type of OS used, programming technologies as well as computing and energy efficiency of IoT ecosystems. For this reason, the hardware architecture of microcontrollers should be selected individually for the specific requirements of IoT solutions.

After analyzing real IoT projects, the following factors can be distinguished when selecting a CPU:

- microprocessor architecture,
- microprocessor software model,
- power consumption.

There are two microprocessor architectures on the market for IoT devices, ×86 and Advanced RISC Machine (ARM). Processors with the ×86 architecture are used on a large scale in personal computers. The advantage of the ×86 architecture is the very high efficiency of single-core and multithreaded operations. Another advantage is the ability to run software programmed for ×86 architecture. In turn, the disadvantage of the architecture is the relatively high power consumption. However, manufacturers offer energy-saving processors dedicated to IoT solutions. The most popular processor architecture for IoT solutions is ARM. The leading solutions in this area are cores based on the architects Cortex-M and Cortex-A. The key advantage of ARM solutions is low power consumption. Depending on the microprocessor lithography, TDP (Thermal Design Power) may vary between 0.5 and 2 W. The limitations of ARM processors include, among others, limited computing performance and lack of full support for virtualization technologies. The use of ARM architecture requires a dedicated OS and software programmed for these microprocessors. It should be noted that the latest microprocessors dedicated to IoT contain hardware systems supporting AI algorithms. In particular, CPUs equipped with programmable systems accelerating ML and Deep Learning (DL) algorithms are being developed. It is worth mentioning that both ×86 and ARM processors are available in various software models. The most advanced IoT-dedicated processors provide a data bus that operates on 64-bit addresses. However, not always 64-bit microprocessors are required in IoT implementations. A significant part of simple IoT end devices still uses popular 8-bit controllers.

Another essential element of the IoT technical architecture is the OS. The OS is responsible for hardware management and provides a runtime environment for the user software. The OS must support the CPU architecture used in the IoT device. In addition, the choice of OS also determines the programming technologies used to implement IoT service software. The published research results show that, in 2020, the most popular OS supporting IoT devices are as follows [19]:

- Linux distributions – 43%,
- FreeRTOS – 35%,
- Windows systems – 31%,
- Zephyr – 8%.

The popularity of Linux is because the kernel code is available under the terms of the GNU GPL. The key factors in favor of using Linux in IoT ecosystems are support for processors of various architectures, including ARM, multitasking, support for the TCP/IP protocol stack, and advanced security mechanisms. There are dedicated Linux distributions specifically designed to support embedded devices. The architecture of Linux distributions dedicated to IoT devices is significantly reduced. The goal is to increase the system's speed on limited hardware resources (limited CPU clock and minimum RAM size) and reduce power consumption, which is especially important on battery-powered IoT devices. It is possible to minimize the energy consumption of Linux-based IoT devices by implementing profiling mechanisms and cyclic operation of devices [20]. The following distributions are used in real IoT projects: Ubuntu Core, Raspberry Pi OS, Debian, and Yocto Project.

IoT also uses real-time operating systems (RTOS). The task of RTOS is to ensure that certain operations are performed within a defined time. RTOS uses a kernel architecture that uses task scheduling algorithms to meet the real-time regime in implementing processes. According to the statistics presented above, the most popular real-time system used in IoT is FreeRTOS. The system is available under the MIT license. FreeRTOS is used to support energy-saving microcontrollers of IoT devices. The system consists of a kernel and many libraries available, thanks to which it is possible to communicate with IoT edge devices and connect with the cloud. FreeRTOS does not contain the advanced kernel mechanisms typical of classic OSs such as Linux or Windows. The system source code has been oriented to the maximum simplicity and speed of task execution. Thanks to this, the system is characterized by low demand for hardware resources and reducing power consumption.

The Zephyr OS is another example of an RTOS system that is popular in IoT projects. The advantage of the Zephyr system is that it supports multiple microprocessor architectures, including ARM and SPARC. In addition, the system is distributed under the Apache License 2.0. The system is designed to support microcontrollers of embedded IoT devices. In addition to the OS kernel, Zephyr also includes device drivers, support for the network protocol stack: TCP/IP, 6LoWPAN, CoAP, IPv4, IPv6, Ethernet, and a file system.

An essential category of OSs used in IoT ecosystems is Windows. There is a version of Windows 10 IoT that is intended for use in embedded systems. The Enterprise edition and the Core edition are available. The Enterprise version is similar to the classic Windows 10 Enterprise with additional locking functions and does not support ARM processors [21]. Examples of Windows 10 IoT Enterprise usage scenarios may include ATMs, medical devices, tablets, and production devices. Windows 10 IoT Core is reduced compared to the Enterprise version but provides ARM architecture support. It is worth adding that a significant advantage of Windows 10 IoT is easy integration with MS Visual Studio development tools, which can be effectively used to create IoT software.

2.5 PROGRAMMING STANDARDIZATION

Programming standardization concerns solutions used to implement software for IoT ecosystems. Programmable microcontrollers and data processing in cloud services pose new requirements for programmers. The key issues that need to be analyzed in the context of IoT are as follows:

- programming languages;
- programming paradigms;
- IDEs.

The distinguished categories are key in terms of software development. Standardization of programming technologies will contribute to the improvement of application programming for IoT ecosystems.

2.5.1 Programming Languages in the IoT Environment

Software implementation is usually the costliest part of IoT projects. According to the TIOBE ranking, the most popular programming languages for May 2021 are C, Python, Java, C++, C# [22]. In turn, according to the results of the Eclipse Foundation research, the most popular programming languages for IoT solutions in 2020 are C, C++, Java, Python, and JavaScript [19]. Therefore, it would seem that the programming of classic desktop applications, web applications, or server logic services is similar to the IoT programming of IoT ecosystem solutions. However, IoT programming often requires the simultaneous use of multiple programming technologies when implementing a single project. The need to use multiple programming languages arises directly from the technical and logical architecture presented in point 4. The microprocessor architecture and the type of OS directly determine the type of programming techniques used.

There are three main components of IoT: sensors and end devices, local gateways, and central service servers (see e.g. [23–25]). The mentioned components are based on different logical and technical architecture, which entails using different programming languages. The most popular programming languages in the field of IoT are C, Java, C++, and Python (see e.g. [19]). Programming languages are constantly being developed to support IoT programming better. The following programming languages should be distinguished:

- JAVA – object-oriented programming language. Thanks to the Java Virtual Machine (JVM) concept, it is possible to minimize the dependence of the source code on the hardware used, which is crucial in implementing software for IoT ecosystems. Java has dedicated programming libraries for mobile devices (see e.g. [26]). The Oracle Java SE Embedded platform facilitates the programming of embedded devices (see e.g. [27]). Java supports software development for the Raspberry Pi architecture based on platforms such as Eclipse Open IoT Stack, IBM Watson IoT, Amazon AWS IoT, and Microsoft Azure IoT (see e.g. [26, 28]). Java is also used to develop back-end server services that are the basis of the IoT cloud architecture.
- C – one of the most popular programming languages, which allows creating software for embedded systems. A significant part of IoT endpoints runs under embedded OSs, often programmed in C. The advantage of software written in C is primarily the speed of source code execution and direct access to hardware interfaces.
- C++ – an object-oriented programming language based on C syntax with an additional abstraction layer for data types. The C++ language enables programming the source code for embedded devices in compact scripts closely adapted to the hardware architecture. The advantage of the C++ language is the ability to implement object-oriented paradigms and effective software development for the Linux OS, which is the basis of many IoT ecosystems.
- Python – a dynamically developing language that can be used in many areas of programming, including software implementation for IoT ecosystems. The language has clear and transparent syntax and a rich collection of

programming libraries. Python enables effective application development for IoT ecosystems, from embedded device development to server-side implementation based on modern RESTful API and WebSockets programming concepts (see e.g. [29]). Popular Raspberry Pi architect and STM32F405RG, ESP8266, ESP32 series microcontrollers support Python [30]. It is worth noting that the MQTT protocol, which is one of the most popular IoT communication protocols, is implemented in Python [30].

- JavaScript – a scripting programming language that is growing in popularity in the IoT environment. The Tessel and Espruino microcontrollers used in the implementation of IoT ecosystems provide support for the JavaScript language ([31, 32]). Provided JavaScript can be used effectively in Raspberry Pi-based ecosystems and Apache server.
- Assembler – a low-level programming language in the CPU machine code. Its advantage is speed and the ability to optimize the code for a specific CPU architecture. In the TIOBE ranking, assembler was promoted from 15th to 8th place in May 2021 compared to 2020 [22]. The growing popularity of assemblers can be explained by the increase in the number of devices embedded in IoT ecosystems on the market.

2.5.2 Programming Paradigms

In addition to programming languages, programming paradigms are also crucial in the field of application development. Programming paradigms define the structure and principles on which an application algorithm is based. Examples of paradigms are imperative programming or object-oriented programming. When analyzing real IoT projects, it is possible to distinguish programming paradigms that are most often used in this environment. In particular, the following specific programming approaches used in the implementation of IoT solutions should be distinguished (see e.g. [25]):

- Dataflow programming – the programming paradigm according to which the source code is oriented toward modeling the dataflow between individual operations that transform data. Dataflow programming can facilitate asynchronous dataflow through multiple nodes that handle data from IoT sensors (see e.g. [33]). In the scientific literature, there are concepts based on the assumptions of distributed dataflow in IoT application programming (see e.g. [34, 35]).
- Event-driven programming – based on the concept in which the control flow in the application depends on events such as user actions or receiving data from IoT input sensors. It is possible to implement a concurrency model and distributed communication (see e.g. [36]). It should be emphasized that recent studies have highlighted possible security problems related to the application of event-driven programming paradigms in the automation of smart houses [37].
- Functional programming – the implementation of the source code is oriented toward functions that are treated as fundamental values (similar to *Int* or *Float* values). When building an algorithm, functions can be passed to

other functions as parameters, and they can be returned as a result of other functions. The research shows that the size and complexity of the source code that processes, collects, and interprets data from IoT sensors can be reduced using functional programming techniques [38].

- Aggregate programming – based on the concept of a computational field being a unified abstraction of construction and inference on large self-organizing networks of devices (see e.g. [39–41]). Aggregate programming enables the implementation of complex distributed services and safe encapsulation, modulation, and composition of services in the IoT environment [41].

The analyzed programming paradigms are not assigned to a specific language. Therefore, it is possible to use different programming paradigms simultaneously.

2.5.3 Integrated Development Environments

An IDE provides development tools for creating and testing source code and compiling it into executable files. There are a large number of IDEs on the market that support different programming languages and paradigms. The research in this chapter focuses on the analysis of development environments that support application programming for IoT ecosystems. In real IoT projects, developers use, among other things, the following environments:

- Microsoft Visual Studio – an environment often used to develop classic desktop or web applications. The environment also supports software development for IoT ecosystems. The available.NET IoT library supports sensors, I/O devices, serial ports, and various other modules (see e.g. [42]). The.NET IoT library enables the applications programming for IoT ecosystems based on the Raspberry Pi, HummingBoard, BeagleBoard, and Pine A64 architecture (see e.g. [42]). Currently, the popularity of Azure IoT Edge is increasing, thanks to which it is possible to implement Docker-compatible execution units (see e.g. [43]).
- PlatformIO – a multi-platform IDE enabling the implementation of software for embedded devices. The environment supports the C and C++ languages. The platform is available as an extension to Microsoft Visual Studio Code. The platform works with a wide range of microcontrollers, in particular, based on the ARM Cortex, AVR, and MSP430.
- Arduino IDE – an environment that supports C and C++ language and software implementation for programmable IoT microcontrollers compatible with Arduino, as well as microcontrollers from other manufacturers. The environment is popular with both amateur and professional applications.
- OpenSCADA – an open implementation of Supervisory Control and Data Acquisition (SCADA) and Human-Machine Interface (HMI) systems. The environment is multi-platform, multi-module, and scalable (see e.g. [44]). The system enables programming the automation of production processes. SCADA systems are used, among others, in industry, energy, transport, and aviation (see e.g. [45]).

The environments highlighted above are mainly solutions for programming embedded devices. Modern IoT ecosystems also include cloud services that store and analyze data from endpoints. The server logic can be implemented in classic programming concepts based on the Service-Oriented Architecture (SOA) and Application Programming Interface (API) paradigms.

The maturity of programming technologies is expressed in the programming environment by the emergence of a programming framework for a specific class of solutions meeting given business requirements. An example of technological maturity can be web application programming frameworks such as Java Spring or Python Django. The framework contains architectural patterns and templates on which to build business logic. On the market, there is an open framework Eclipse Kura based on Java language, which offers API access to IoT hardware interfaces (see e.g. [46]). In addition to market solutions, there are also scientific studies on the programming frameworks based on the server-less architecture (see e.g. [47]). Thus, improving programming frameworks for IoT ecosystems is a crucial direction in the development of programming tools.

2.6 COMMUNICATION STANDARDS

Communication between IoT devices is the foundation for the functioning of the entire IoT ecosystem. IoT ecosystems can be embedded in the structure of classical computer networks. Communication in computer networks uses devices such as bridges, hubs, switches, and routers. Standardization of network communication requires the use of appropriate transmission protocols. Data transmission for IoT ecosystems uses classic protocol stacks in the transport and network layers used in computer networks, especially TCP/IP, UDP/IP.

The unique requirements of IoT systems in communication have led to the creation of transmission standards and protocols dictated for IoT. One of the basic requirements is the functioning of IoT ecosystems in a wireless environment. The second essential requirement for communication systems in IoT ecosystems is to ensure the interoperability of devices and software. Interoperability is reflected in the exchange of services and information between devices and software from different vendors. Many institutions are developing technical specifications and test procedures to ensure the interoperability of IoT ecosystems. Communication standards can be analyzed in terms of the area covered by the computer network. IoT systems can operate in local and wide area networks (WAN). Typical network standards such as IEEE 802.11, NFC, and Bluetooth (IEEE 802.15.1) are used in wireless local area networks (WLAN). Nevertheless, research is currently underway on the development of modern energy-saving communication standards, such as:

- Z-Wave – the protocol that uses low-energy radio waves to communicate with IoT devices in a mesh network concept. An essential feature of the Z-Wave specification is the support for the interoperability of IoT ecosystems through the exchange of information and cooperation of end devices and software. Thanks to the mesh structure of the Z-Wave network, it is possible to connect to an IoT device through another device if it is not

possible to establish a connection directly. The transmission speed varies in the range of 40–100 kbps [48].
- Zigbee – a specification based on the wireless standard IEEE 802.15.4 in the 2.4-GHz band and enables implementing a dynamic and nonhierarchical device structure in IoT ecosystems. The protocol is characterized by a wide range (up to 100 m indoors), a throughput of up to 250 kbps, and low power consumption [49]. In terms of security requirements, the transmission is encrypted with the AES algorithm with a 128-bit key. The specification is widely used to implement intelligent facilities, home appliance automation, alarm systems, and personal wireless networks (WPAN).

IoT ecosystems can also function in WANs. WANs dedicated to IoT are characterized by data transmission over distances exceeding 500 m and low power consumption. These types of wireless WANs are referred to as LPWAN (low-power wide area network). In terms of the essential LPWAN standards that can be effectively used in IoT, the following should be distinguished:

- LoRaWAN – a specification that ensures record energy efficiency and long-distance device communication, so it is used in battery-powered IoT devices and even in devices using photovoltaics. The advantage of the technology is using a licensed frequency band, which means no need to pay license fees. Typical ranges achieved in LoRaWAN networks are measured in kilometers, while the transmission capacity varies between 250 bps and 50 kbps [50]. Energy efficiency, no license fees, and long-range means that the standard is suitable for implementing IoT in the areas of Smart Cities (traffic control, lighting control, measurement systems), Smart Agriculture, and Industry 4.0. It should be noted that there are scientific studies that highlight the possible limitations of LoRaWAN in guaranteeing real-time operations (see e.g. [51]).
- NB-IoT – a standard based on the generalized LTE protocol, which requires a licensed frequency band. Typical maximum bit rates are 234.7 kbps (download) and 204.8 kbps (upload), respectively [52]. Several advantages of NB-IoT technology are indicated in terms of ensuring QoS, low transmission delay, high reliability, and wide network coverage [52].
- LTE-M – a technology based on LTE transmission that ensures fast communication between M2M applications and IoT devices. Depending on the version of the LTE-M standard, it provides transmission at the level of 4 Mbps (download) and 7 Mbps (upload), respectively [53]. The standard is used in IoT ecosystems that require higher link bandwidth.

It is worth noting that the development of the 5G network is another research area in implementing this technology in the WANs IoT networks. Moreover, it should be added that the 6LoWPAN protocol is available in the application layer, which implements the mechanisms of encapsulating IPv6 packets on devices with limited computing capabilities and low power consumption.

2.7 CYBERSECURITY STANDARDIZATION

Due to the dynamic expansion of the IoT concept into various aspects of human life, the cybersecurity of IoT technology is one of the key factors that should be standardized. IoT cybersecurity is a complex research area. Standardization of IoT cybersecurity should be carried out systemically and holistically. The development of IoT security standards is the focus of many international organizations. IoT cybersecurity system standardization should consider protecting users, data, applications, and networks (see e.g. [54]). IoT cybersecurity requirements include, among other things [55]:

- technical factors – detection techniques, communication methods, network technologies;
- protection and security – information confidentiality, transmission security, privacy protection;
- business problems – business models and business processes.

Currently, the following standards are available that apply to IoT security:

- Security TS-0003 – the specification includes, among other things, solutions in the field of security of M2M systems, the security architecture model, service security, authorization mechanisms, and security framework [56].
- Security TS-0008 – the specification describes the oneM2M communication protocol through "RESTful CoAP binding" [57].
- End-to-End Security and Group Authentication TR-0012 – the technical report describes the functions and security mechanisms for oneM2M. The document includes, among other things, use cases, threat intelligence, high-level architecture, general requirements, and end-to-end security and group authentication procedures [58].
- ISO 31000 – a group of risk management standards. The ISO 31000: 2018 standard was introduced in February 2018 and provided a universal model for designing, implementing, and maintaining risk management processes [59].

In addition to the standards mentioned above, the holistic standardization of IoT cybersecurity should also include the basics of security science. Based on the system analysis, two views of security can be distinguished [60]:

- System security is considered a property (feature) of the tested object, characterized by resistance to the emergence of threats, with the focus being on vulnerabilities to security incidents.
- System security is defined in terms of the ability to protect the internal values (resources) of an object from threats.

Due to the limited computing power of microcontrollers, it is crucial to develop lightweight cryptographic algorithms dedicated to IoT solutions (see e.g. [61]). Another important research direction is implementing Blockchain technology in IoT ecosystems (see e.g. [62]). The available research in the area of cybersecurity indicates that the human factor and technical requirements are critical factors in the security of information systems (see e.g. [63, 64]). The risks associated with attack vectors on IoT ecosystems using social engineering can be reduced through systematic social campaigns and training. In safety education, it is possible to effectively use new technologies (see e.g. [65]).

Considering the above, the future development of IoT cybersecurity standards should take into account the foundations of security science, technical sciences, and management sciences [54].

2.8 A FRAMEWORK FOR SYSTEMIC IoT STANDARDIZATION

System standardization of IoT should cover, in particular, the architectural area, programming technologies, network communication, and solutions related to ensuring cybersecurity. Based on the analyses of real IoT projects, current solutions in selected areas are discussed. Below are the results of the research on the open challenges and future directions of standardization. The findings are presented in Table 2.1.

By synthesizing the conducted research, a framework for systemic IoT standardization was developed, taking into account the standardization objectives. A proposal for a layered standardization model is shown in Figure 2.2.

The proposed model considers the business logic and user requirements, which constitute the basic guidelines for improving the quality of IoT ecosystems. Standardization in the distinguished layers of the model is aimed at ensuring the interoperability, compatibility, scalability, and cybersecurity of IoT ecosystems.

2.9 CONCLUSIONS

IoT standardization is an open challenge in the area of computer science, security sciences, and management. Providing added value through the implementation of business logic and user requirements, as well as the multitude of perspectives analyzed in the chapter, means that standardization in the IoT area must be carried out holistically and systemically. The proposed layered model helps ensure the interoperability, compatibility, scalability, and cybersecurity of IoT ecosystems. In the architectural layer, the key challenges are related to ensuring the compatibility of devices, OSs, and services provided by different vendors. In the area of programming, it is postulated to develop IDEs dedicated to programming IoT applications. It is also crucial to develop an open SOA and API-oriented IoT programming framework, including a RESTful API. In network communication, the key is to standardize 5G technology in WAN IoT networks. In turn, future directions in cybersecurity include the implementation of Blockchain technology in the area of the IoT environment and the implementation of lightweight encryption algorithms dedicated to low-power processors.

TABLE 2.1
Open Challenges and Future Directions of IoT Standardization

	Open Challenges	Future Directions
Architecture	• Incompatibility of devices from different manufacturers • Deficiencies in the interoperability of IoT ecosystems in the field of data exchange and service sharing • Difficulties in updating device operating systems and control software • The low computing power of embedded systems • Relatively high power consumption by processors based on ×86 architects	• Development of standards ensuring the implementation of the company's business logic through the use of IoT solutions • Development of universal operating systems capable of operating with various IoT hardware architectures • Development of intelligent sensors capable of preprocessing data at the perception stage • Reduce power consumption by developing CPU architecture standards and logical models for their use
Programming	• The need for dedicated programming for various hardware architectures of processors, microcontrollers, and operating systems • Frequent need to program firmware in low-level programming languages • Problems in processing heterogeneous Big Data sets generated by IoT end sensors	• Development of an open-source programming framework dedicated to software implementations for IoT, taking into account SOA and API paradigms, including RESTful API • Development of IDEs supporting programming for various IoT hardware and software architectures
Communication	• Low transmission bandwidth in unlicensed frequency bands • Potentially possible restrictions in implementing real-time systems with LoRaWAN specification (see e.g. [51]) • Potential problems in ensuring compatibility with classic network solutions based on the TCP/IP stack	• Standardization of the use of 5G networks in WAN IoT solutions • Development of network communication standards focusing on transmission speed, ensuring QoS, low latency, reliability, and wide network coverage
Cybersecurity	• Standardization deficiencies or limited guidelines for Internet cybersecurity management, e.g. risk management • Limitations in the implementation of transmission encryption in the case of limited processor power • Problems related to the security of device authorization • Threats in the transport layer, e.g. MiTM attacks, network message spoofing, attacks on the routing path, DoS attacks	• Development of IoT cybersecurity management standards based on security science • Development of energy-efficient processors implementing hardware support for the AES encryption algorithm • Development of lightweight encryption algorithms adapted to the low performance of microcontrollers • Implementation of Blockchain technology in the IoT environment

Source: Own work.

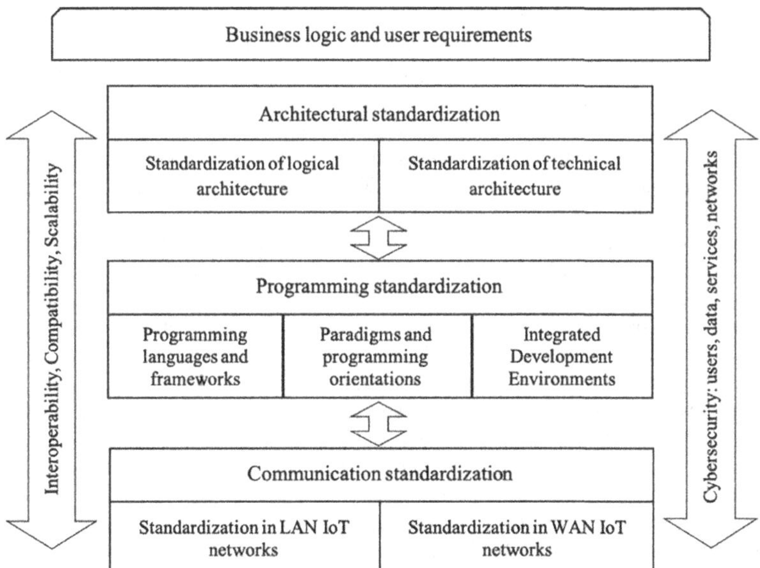

FIGURE 2.2 A framework for systemic IoT standardization.

Source: Own work.

The open challenges identified in this chapter constitute future research directions. Particularly noteworthy are research on the processing of heterogeneous Big Data sets generated by IoT sensors and research related to the multidimensional reduction of power consumption by IoT ecosystems.

REFERENCES

1. Cambridge Dictionary. Available online: https://dictionary.cambridge.org/pl/dictionary/english/standardization (Accessed 21 April 2021).
2. Wiegmann, P.M., Vries, H.J., Blind, K. (2017). Multi-mode standardisation: A critical review and a research agenda. *Research Policy*. Volume 46, Issue 8, pp. 1370–1386. Elsevier. https://doi.org/10.1016/j.respol.2017.06.002.
3. Xie, Z., Hall, J., Ian, P., McCarthy, I.P., Skitmore, M., Shen, L. (2016). Standardization efforts: The relationship between knowledge dimensions, search processes and innovation outcomes. *Technovation*. Volumes 48–49, pp. 69–78. Elsevier. https://doi.org/10.1016/j.technovation.2015.12.002.
4. Lorenz, A., Raven, M., Blind, K. (2019). The role of standardization at the interface of product and process development in biotechnology. *The Journal of Technology Transfer*. Volume 44, Issue 4, pp. 1097–1133. Springer. https://doi.org/10.1007/s10961-017-9644-2.
5. Zarzycka, E., Dobroszek, J., Lepistö, L., Moilanen, S. (2019). Coexistence of innovation and standardization: Evidence from the lean environment of business process outsourcing. *Journal of Management Control*. Volume 30, Issue 3, pp. 251–286. Springer. https://doi.org/10.1007/s00187-019-00284-x.
6. Jovanović, B. (2021). DataProt. Internet of Things statistics for 2021 – Taking Things Apart. Available online: https://dataprot.net/statistics/iot-statistics/ (Accessed 16 April 2021).

7. Manyika, J., Chui, M., Bisson, P., Woetzel, J., Bughin, J., Aharon, D. (2015). *The Internet of Things: Mapping the Value Beyond the Hype*. McKinsey Global Institute.
8. Edquist, H., Goodridge, P., Haskel, J. (2021). The Internet of Things and economic growth in a panel of countries. *Economics of Innovation and New Technology*. Volume 30, Issue 3, pp. 262–283. Taylor & Francis Group. https://doi.org/10.1080/10438599.2019.1695941.
9. Banafa, A. (2016). IEEE. IoT Standardization and Implementation Challenges. Available online: https://iot.ieee.org/newsletter/july-2016/iot-standardization-and-implementation-challenges.html (Accessed 15 May 2021).
10. World Economic Forum. (nd). Shaping the Future of the Internet of Things and Urban Transformation. Available online: https://www.weforum.org/platforms/shaping-the-future-of-the-internet-of-things-and-urban-transformation (Accessed 11 June 2021).
11. European Commission. (nd). How to Unleash the full potential of the Internet of Things. Available online: https://cordis.europa.eu/article/id/124539-how-to-unleash-the-full-potential-of-the-internet-of-things/pl (Accessed 11 June 2021).
12. Sienkiewicz, P. (1994). *Analiza systemowa. Podstawy i zastosowania*. Warsaw: Wydawnictwo Bellona.
13. Burhan, M., Rehman, R.A., Khan, B., Kim, B.S. (2018). IoT elements, layered architectures and security issues: A comprehensive survey. *Sensors*. Volume 18, Issue 9, pp. 2796. https://doi.org/10.3390/s18092796
14. Said, O., Masud, M. (2013). Towards Internet of Things: survey and future vision. *International Journal of Computer Networks*. Volume 5, Issue 1, pp. 1–17.
15. Liu, J., Li, X., Chen, X., Zhen Y., Zeng, L. (2011). Applications of Internet of Things on smart grid in China. *13th International Conference on Advanced Communication Technology (ICACT2011)*. pp. 13–17. IEEE. Seoul. http://dx.doi.org/10.11591/telkomnika.v12i2.4178.
16. Wu, M., Lu, T. J., Ling, FY., Sun, J., Du, H. (2010). Research on the architecture of Internet of Things. 3rd International Conference on Advanced Computer Theory and Engineering (ICACTE). pp. V5-484–V5-487. IEEE. Chengdu. https://doi.org/10.1109/ICACTE.2010.5579493.
17. Nord, J.H., Koohang, A., Paliszkiewicz, J. (2019). The Internet of Things: Review and theoretical framework. *Expert Systems With Applications*. Volume 133, pp. 97–108. Elsevier. https://doi.org/10.1016/j.eswa.2019.05.014.
18. International Telecommunication Union. Y.4000: Overview of the Internet of things. Retrieved from: https://www.itu.int/rec/T-REC-Y.4000/en (Accessed 21 May 2021).
19. Ecplipse Fundation. 2020 IoT Developer Survey Key Findings. Retrieved from: https://f.hubspotusercontent10.net/hubfs/5413615/2020%20IoT%C2%A0Developer%20Survey%20Report.pdf (Accessed 29 May 2021).
20. Amirtharaj, I., Groot, T., Dezfouli, B. (2020). Profiling and improving the duty-cycling performance of Linux-based IoT devices. *Journal of Ambient Intelligence and Humanized Computing*. Volume 11, pp. 1967–1995. https://doi.org/10.1007/s12652-019-01197-2.
21. Microsoft. (2018). An Overview of Windows 10 IoT. Available online: https://docs.microsoft.com/en-us/windows/iot-core/windows-iot (Accessed 19 May 2021).
22. Tiobe. (2021). TIOBE Index for May 2021. Available online: https://www.tiobe.com/tiobe-index/ (Accessed 21 May 2021).
23. Gubbi, J., Buyya, R., Marusic, S., Palaniswami, M. (2013). Internet of Things (IoT): A vision, architectural elements, and future directions. *Future Generation Computer Systems*. Volume 29, Issue 7, pp. 1645–1660. Elsevier. https://doi.org/10.1016/j.future.2013.01.010.
24. Intersog. (2017). 12 Popular Programming Languages For IoT Development In 2017. Available online: https://medium.com/@Intersog/12-popular-programming-languages-for-iot-development-in-2017-b8bf6ab5aef3 (Accessed 11 June 2021).
25. Devopedia. (2020). Programming for IoT. Available online: https://devopedia.org/programming-for-iot (Accessed 11 June 2021).

26. Xiao, P. (2019). *Practical Java Programming for IoT, AI, and Blockchain*. Indianapolis, IN: John Wiley & Sons.
27. Oracle. (nd). Oracle Java SE Embedded. Available online: https://www.oracle.com/java/technologies/javase-embedded/javase-embedded.html (Accessed 15 May 2021).
28. Chin, S., Weaver, J. L. (2015). *Raspberry Pi with Java: Programming the Internet of Things (IoT)*. McGraw-Hill. New York: Oracle Press.
29. Smart, G. (2020). *Practical Python Programming for IoT*. Birmingham: Packt Publishing.
30. Svitla. (nd). Internet of Things with Python. Available online: https://svitla.com/blog/internet-of-things-with-python (Accessed 22 May 2021).
31. Tessel. (nd). Tessel 2. Available online: https://tessel.io (Accessed 24 May 2021).
32. Espruino. (nd). Bangle.js, the Hackable Smart watch. Available online: http://www.espruino.com (Accessed 26 May 2021).
33. Tanaka, K., Tsujino, C., Maeda, H. (2020). IoT Software by Dataflow Programming in Mruby Programming Environment. In: Gervasi, O., Murgante, B., Misra, S., Garau, C., Blečić, I., Taniar, D., Apduhan, B.O., Rocha, A.M., Tarantino, E., Torre, C.M., Karaca, Y. (eds). Computational Science and Its Applications – ICCSA 2020. Lecture Notes in Computer Science. Volume 12252. Springer, Cham. https://doi.org/10.1007/978-3-030-58811-3_15.
34. Giang, N.K., Blackstock, M., Lea, R., Leung, V.C.M. (2015). Developing IoT applications in the Fog: A Distributed Dataflow approach. 2015 5th International Conference on the Internet of Things (IOT). pp. 155–162, https://doi.org/10.1109/IOT.2015.7356560.
35. Giang, N.K., Blackstock, M., Lea, R., Victor C. M Leung, V.C.M. (2015). Distributed Data Flow: a Programming Model for the Crowdsourced Internet of Things. In Proceedings of the Doctoral Symposium of the 16th International Middleware Conference. Association for Computing Machinery, New York, NY, Article 4, 1–4. https://doi.org/10.1145/2843966.2843970.
36. Cutsem, T.V., Mostinckx, S., Boix, E.G., Dedecker, J., Meuter, W.D. (2007). AmbientTalk: Object-oriented Event-driven Programming in Mobile Ad hoc Networks. XXVI International Conference of the Chilean Society of Computer Science. pp. 3–12. https://doi.org/10.1109/SCCC.2007.12.
37. Zhang, Q., Wang, X., Shen, S., Zhang, S., Bu, L., Li, X. (2019). Automated configuration, simulation and verification platform for event-driven home automation IoT system. *Chinese Journal on Internet of Things*. 2019, Volume 3, Issue: 3, pp. 90–101.
38. Haenisch, T. (2016). Case study on using functional programming for Internet of Things applications. *Athens Journal of Technology & Engineering*. Volume 3, Number 1, pp. 29–38. Retrieved from: https://www.athensjournals.gr/technology/2016-3-1-2-Haenisch.pdf (Accessed 11 June 2021).
39. Beal, J., Viroli, M. (2014). Building Blocks for Aggregate Programming of Self-Organising Applications. IEEE Eighth International Conference on Self-Adaptive and Self-Organizing Systems Workshops, pp. 8–13. https://doi.org/10.1109/SASOW.2014.6.
40. Beal, J., Viroli, M. (2016). Aggregate Programming: From Foundations to Applications. In: Bernardo M., De Nicola R., Hillston J. (eds). *Formal Methods for the Quantitative Evaluation of Collective Adaptive Systems. SFM 2016*. Lecture Notes in Computer Science. Volume 9700. Springer, Cham. https://doi.org/10.1007/978-3-319-34096-8_8.
41. Beal, J., Pianini, D., Viroli, M. (2015). Aggregate programming for the Internet of Things. *Computer*. Volume: 48, Issue: 9, pp. 22–30. IEEE. https://doi.org/10.1109/MC.2015.261
42. Microsoft. (nd). .NET IoT Apps. Available online: https://dotnet.microsoft.com/apps/iot (Accessed 29 May 2021).
43. Microsoft. (2019). What is Azure IoT Edge. Available online: https://docs.microsoft.com/en-us/azure/iot-edge/about-iot-edge?view=iotedge-2020-11 (Accessed 29 May 2021).
44. OpenSCADA. (nd). Available online: http://oscada.org/main/about-the-project/ (Accessed 5 June 2021).

45. Prokhorov, A.S., Chudinov, M.A., Bondarev, S.E. (2018). Control systems software implementation using open source SCADA-system OpenSCADA. IEEE Conference of Russian Young Researchers in Electrical and Electronic Engineering (EIConRus), pp. 220–222, doi: https://doi.org/10.1109/EIConRus.2018.8317069.
46. Eclipse Foundation. (nd). Kura. The extensible open source Java/OSGi IoT Edge Framework. Available online: https://www.eclipse.org/kura/ (Accessed 9 June 2021).
47. Nakagawa, I., Hiji, M., Esaki, H. (2015). Dripcast – Architecture and implementation of server-less java programming framework for billions of IoT devices. *Journal of Information Processing*. Volume 23, Issue 4, pp. 458–464. https://doi.org/10.2197/ipsjjip.23.458.
48. Silicon Laboratories. (nd). Safer. Smarter. Z-wave. Available online: https://www.z-wave.com/learn (Accessed 11 June 2021).
49. Connectivity Standards Alliance. (nd). Zigbee. Available online: https://zigbeealliance.org/solution/zigbee/ (Accessed 10 June 2021).
50. LoRa Alliance. (2015). A technical overview of LoRa and LoRaWAN. Retrieved from: https://lora-alliance.org/wp-content/uploads/2020/11/what-is-lorawan.pdf (Accessed 10 June 2021).
51. Adelantado, F., Vilajosana, X., Tuset-Peiro, P., Martinez, B., Melia-Segui, J., Watteyne, T. (2017). Understanding the Limits of LoRaWAN. IEEE Communications Magazine. Volume 55, Number 9, pp. 34–40. https://doi.org/10.1109/MCOM.2017.1600613.
52. Sinha, R.S., Wei, Y., Hwang, S-H. (2017). A survey on LPWA technology: LoRa and NB-IoT. *ICT Express*. Volume 3, Issue 1, pp. 14–21. Elsevier. https://doi.org/10.1016/j.icte.2017.03.004.
53. GSMA. (2019). LTE-M Deployment Guide to Basic Feature Set Requirements. Retrieved from: https://www.gsma.com/iot/wp-content/uploads/2019/08/201906-GSMA-LTE-M-Deployment-Guide-v3.pdf (Accessed 2 June 2021).
54. Szczepaniuk, H., Szczepaniuk, E.K. (2021). Cybersecurity Management within the Internet of Things. In Sharma, S.K., Bhushan, B., Debnath, N.C. (Eds.). IoT Security Paradigms and Applications: Research and Practices. Boca Raton: Taylor & Francis Group.
55. Li, S., Xu, L.D. (2019). *Securing the Internet of Things*. Cambridge: Elsevier.
56. oneM2M. (2018). Technical Specification TS-0003-V2.12.1. Security Solutions. Retrieved from: https://www.onem2m.org/images/files/deliverables/release2a/ts-0003-security_solutions-v_2_12_1-.pdf (Accessed 6 June 2021).
57. oneM2M. (2016). Technical Specification TS-0008- V-1.3.2. CoAP Protocol Binding. Retrieved from: https://www.onem2m.org/images/files/deliverables/release2a/ts-0003-security_solutions-v_2_12_1-.pdf (Accessed 6 June 2021).
58. ETSI, oneM2M. (2016). Technical Report TR-0012 version 2.0.0. End-to-End Security and Group Authentication. Retrieved from: https://www.etsi.org/deliver/etsi_tr/118500_118599/118512/02.00.00_60/tr_118512v020000p.pdf (Accessed 8 June 2021).
59. ISO. (2018). ISO 31000: 2018 Risk Management. Available online: https://www.iso.org/obp/ui/#iso:std:iso:31000:ed-2:v1:en (Accessed 9 June 2021).
60. Sienkiewicz, P. (2010). Systems analysis of security management. *Scientific Journals Maritime University of Szczecin*. 24(96), pp. 93–99.
61. Sharma, S.K., Bhushan, B., Debnath, N.C. (2020). *Security and Privacy Issues in IoT Devices and Sensor Networks*. London, San Diego: Academic Press. Elsevier.
62. Sharma, S.K., Bhushan, B., Khamparia, A., Astya, P.N., Debnath, N.C. (Eds.). (2021). *Blockchain Technology for Data Privacy Management*. Taylor & Francis Group. https://doi.org/10.1201/978100313339.
63. Szczepaniuk, E.K., Szczepaniuk, H., Rokicki, T., Klepacki, B. (2020). Information security assessment in public administration. *Computers & Security*. Volume 90. https://doi.org/10.1016/j.cose.2019.101709.
64. Szczepaniuk, E.K. (2016). *Bezpieczeństwo struktur administracyjnych w warunkach zagrożeń cyberprzestrzeni państwa*. Warsaw: AON.
65. Gawlik-Kobylińska M. (2016). *Nowe technologie w edukacji dla bezpieczeństwa*. Warsaw: Rozpisani.

3 A Node Reduction Technique for Trojan Detection and Diagnosis in IoT Hardware Devices

Sree Ranjani Rajendran
University of Florida, Gainesville, Florida, USA

Nirmala Devi and Jayakumar M
Amrita Vishwa Vidyapeetham, Ettimadai, India

CONTENTS

3.1 INTRODUCTION

Internet of things (IoT) is the interconnection of a large number of devices that generate volume of data. The data obtained from the devices like sensors, actuators and nodes are processed into useful actions in applications, including home and building automation, intelligent transportation and connected vehicles, industrial automation,

DOI: 10.1201/9781003219620-3

smart healthcare and smart cities. These connected devices are used as data collectors. The information collected with these devices can aid criminals in stealing the identity of user, accessing the data, like name, age, health data and location. These devices used in IoT are to be secured at both hardware and software levels. However, the secure hardware is mostly required to build a secured protocol. However, threats in hardware may lead a way to many new attacks both at hardware and software levels [1]. The authors in [2] discuss the idea of smart city along with the notable implementations and depict different security and privacy issues with a solid understanding. In [3, 4], the comprehensive review on how to remodel blockchain to the specific IoT needs has been discussed, in order to develop blockchain-based IoT (BIoT) applications, and aims to shape a coherent picture of the current state-of-the-art efforts.

Nowadays, a system-on-chip (SoC), equipped with different types of hardware resources, is used widely in IoT. Memory blocks, processors, coprocessors and their interfaces are included in these SoCs to realize high-performance computing systems. The high-complexity design and shortened time-to-market had driven the modern design flow horizontally. The design process is expedited with the *third-party intellectual properties (3PIPs)* in the forms of soft, firm and hard *intellectual properties (IPs).* In such a situation, a secured hardware platform is necessary for any architecture to perform its operation in applications like defense, health monitoring, and IoT. Researchers have presented a considerable amount of work on Trojan detection [5, 6], which has been broadly categorized into two: Trojan detection by *side-channel analysis (SCA)* approaches and logic testing scheme [7]. The Trojan detection approaches are enhanced by runtime monitoring systems, in which an extra module is embedded into the original design to continuously monitor the circuit performance or any side-channel signal behavior [8].

Hardware Trojan (HT) detection through SCA is challenging and this forms the basis for the proposed work. The existing HT detection schemes may incorrectly detect the effect of process variations (PV) as the Trojan and this is because of referring the SCA parameters to the golden chip. In addition, in some self-referencing schemes of Trojan detection, the portion of the design is compared with the similar part of the same design, which leads to intra-die PV. The computational time required to detect HT is more in the existing schemes, since the algorithm will check each and every gate of the circuit. However, the adversary will embed these stealthy Trojans to the nets with low toggling, so that Trojan activation probability is reduced. Furthermore, the nets in the noncritical paths of the design that have low toggling are to be studied and investigated.

In this proposed scheme, the limitations of the existing HT detection schemes are overcome by identifying specific low toggling gates. The circuit is analyzed to identify such low transition nets either by the transition probability (TP) analysis or the observability analysis of each node. The dynamic power measurements of these specific gates are observed to generate corresponding power signatures. Since the power signatures of important gates alone are considered for Trojan detection and diagnosis, it reduces the time complexity with minimum power measurements. In order to avoid the false-positive (FP) Trojan detection, the dynamic power signatures of the selective gates are measured at different timeframes. The inconsistency in the scaling factor of the measured power parameters at different time instance will detect the presence of Trojan without referring to any golden chip. Hence, the proposed scheme will nullify the effect of PV that would affect the side-channel parameters with reduced computational time.

The rest of the chapter is organized as follows. Section 3.2 discusses the related works concerning various security issues of IoT devices addressed in literature. The proposed node reduction technique for Trojan detection and diagnosis method is detailed in Section 3.3. We apply the proposed testing method to test several benchmark circuits, and the results are reported and analyzed in Section 3.4 and Section 3.5, respectively. Finally, Section 3.6 summarizes the findings.

3.2 BACKGROUND WORK

Various security issues in IoT devices are discussed in Refs. [1–4]. The author address the hardware threats that are most vulnerable than software threats. HTs are most vulnerable and it causes various threats in hardware and software levels. HT detection and prevention schemes are classified based on logical testing, side-channel parameter analysis, Trojan prevention schemes and circuit hardening. Taxonomy of HT detection and their prevention techniques are discussed [5, 6, 9]. This scope of this proposed work is focused toward SCA and logical testing for HT detection; hence, the review concentrates only on these two techniques. Logical testing is a process of comparing the output of the chip with the output of the golden chip. This process is difficult due to a large number of circuit inputs and their internal states. SCA is the basic scheme to detect the presence of the Trojan in the circuit. Govindan and Chakraborty [10] proposed an effective logic testing algorithm to detect relatively small Trojans. This scheme of Trojan detection based on the reduction of number test patterns using genetic algorithm and Boolean satisfiability will detect only the malfunction based on weighted random pattern-based test generation technique.

Salmani, in Refs. [11–13], presented an overview of various HT detection techniques based on test generation after circuit manufacturing. Amelian and Borujeni [14] proposed a path delay-based HT detection. However, identifying the path affected by the Trojan after fabrication is critical and it thus adds a time complexity in testing phase. In Ref. [15], Hiramoto and Ohtake proposed a design verification technique to detect the HT to improve the security of integrated circuits (ICs). Among the proposed Trojan detection and diagnosis schemes, the schemes that overcome the PV are considered important. In Ref. [16], the effect of PV is reduced by using supply current to explore signal processing techniques. A transient power-supply-based approach was proposed in Ref. [17], where multiple power port current signals are measured and a characterization of process noise is used for Trojan detection. Wei and Potkonjak [18] had extended Trojan detection and diagnosis with the self-consistency approach. They applied variable elimination by reusing the fixed set of power measurements.

Du et al. [29] proposed a self-referencing side-channel approach to improve a Trojan detection by the vector generation algorithm. Comparing the transient current signature of one region with that of the others nullifies the effect of PV. Narasimhan et al. [22] eliminated the effect of PV by comparing the current signature at different timeframes. In Ref. [30], Salmani proposed a controllability and observability-analysis-based Trojan detection scheme in gate-level netlist without referring to any golden chip. The unsupervised clustering analysis is used to distinguish Trojan signals from the genuine signals. Although the scheme is reference free, it requires controllability and observability analysis to cluster the signals. Devi et al. [31] proposed a malicious circuit detection technique using TP to identify a specific node. However, plausible Trojan detection scheme, proposed in Ref. [32], uses frequency analysis to increase the precision of Trojan detection. Recently, machine-learning-based Trojan detection techniques are proposed [33] to improve the detection accuracy.

The summary of the existing HT detection schemes and their limitations is described in Table 3.1. Among all the self-referencing approaches, a new approach with a minimum number of power measurements to reduce the time complexity of Trojan detection and diagnosis with the nullified inter-die and intra-die PVs is necessary. This motivates the need for selecting specific nodes for Trojan detection and diagnosis. Thus, the research adds another dimension to the direction of thinking; the proposed work analyzes only the critical nodes that lead to the reduction in computation complexity and increased detection rate.

Our main technical contributions of the proposed work are summarized as follows:

- A scalable golden chip-free scheme of Trojan detection and diagnosis.
- A timeframe-based self-referencing scheme that nullifies the effect of PV.
- Segmentation improves detection rate by minimizing false negative (FN) and FP.
- More importantly by examining only the specific node of interest,
    - The number of power measurements is reduced and this reduces the time complexity of Trojan detection and diagnosis.
- Hardware setup for the validation of HT detection and diagnosis is done on Spartan-6 field-programmable gate arrays (FPGA) trainer board and Mixed Signal Oscilloscope (MSO).

The Trojan models used in this scheme are described in detail with their appropriate malfunction. The chosen Trojans are time bombs, so that they will remain stealthy for a long time and they will produce malicious behavior after certain time periods as shown in Figure 3.1. These Trojans are inserted at low activity nodes and their payloads will change the logic of the intermediate nets and need not the circuit functionality. Thus, they are challenging to get detected by the conventional logic testing-based Trojan detection schemes.

The Trojan model with a corresponding truth table, which describes the triggering state and malfunction creation, is shown in Figure 3.1. The Trojan model in Figure 3.1(a) will remain stealthy up to a specific clock cycle and produce malfunction at the intermediate nodes of the circuit under test (CUT), whereas Trojan shown in Figure 3.1(b) will be triggered after a specified time window. The Trojan model shown in Figure 3.1(c) will produce a malfunction only when they are triggered by *01* input at the specified clock pulse. These Trojans are inserted in ISCAS'85 benchmark circuits at the rarely toggling nodes, such that their triggering will be at the low observable nodes and their payloads will be at low controllable nodes. The proposed scheme will

TABLE 3.1
Summary and Limitations of Some of the Existing HT Detection Schemes

Author and Year	Test Modality	Detection Method	Referring to Golden Chip	Disadvantages
Agrawal et al. [16]	Transient power	Kullback-Leibler (KL) distance	Yes	Reduction in detection sensitivity with increasing process variations
Jin et al. [19]	Delay	Path delay analysis	Yes	Delay due to process variation may be predicted as a Trojan effect
Potkonjak et al. [20]	Gate-level characterization (GLC)	Static power and circuit switching activity	Yes	GLC for each gate will have high computational complexity
Narasimhan et al. [21]	Intrinsic relationship between dynamic current (I_{DDT}) and maximum operating frequency (F_{max})	Vector generation approach for I_{DDT} measurement	No	Input vector dependent
Narasimhan et al. [22]	Current signature analysis	Side-channel signature analysis	No	Input vector dependent
Bazzazi et al. [23]	Functional	Logic testing	No	The logical relationship between the candidate nodes of the Trojan-free and Trojan-infected modes
Bao et al. [24]	Reverse engineering (RE)	K-means clustering approach	No	RE is a very costly process, which consumes lots of time and intensive manual effort
Chen et al. [25]	Statistical learning	Under-determined linear system by a sparse gate profiling technique	No	Detects HTs inserted during fabrication
Tehranipoor et al. [26]	Verification	Information flow security verification	–	–
Mishra and Lyu [27]	Delay	Side-channel analysis	–	–
Forte, et al. [28]	Degradation of power supply rejection ratio	Side-channel analysis	–	–

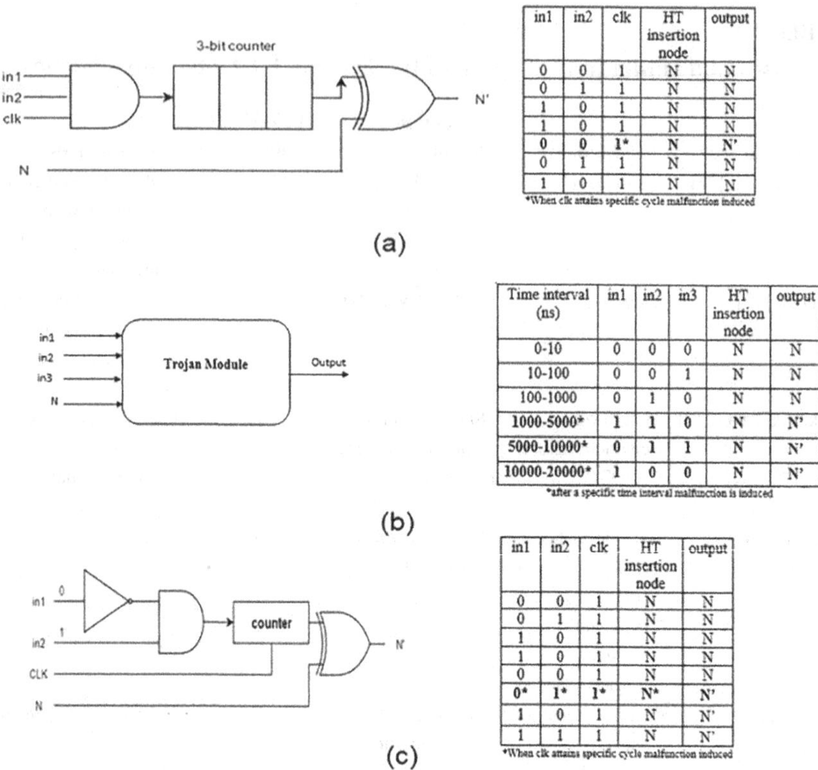

(a)

in1	in2	clk	HT insertion node	output
0	0	1	N	N
0	1	1	N	N
1	0	1	N	N
1	0	1	N	N
0	**0**	**1***	**N**	**N'**
0	1	1	N	N
1	0	1	N	N

*When clk attains specific cycle malfunction induced

(b)

Time interval (ns)	in1	in2	in3	HT insertion node	output
0-10	0	0	0	N	N
10-100	0	0	1	N	N
100-1000	0	1	0	N	N
1000-5000*	**1**	**1**	**0**	N	**N'**
5000-10000*	**0**	**1**	**1**	N	**N'**
10000-20000*	**1**	**0**	**0**	N	**N'**

*after a specific time interval malfunction is induced

(c)

in1	in2	clk	HT insertion node	output
0	0	1	N	N
0	1	1	N	N
1	0	1	N	N
1	0	1	N	N
0	0	1	N	N
0*	**1***	**1***	**N***	**N'**
1	0	1	N	N'
1	1	1	N	N'

*When clk attains specific cycle malfunction induced

FIGURE 3.1 Trojan Model with corresponding truth table: (a) Trojan-1: Produce malfunction after specified clock cycles; (b) Trojan-2: Produce malfunction at a specified time window; (c) Trojan-3: Create malfunction when triggered by 01 after specified clock cycle.

detect these time bomb Trojans through consistency checking of power signatures at different timeframes and during the runtime by means of continuous monitoring to improve the reliability of the design. Three types of Trojan models considered in the proposed scheme are briefly discussed and their effects are analyzed on the ISCAS'85 benchmark circuits.

3.3 THE PROPOSED SCHEME

The shortcomings of the techniques discussed in the previous section made the authors propose a novel self-referencing Trojan detection method to overcome the current obstacles. The proposed scheme is characterized as below:

1. It is suitable to implement the Trojan detection during the test phase in applications like application-specific ICs (ASICs), microprocessors or digital signal processors (DSPs) or FPGA circuits.

2. It is applicable to both combinational and sequential circuits. In this work, combinational circuits are considered and the same is ideal for sequential circuits.
3. This method is implemented to detect three types of Trojans discussed in the above section.
4. It is ideal for detecting malfunction-creating functional Trojans.

Figure 3.2 illustrates the proposed scheme in detail. It involves identification of specific nodes either by the TP analysis [34] or observability analysis [35] of CUT and their power measurements are compared at different time windows to check whether they are consistent or not.

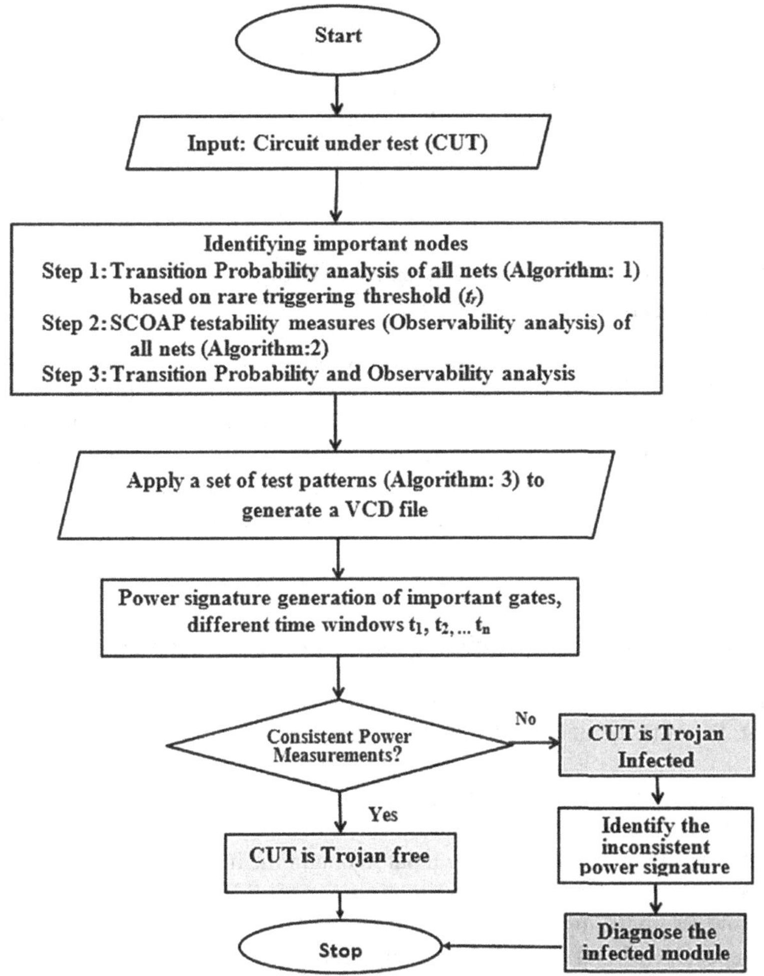

FIGURE 3.2 Proposed Scheme of HT detection and diagnosis.

3.3.1 Identification of Important Nodes

The first step in the proposed scheme is the identification of important nodes to reduce the computational complexity of the large circuits. The node identification is first done with TP analysis used in Ref. [31]. Furthermore, to reduce the time complexity, the testability measure is used by calculating the controllability and observability measures of all the gates in the circuit. Then the proposed scheme will consider the gates with both low observability and TP for power signature generation.

To reduce the complexity of computation, the proposed technique will monitor only the power signatures of the *low transition probability* (LTP) as "the gates of interest". Whereas, the malfunction created in the *medium transition probability* (MTP) and *high transition probability* (HTP) are propagated toward the output and they are easily detected by the conventional methods of Trojan detection. Following the TP measurements, testability measures [35], namely, controllability and observability, are computed. The difficulty of setting each node to *0 or 1* by assigning to the circuits primary inputs is known as controllability, whereas the difficulty in observing the circuit nodes is known as observability. The gates *observability* values are calculated to identify the nodes of interest for power signature analysis.

3.3.2 Apply the Selected Test Patterns to Toggle the Specific Node of Interest

The proposed scheme requires a set of test vectors to measure a dynamic power of any CUT. Hence, the test vector applied for power signature extraction plays a major role in Trojan detection and diagnosis process. The randomly generated test patterns need not always trigger the rarely toggling nodes. Hence, an algorithm is developed to obtain a set of patterns that toggle the specific gates of interest. Once the important gates are identified, the set of vectors to toggle these gates are obtained from the Algorithm 3.1.

Algorithm 3.1: Test set generation to toggle the specific gates of CUT

Input: CUT with the specific node of interest
Output: Set of test patterns to trigger the gates of interest

1. Read the CUT (.v) file as the input and identify the node of interest
2. Initialize the specific node value to 0
3. Backtrack toward the primary input to obtain the input pattern set
4. While backtracking determine implications of all PIs
5. Now assign the logic of specific node to 1
6. Repeat step 3 and 4 to obtain the input vectors
7. Repeat the steps 1–6 to obtain a set of input patterns for all gates of interest

3.3.3 Power Signature Generation of Important Gates at Different Time Windows

The power signatures of the specific gates of interest are generated for the selected set of input patterns at different timeframes as shown in Algorithm 3.2. Then the average power supply consumed by the important gates is compared for different time instances to confirm whether the gates in the group of interest are maintaining the same power throughout the process. In order to increase the Trojan sensitivity, this approach will be repeated by hierarchically considering the cells around the gates of interest. The relative difference of power signatures for each region is clearly observed and this eliminates intra-die systematic variation by using adjacent power signatures of the same die. On the other hand, by considering switching activity of many gates, this scheme will also eliminate the intra-die random variations. Simulation results show that the random deviations in power supply can be effectively eliminated by considering the switching of group of logic gates.

For every high activity vector, the dynamic switching power supply is larger than the power consumed by tiny Trojans. For large circuits, the variation in the measured power value due to PV is large and it differs from chip to chip. Hence, the proposed approach enhances the Trojan detection by obtaining the power signatures at low activity nodes. This scheme is referred to self-referencing, since the power signatures of low activity nodes are examined at different timeframes without referring to any other region of the same chip or golden chip.

Algorithm 3.2: Power signature extraction. Generate the power signature for the gates of interest

Input: netlist (.v) and a set of input patterns obtained from Algorithm 3.1
Output: Power signatures of the important gates

1. Read the netlist file (.v)
2. Create a value change dump (.vcd) file after compilation
3. Read the. vcd file
4. For a time window 1 (TW1)
5. //Create a power group for the important gates using a syntax
    - create_power_group -name g_1 [get_cells "gate_0 gate_1 gate_2 gate_3 gate_4"]
    - report_power -groups g_1 > g_1_tw1.power
6. Generate.fsdbfile for Waveform View
7. Generate power signatures
8. **end for**
9. Repeat step 4 for different time windows
10. **for** a group of important gates
11. //Check the consistency of power at different time instances
12. **if** (power at TW_1, TW_2, TW_3.... TW_n are consistent) **then**
13. CUT is Trojan free

```
else
Trojan Infected
end if
end for
```

3.3.4 Consistency Analysis of the Measured Power

In this process, the power signatures of the group of cells around the gates of interest are compared to make the detection and diagnosis. In the first step, the TP analysis of the CUT is done and the important gates are identified for consistency-based Trojan detection and diagnosis. From the TP-based analysis, the gates are categorized as *LTP, MTP and HTP* gates. If the discrepancy in power signature is examined at any stage of computation, then the algorithm will stop the process and report that the design is infected.

The proposed scheme primarily considers the gates of low controllability in addition to TP, in order to reduce the time complexity of the design to a large extent. To improve reliability of detection by minimizing FP and FN, the next stage considers both LTP and *LO* gates for power signature analysis. The proposed scheme will pinpoint the Trojan location exactly; since only a small group of cells are considered the node of interest, it is easy to diagnose the Trojan with the same effort of Trojan detection.

3.4 EVALUATION

The proposed scheme was tested and validated on the benchmark circuits. The scheme uses visual studio tool to identify the specific nodes of interest. Synopsys tool is used to generate the power signatures and the entire procedure is implemented using the Digilent Anvyl Spartan-6 FPGA trainer board. The experimental setup used is shown in the results and analysis section with neat figures. Evaluation is done by comparing with the existing transition-probability-based node reduction technique [31] and the performance measure is examined by the metrics like detection accuracy, detection complexity reduction and time complexity reduction.

Since the Trojan remains stealthy most of the time, the static power consumption is negligible and the dynamic power plays a role during the node toggling. Hence, in the proposed scheme, the dynamic power signatures are considered side-channel parameters for Trojan detection and they diagnose at specific nodes of the circuit. This improves the detection accuracy with reduced time complexity, which forms the main objective of this work. This section includes the results of node reduction by identifying important gates with TP analysis or the testability measures. The selection of test vectors used to measure dynamic power at different timeframes is also discussed. The proposed scheme is evaluated with the metrics of detection accuracy, detection complexity reduction and time complexity reduction. These are the major findings addressed in the below mentioned sections.

TABLE 3.2
Grouping the Gates of c499 Based on TP Estimation and LO

Group of Node	LTP ($0 < \tau/2$)	MTP ($\tau/2 < \tau$)	HTP ($\tau < 1$)	LO gates ($0 < \theta/2$)	LTP + LO gates
No. of Gates	50	40	112	40	50

Notes:
TP threshold $\tau = 0.167085$
Observability threshold $\theta = 158$

3.4.1 Simulation Results of Node Identification

The generated TP will estimate the TP at each node in the circuit. The LTP, MTP and HTP gates, which are the gates of interest for Trojan detection, are identified. Here, the *TP threshold* (τ) *is 0.167085* and hence the range of node identification is $(0 < (\tau/2)0.083542 < (\tau)\ 0.167085 < 1)$. Table 3.2 shows the number of LTP, MTP, HTP, LO and LTP + LO gates of c499 circuit. The *202* gates are categorized as *50* LTP gates, *40* MTP gates and *112* HTP gates, respectively. First priority is given to the cells in LTP, followed by MTP and HTP cells for power estimation (only if necessary). Thus, the TP estimation will reduce the complexity of the Trojan detection for the large designs. In order to reduce the algorithm complexity, the testability measures are further used to calculate the controllability and observability values of each signal/node in the circuit. The signals with low observable values are identified for power signature extraction. This will further reduce the number of gates considered for SCA and thus the time of computation and the number of power measurements are considerably reduced.

3.4.2 Results for Test Vector Generation

When the specific node of any circuit is given as input to the test vector generation algorithm, it will generate all possible input vectors that trigger. The test vector, toggling the gates of interest based on TP analysis and testability measures for c499, is shown in Table 3.3. These test patterns are applied to trigger the specific node and the corresponding dynamic power signatures are generated at different timeframes.

TABLE 3.3
Test Vector to Toggle the Specific Gates of c499

Name of the Specific Gate	Toggling Test Vectors
Gate_138, Gate_139, Gate_153	000011110000011110101010111011101000000000 & 11111111000011110101010111011101000000000
Gate_149, Gate_164	0101010111011101000011110000111000000000 & 01010101110111011111111100001111000000000

It has to be noted that a common vector toggles many important gates and hence the test set used for power measurements is made compact.

3.4.3 Power Signature Analysis at Different Time Windows

The dynamic power of Trojan-infected ISCAS'85 benchmark circuits at different time instances is measured. Once the power measurements are done for the same set of test vectors, the consistency in the measured power signatures is analyzed at regular time intervals. If there is any discrepancy among the signatures, the presence of Trojan is ensured. Once the presence of Trojan is confirmed from the power measurement, the algorithm will scan the cells with inconsistent power signatures to isolate the location of Trojan. Figure 3.3 shows the comparison effect of three types of Trojan considered in this work at various timeframes in c1355 circuit. Figure 3.4 shows inconsistent power signatures after a timeframe of *1000s* in c1355 circuit infected by the Trojan-1, whereas the Trojan-free c1355 circuit has a consistent, power signature at all timeframes.

Figure 3.5 shows the discrepancy in power signature between a timeframe of *1000* and *5000s* at MTP gates due to Trojan-2 in c1355 based on TP analysis. Thus, when a malfunction is triggered, the specific gates will have inconsistent power for the same toggling test vectors, and if triggering is withdrawn, the power consumption becomes consistent for a normal node function. Here, the algorithm first computes the consistency analysis on LTP gates and it is consistent, so the algorithm computes for MTP gates, and if they are also consistent, then the HTP gates are examined for consistency at the different time windows. It is seen that there exists inconsistency at the HTP gates of c1355 circuit. Thus, to check all gates of interest, the computational time of TP-based analysis is high. The proposed algorithm is executed on ISCAS'85 benchmark circuits at different timeframes for the same set of triggering test vectors. It is examined that the Trojan-2 shows a discrepancy in power signature at the

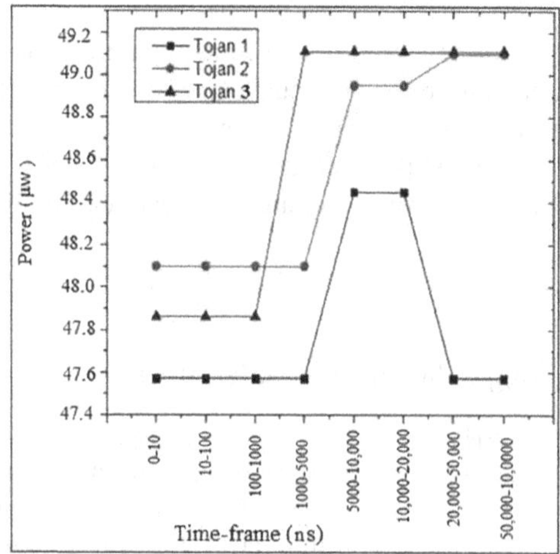

FIGURE 3.3 Trojan effects on c1355 at different time windows.

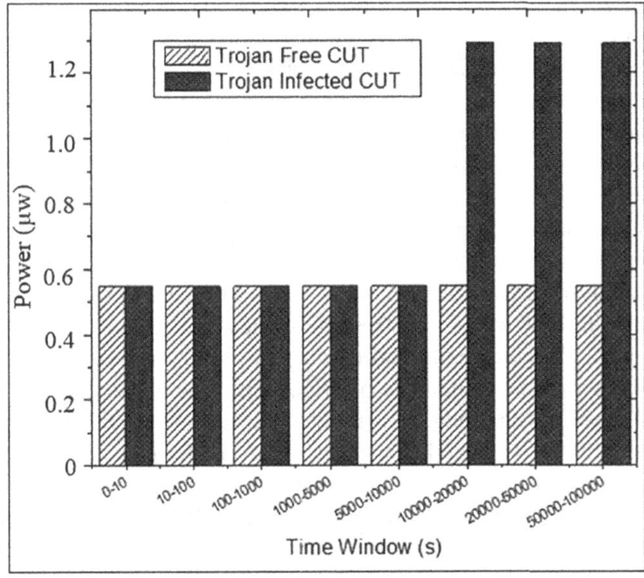

FIGURE 3.4 Inconsistent power signatures after a timeframe of 1000s due to Trojan-1 in c1355.

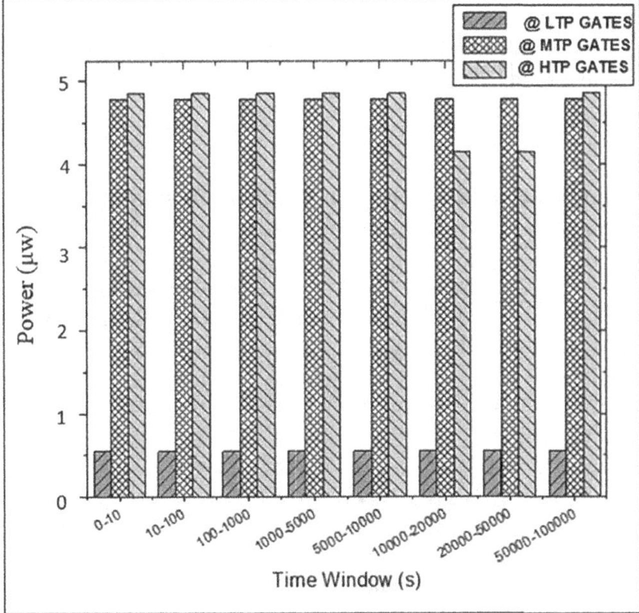

FIGURE 3.5 Inconsistent power signatures between a timeframe of 1000–5000s due to Trojan-2 in c1355.

timeframe of *10000–20000s and 20000–50000s*; on other timeframes, there is a consistent power measurement; this is because Trojan-2 will create a malfunction at a specific timeframe and retain the original circuit function after the particular time period. Whereas Trojan-3 has its power discrepancy in timeframe between *1000 and 5000s*, because once the Trojan-3 is triggered, it creates malfunction continuously. The main objective of the proposed scheme is to detect and diagnose the HT with a minimum power measurement and short time for computation even for large circuits. This is achieved by considering the specific gates as important gates among the millions of gates in a circuit. The proposed scheme identifies the gates of interest either through TP analysis or computing SCOAP testability measures of any circuit or the combination of both TP analysis and testability measures. By identifying the specific gates of interest, test complexity and number of power measurements are reduced and this reduces the time complexity of Trojan detection and diagnosis.

Table 3.4 describes the reduction in detection complexity by the TP, testability measure and both TP + testability measure analysis. By considering only a group of gates among the millions of gates for power signature extraction will reduce the algorithm complexity to a great extent. By applying TP analysis even for a large circuit like c5315, among *2307* gates, only *1980* gates are considered for SCA and thus it reduces the algorithm complexity of power signature extraction by *14.17%*. This is obtained by considering the LTP, MTP and HTP gates as gates of interest, whereas the gates with zero TPs are not considered as they remain untoggled for any set of input patterns, and by removing these untoggling gates, the computational time is highly reduced.

In some of the circuits like c499, c2670, c3540, c5315 and c6288, there are common gates of LTP and LO, which will further reduce the computational complexity. In circuits, like c17, c499 and c6288, the gate of specific interest reduction is *0%* because, in these circuits, there do not exist any untoggled gates, and hence, in these

TABLE 3.4
Detection Complexity Reduction (ΔD) Using TP, Testability Measure and LTP + LO Gates

Circuits	Number of Gates	TP Gates				LO Gates	LTP + LO Gates	ΔD with TP Gates (%)	ΔD with LO Gates (%)	ΔD with LTP + LO Gates (%)
		LTP	MTP	HTP	Total TP Gates					
c17	6	2	2	2	6	3	5	0	50.00	16.66
c432	160	45	9	45	99	27	72	38.12	83.12	55.00
c499	202	50	40	112	202	40	50	0	80.19	75.24
c1355	546	79	63	40	182	168	247	66.66	69.23	54.76
c1908	880	115	52	239	406	161	276	53.86	81.70	68.63
c2670	1269	504	152	395	1051	297	704	17.17	76.59	44.52
c3540	1669	814	84	338	1236	203	901	25.94	87.83	46.02
c5315	2307	773	476	731	1980	821	1211	14.17	64.41	47.51
c6288	2416	15	334	2067	2416	574	584	0	76.24	75.62
Avg.	*1051*	*266*	*135*	*441*	*842*	*255*	*450*	*24*	*74*	*53.79*

circuits, the total number of important gates is equal to the original number of gates in the circuit. An average of all ISCAS'85 circuits is considered, such that for an average of *1051* gates, *detection complexity* (ΔD) is reduced to *24%* with TP gates, *74%* with LO gates and *53.79%* with LTP + LO gates. The algorithm attains a high ΔD by considering LO gates, but in order to obtain a high detection accuracy without any FP and FN detection rates, LTP + LO gates are considered for the Trojan detection. The analysis performed for all the three Trojans is considered in this work separately. LTP + LO-based gate selection for Trojan detection and diagnosis has better detection accuracy even for a randomly inserted Trojans than the other two schemes. Detection accuracy is chosen as an evaluating metric by inserting Trojan models at random nodes, at low activity nodes and medium activity nodes as in Equation (3.1). Whereas, *true positive* (TP) is the number of Trojans identified correctly, FP is the number of Trojans incorrectly identified, *true negative* (TN) is the number of Trojan-free circuits identified correctly and FN is the number of incorrectly identified as Trojan-free circuits.

Figure 3.6 shows the detection accuracy analysis of the proposed algorithm on three different locations. The analysis is performed for all the three types of Trojans considered in this work separately. In case of randomly inserted Trojans, the detection accuracy is relatively small while compared to other two cases. This is because the Trojan may be embedded in the zero toggling gates, which do not affect the manifestational properties of the design. It is observed that the LO-based gate selection for the Trojan detection and diagnosis has better detection accuracy even for a randomly inserted Trojans than the TP-based gate selection scheme; this is because the proposed algorithm identifies the Trojan-infected and -noninfected circuits without referring to any golden reference.

$$\text{Detection Accuracy (\%)} = \frac{\text{TP} + \text{TN}}{\text{TP} + \text{TN} + \text{FP} + \text{FN}} \times 100 \tag{3.1}$$

Hence, to reduce the detection complexity further, the testability measures of ISCAS'85 benchmark circuits are computed and tabulated in Table 3.4. Here the total number of TP gates of some circuits is not equal to the total gates in the circuit;

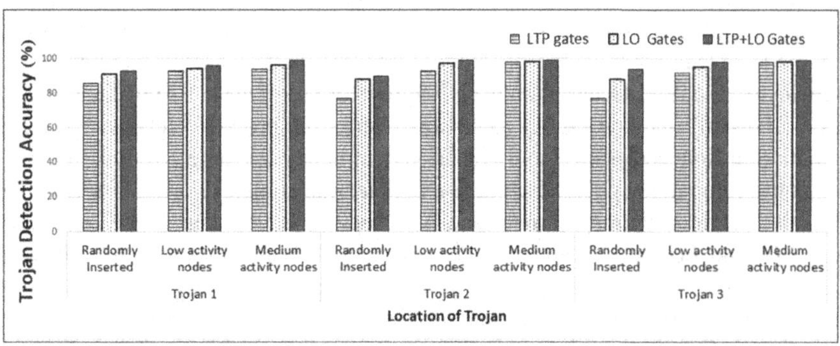

FIGURE 3.6 Detection accuracy analysis of the proposed algorithm at three different locations.

TABLE 3.5
Time Complexity Reduction (Δt) Using TP, Testability Measure and LTP + LO Gates

		Computation Time (μs)				Time Complexity (Δt) Reduction (%)		
Circuit	Gates	Conventional Method	TP	LO	LTP + LO	LTP	LO	LTP + LO
c17	6	192	179	174	174	6.77	9.38	9.38
c432	160	267	139	126	132	47.94	52.81	50.56
c499	202	293	155	125	145	47.10	57.34	50.51
c1355	546	386	186	145	176	51.81	62.44	54.40
c1908	880	423	139	131	137	67.14	69.03	67.61
c2670	1269	444	149	146	149	66.44	67.12	66.44
c3540	1669	489	155	149	153	68.30	69.53	68.71
c5315	2307	513	154	136	142	69.98	73.49	72.32
c6288	2416	484	137	149	155	71.69	69.21	67.98
Avg.	*1051*	*388*	*155*	*142*	*151*	*55*	*60*	*54.6*

it is because the node selection algorithm will exclude the redundant gates while processing. It is observed that, for a c5315 circuit among the *2307* gates, only exists there *821* low observability gates, and this results in reducing the detection complexity by *64.41%*. Thus, the algorithm reduces computation by identifying important gates using testability measures. In order to get a zero FP and FN Trojan detection rates, the proposed scheme is further improvised by considering both LTP and LO gates as gates of interest and it is clear from Table 3.4 that the detection complexity reduction obtained was *30.90%* for c5315 circuit. Since the proposed algorithm considers only specific gates for SCA, time complexity is highly reduced than the conventional method.

Table 3.5 compares the reduction in time complexity on the benchmark circuits for the proposed algorithm and conventional method of power signature analysis. It is observed that, for an average of *1051* gates, the time required for the conventional scheme is *388ms*, whereas the proposed TP-based analysis has a reduction in time complexity of *55%*, LO-based analysis acquires *60%* and LTP + LO-based analysis has a *54.6%* of time complexity reduction. Hence, the proposed scheme has a high reduction in time complexity and detection complexity for large circuits. Figure 3.7 shows the comparison chart of reduction in detection complexity and time complexity with detection accuracy of the proposed scheme. It is clear that, while considering both *LTP* and *LO* gates, comparatively high detection accuracy is achieved with a reduced *time complexity (ΔT)*. Although the detection complexity reduction is more than the other two-gate selection schemes, this is to test all possible gates for power signature analysis. Thus, among the three-gate selection schemes, the LTP + LO gates for the proposed scheme give a better detection accuracy with acceptably minimal time complexity.

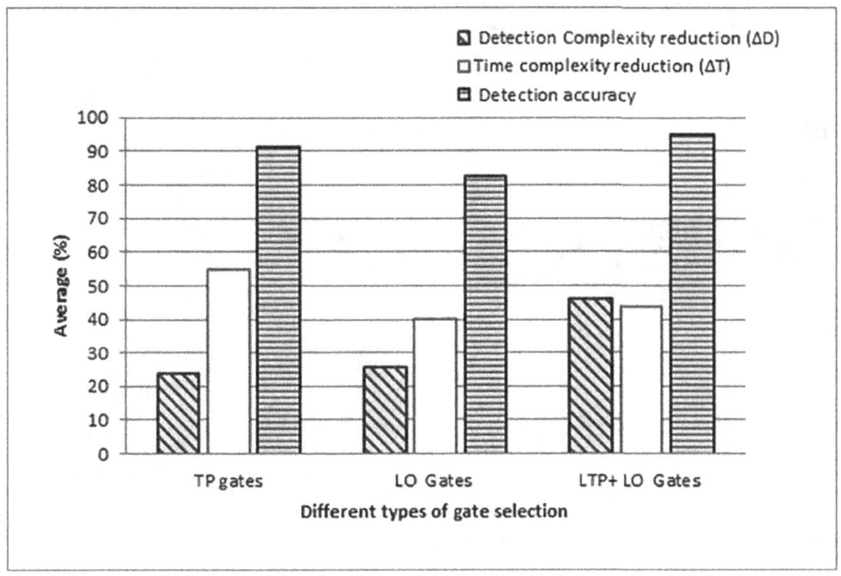

FIGURE 3.7 Decision analysis on ISCAS'85 circuits.

3.5 EVALUATION OF THE PROPOSED NODE SELECTION SCHEME

The proposed timeframe-based self-referencing for HT detection and diagnosis scheme is evaluated on ISCAS'85 benchmark circuits and the percentage of node reduction is compared with the existing transition-probability-based node reduction technique as shown in Figure 3.8. It is observed that the proposed scheme identifies less number of nodes than the existing scheme. The proposed algorithm reduces the

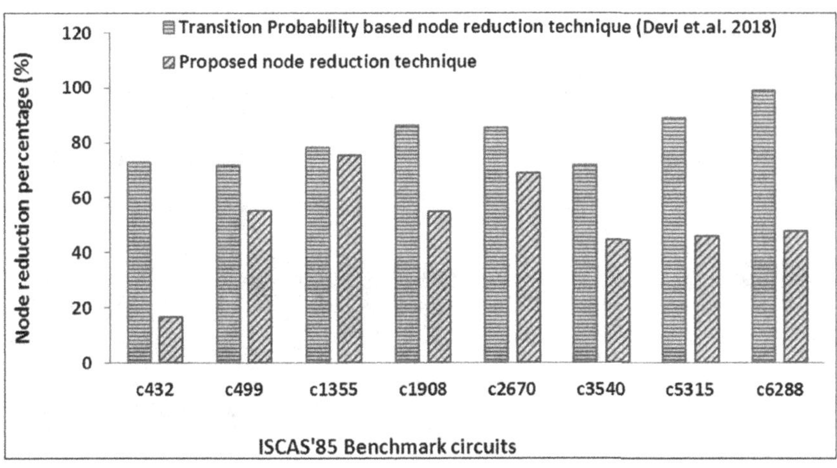

FIGURE 3.8 Comparison of proposed scheme with TP-based node reduction scheme.

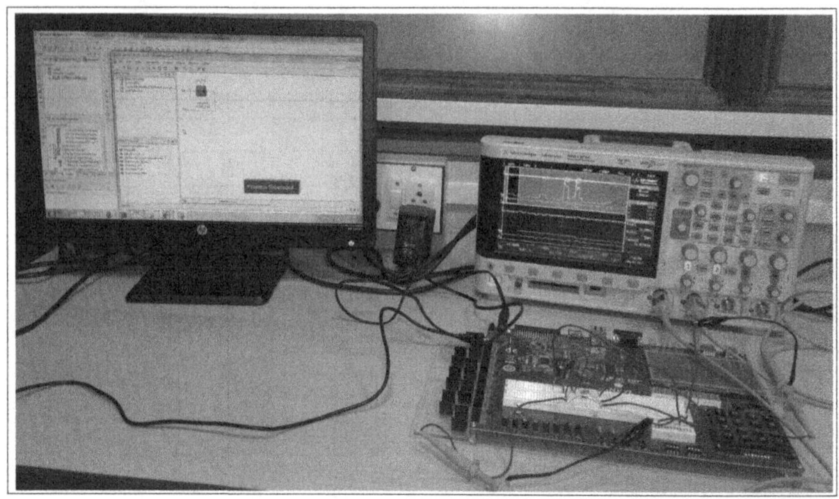

FIGURE 3.9 Hardware setup for self-referencing Trojan detection and diagnosis.

computational complexity of the scheme as well as time complexity of an average of *54.6%*. The above observation justifies the improvement of the proposed scheme over the transition-probability-based node reduction technique [31].

3.5.1 Implementation of the Proposed Scheme

The timeframe-based HT detection and diagnosis is implemented on the Anvyl Spartan-6 FPGA trainer board, as shown in Figure 3.9. Figure 3.10 shows the power signature generation model designed to analyze the differential power at different timeframes. The experimental setup includes a Spartan-6 FPGA, a MSO and an analog module to measure a differential voltage.

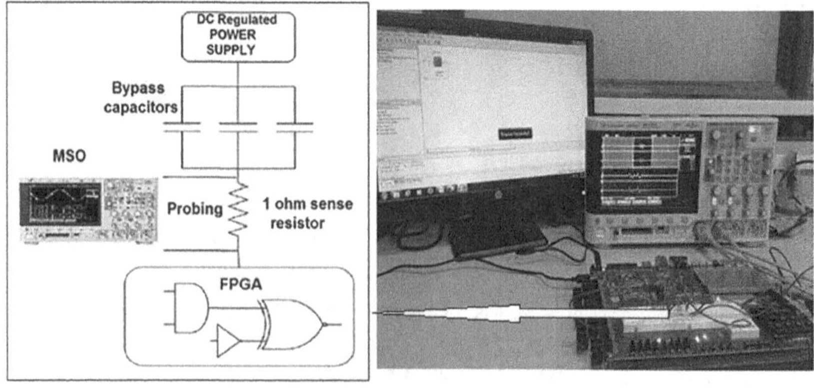

FIGURE 3.10 Power signature module in the hardware setup.

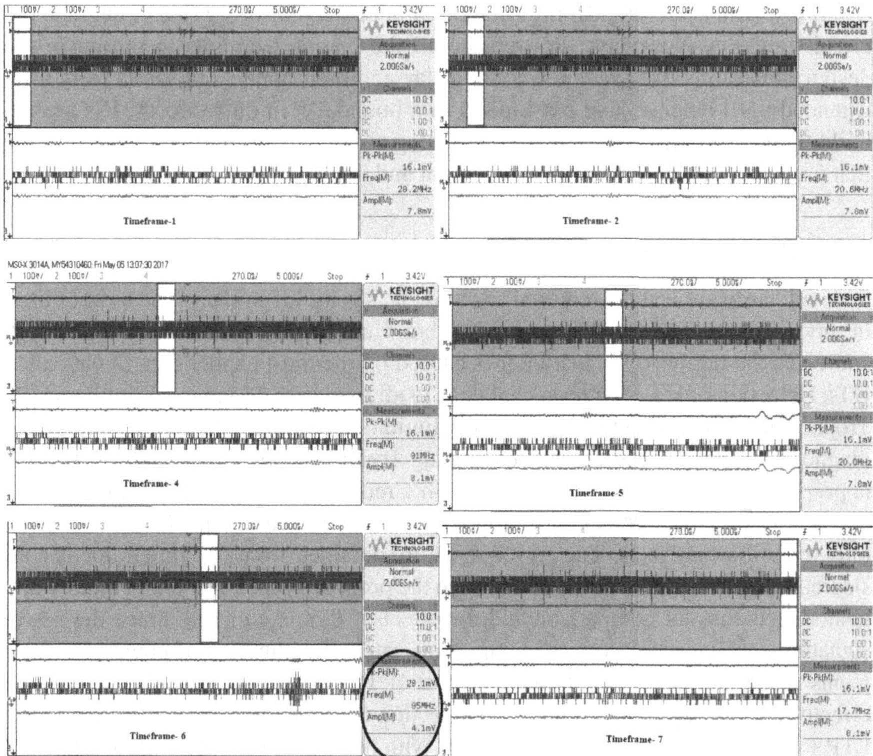

FIGURE 3.11 Peak-to-peak voltage difference and amplitude at different timeframes of c5315 circuit.

A bypass capacitor in an analog design will remove the alternative current (AC) noises in the direct current (DC) signal so that a pure DC signal is obtained without any AC ripple. From the voltage difference across the shunt resistor, the power signatures are extracted at different timeframes and their consistency is studied. In Figure 3.11, the consistency analysis of differential voltage at various timeframes (1,2,4,5,6 &7) and the peak-to-peak voltage difference, amplitude and its corresponding frequency captured in the oscilloscope are shown. It is observed that, in the timeframe 6, the peak-to-peak voltage difference, amplitude and its frequency are inconsistent and this ensures the presence of Trojan in the CUT.

3.5.2 Outcome of This Work

Experimental and simulation results have been analyzed to assess the outcomes of the proposed work. Our results have shown the scope and context of the HT detection and diagnosis without referring to the golden chip. Since the power signatures are checked for consistency at different timeframes, the proposed self-referencing scheme is free from false-positive and false-negative results of Trojan detection. The HT in digital and analog circuits will create a malfunction to destroy the circuit reliability, leak secret logical information, produce denial-of-service or degrade the circuit function. Nowadays, ICs replace all the complex circuits to minimize the design constraints like area, power and delay and their application starts from toys, extended until defense, and not limited to a boundary. In other words, ICs are the building blocks of all digital as well as analog circuits to ensure the trustworthiness that proposed self-referencing HT detection and diagnosis, which is necessary.

3.6 CONCLUSION

Secured hardware is most required to build any strong IoT system. The devices used in any IoT system should be secured to build any smart environment. This work addresses the hardware security issues caused by the most challenging threat called HTs. Detection of HT is an open challenge to all hardware designers. The FP and FN Trojan detections due to PV are the major limitation in the existing HT detection and diagnosis schemes. Hence, a self-referencing Trojan detection and diagnosis scheme is proposed to nullify the effect of PV with a reduced computation complexity. The complexity of Trojan detection in large circuits is reduced to *54.6%* by considering only the important gates. The power signature obtained from one timeframe is considered a reference for the other, and hence the proposed scheme does not require any golden signature for comparison and the effect of PVs is nullified. Since the power signatures are measured only for important gates, the numbers of power measurements are reduced and this reduces the cost of Trojan detection and diagnosis more effectively. The algorithm attains high detection coverage of *99%,* while considering LTP + LO gates as critical gates of interest with the reduced average time complexity of *54.6%* on ISCAS'85 benchmark circuits. Whereas, the Trojan detection complexity is highly reduced to *53.79%* while considering important gates as a node of interest.

ACKNOWLEDGMENT

This work was supported by the Defence Research and Development Organization (DRDO), New Delhi, "ERIP/ER/1503187/M/01/1582".

REFERENCES

1. Rajendran, S. R. (2020). Security Challenges in Hardware Used for Smart Environments. *Internet of Things and Secure Smart Environments: Successes and Pitfalls* (pp. 363–381). doi: 10.1201/9780367276706-9
2. Haque, A. B., Bhushan, B., & Dhiman, G. (2021). Conceptualizing smart city applications: requirements, architecture, security issues, and emerging trends. Expert Systems. *2021* (pp. 1–23).
3. Bhushan, B., Sahoo, C., Sinha, P., & Khamparia, A. (2021). Unification of blockchain and Internet of Things (BIoT): requirements, working model, challenges and future directions. Wireless Networks, 27(1), 55201390.
4. Saxena, S., Bhushan, B., & Ahad, M. A. (2021). Blockchain based solutions to secure IoT: background, integration trends and a way forward. Journal of Network and Computer *Applications*, 181, 103050.
5. Ranjani, R. S., & Devi, M. N. (2017). Malicious hardware detection and design for trust: an analysis. Elektrotehniski Vestnik, 84(1/2), 7.

6. Rajendran, S. R., Mukherjee, R., & Chakraborty, R. S. (2020, November). SoK: Physical and Logic Testing Techniques for Hardware Trojan Detection. In Proceedings of the 4th ACM Workshop on Attacks and Solutions in Hardware Security (pp. 103–116).
7. Tehranipoor, M., & Koushanfar, F. (2010). A survey of hardware trojan taxonomy and detection. IEEE Design & Test of Computers, 27(1). doi:10.1109/MDT.2010.7.
8. Li, M., Davoodi, A., & Tehranipoor, M. (2012, March). A sensor-assisted self-authentication framework for hardware Trojan detection. In Design, Automation & Test in Europe Conference & Exhibition (DATE), *2012* (pp. 1331–1336). IEEE. doi: 10.1109/DATE.2012.6176698.
9. Chakraborty, R. S., Narasimhan, S., & Bhunia, S. (2009, November). Hardware Trojan: threats and emerging solutions. In High Level Design Validation and Test Workshop, 2009. HLDVT 2009. IEEE International (pp. 166–171). IEEE. doi: 10.1109/HLDVT.2009.5340158.
10. Govindan, V., & Chakraborty, R. S. (2018). Logic Testing for Hardware Trojan Detection. In *The Hardware Trojan War* (pp. 149–182). Springer, Cham. doi:10.1007/978-3-319-68511-3_7.
11. Salmani, H. (2018). Trusted Testing Techniques for Hardware Trojan Detection. In *Trusted Digital Circuits* (pp. 109–119). Springer, Cham. doi: 10.1007/978-3-319-79081-7_8.
12. Salmani, H., & Tehranipoor, M. (2012). Layout-aware switching activity localization to enhance hardware Trojan detection. IEEE Transactions on Information Forensics and Security, 7(1), 76–87. doi: 10.1109/TIFS.2011.2164908.
13. Salmani, H., Tehranipoor, M., & Plusquellic, J. (2012). A novel technique for improving hardware trojan detection and reducing trojan activation time. IEEE Transactions on Very Large Scale Integration (VLSI) Systems, 20(1), 112–125. doi: 10.1109/TVLSI.2010.2093547.
14. Amelian, A., & Borujeni, S. E. (2018). A side-channel analysis for hardware trojan detection based on path delay measurement. Journal of Circuits, Systems and Computers, 27(09), 1850138. doi: 10.1142/S0218126618501384.
15. Hiramoto, Y., & Ohtake, S. (2018, July). A method of hardware-Trojan detection using design verification techniques. In *Conference on Complex, Intelligent, and Software Intensive Systems* (pp. 978–987). Springer, Cham. doi:10.1007/978-3-319-93659-8_91.
16. Agrawal, D., Baktir, S., Karakoyunlu, D., Rohatgi, P., & Sunar, B. (2007, May). Trojan detection using IC fingerprinting. In Security and Privacy, 2007. SP'07. IEEE Symposium on (pp. 296–310). IEEE. doi: 10.1109/SP.2007.36.
17. Rad, R., Plusquellic, J., & Tehranipoor, M. (2010). A sensitivity analysis of power signal methods for detecting hardware Trojans under real process and environmental conditions. IEEE Transactions on Very Large Scale Integration (VLSI) Systems, 18(12), 1735–1744. doi: 10.1109/TVLSI.2009.2029117.
18. Wei, S., & Potkonjak, M. (2014). Self-consistency and consistency-based detection and diagnosis of malicious circuitry. IEEE Transactions on Very Large Scale Integration (VLSI) Systems, 22(9), 1845–1853. doi: 10.1109/TVLSI.2013.2280233.
19. Jin, Y., & Makris, Y. (2008, June). Hardware Trojan detection using path delay fingerprint. In Hardware-Oriented Security and Trust, 2008. HOST 2008. IEEE International Workshop on (pp. 51–57). IEEE. doi: 10.1109/HST.2008.4559049.
20. Potkonjak, M., Nahapetian, A., Nelson, M., & Massey, T. (2009, July). Hardware Trojan horse detection using gate-level characterization. In Proceedings of the 46th Annual Design Automation Conference (pp. 688–693). ACM. doi: 10.1145/1629911.1630091.
21. Narasimhan, S., Du, D., Chakraborty, R. S., Paul, S., Wolff, F., Papachristou, C., and Bhunia, S. (2010, June). Multiple-parameter side-channel analysis: a non-invasive hardware Trojan detection approach. In Hardware-Oriented Security and Trust (HOST), 2010 IEEE International Symposium on (pp. 13–18). IEEE. doi: 10.1109/HST.2010.5513122.

22. Narasimhan, S., Wang, X., Du, D., Chakraborty, R. S., & Bhunia, S. (2011, June). TeSR: a robust temporal self-referencing approach for hardware Trojan detection. In Hardware-Oriented Security and Trust (HOST), 2011 IEEE International Symposium on (pp. 71–74). IEEE. doi:10.1109/HST.2011.5954999.
23. Bazzazi, A., Shalmani, M. T. M., and Hemmatyar, A. M. A. (2017). Hardware Trojan detection based on logical testing. Journal of Electronic Testing, 33(4), 381–395. doi: 10.1007/s10836-017-5670-0.
24. Bao, C., Forte, D., and Srivastava, A. (2016). On reverse engineering-based hardware Trojan detection. IEEE Transactions on Computer-Aided Design of Integrated Circuits and Systems, 35(1), 49–57. doi: 10.1109/TCAD.2015.2488495.
25. Chen, X., Wang, L., Wang, Y., Liu, Y., and Yang, H. (2017). A general framework for hardware trojan detection in digital circuits by statistical learning algorithms. IEEE Transactions on Computer-Aided Design of Integrated Circuits and Systems, 36(10), 1633–1646. doi: 10.1109/TCAD.2016.2638442.
26. M. Tehranipoor, A. Nahiyan, D. Forte, Hardware trojan detection through information flow security verification, published on November 14, 2019.
27. Prabhat Mishra and Yangdi Lyu, Delay-based Side-channel Analysis for Trojan Detection, U.S. Utility Patent Application Serial No. 17/085,213, filed October 30, 2020.
28. D. Forte, S. Chowdhury, F. Ganji, N. Maghari, Detection of recycled integrated circuits and system-on-chips based on degradation of power supply rejection ratio, published on March 18, 2021.
29. Du, D., Narasimhan, S., Chakraborty, R. S., & Bhunia, S. (2010, August). Self-referencing: a scalable side-channel approach for hardware Trojan detection. In *International Workshop on Cryptographic Hardware and Embedded Systems* (pp. 173–187). Springer, Berlin, Heidelberg. doi: 10.1007/978-3-642-15031-9_12.
30. Salmani, H. (2017). *COTD: Reference-free Hardware Trojan Detection in Gate-level Netlist.* Howard University Washington United States. doi: 10.1109/TIFS.2016.2613842.
31. Devi, N. M., Jacob, I. S., Ranjani, S. R., & Jayakumar, M. (2018). Detection of malicious circuitry using transition probability based node reduction technique. *TELKOMNIKA*, 16(2), 573–579. doi:10.12928/TELKOMNIKA.v16i2.6812.
32. Maruthi, V., Balamurugan, K., & Mohankumar, N. (2020, July). Hardware trojan detection using power signal foot prints in frequency domain. In 2020 International Conference on Communication and Signal Processing (ICCSP) (pp. 1212–1216). IEEE.
33. Gayatri, R., Gayatri, Y., Mitra, C. P., Mekala, S., & Priyatharishini, M. (2020, June). System level hardware trojan detection using side-channel power analysis and machine learning. In 2020 5th International Conference on Communication and Electronics Systems (ICCES) (pp. 650–654). IEEE.
34. Rabaey, J. M., Chandrakasan, A. P., & Nikolic, B. (2002). *Digital Integrated Circuits* (Vol. 2). Englewood Cliffs, NJ: Prentice hall. Retrieved from: https://web1.unirc.it/documentazione/materiale_didattico/599_2008_90_1324.pdf.
35. Bushnell, M., & Agrawal, V. (2004). *Essentials of Electronic Testing for Digital, Memory and Mixed-Signal VLSI Circuits* (Vol. 17). Springer Science & Business Media. Retrieved from: https://mtechlib.files.wordpress.com/2016/07/essentials-of-electronic-testing-for-digital-memory-and-mixed-signal-vlsi-circuits.pdf.

4 Deep-Learning-Empowered Edge Computing-Based IoT Frameworks

Mithra Venkatesan, Anju V. Kulkarni, and Radhika Menon
Dr. D. Y. Patil Institute of Technology, Pune, India

CONTENTS

4.1 INTRODUCTION

Internet of Things (IoT) has emerged to be multidisciplinary encompassing different technologies, varied capabilities of devices covering wide spectrum of applications. The robust architecture involves interconnection across physical and virtual things based on present and ever evolving communication technologies. Chen et al. [1] detailed various standard IoT frameworks developed covering design issues in terms of IoT hardware and software components involving different application domains

DOI: 10.1201/9781003219620-4

such as smart cities, healthcare, agriculture etc. Tournier et al. [2] discussed the increased usage and proliferation of IoT devices resulting in huge volume of content in terms of image, video, audio and speech. Dung and Nguyen [3] illustrated deep neural network-based machine learning models that are widely being used for making predictions or forecast involving huge volume of data. Almiania et al. [4] specified the application of deep learning models to IoT devices and performance enhancement of complex sensing and recognition tasks with increased interactions between human and the physical environment around. Deep learning models are capable of extracting data with precision from IoT devices under complex environmental conditions. With emergence of cloud computing, massive datasets were stored in cloud and processing of datasets was being done. However, the processing time and latency can be further reduced with the emergence of edge computing. Shi et al. [5] explored nuances on how computation and servicing have moved closer to the consumers on the edge and edge computing is useful. Amanullaha et al. [6] discussed the applicability of deep learning models with IoT and information centric networking where all these features combined together will form a useful amalgamation. This can find application in a variety of domains such as healthcare, agriculture, assisted living, asset tracking, surveillance and many more.

The prediction of improved performance metrics because of the convergence of these three techniques or technologies namely IoT, edge computing and deep learning models have been the main motivation for work in this chapter. Furthermore, the scope of its application in a variety of domain has also been a contributory factor for work.

The major contributions of the paper have been the development of an IoT framework powered by deep learning and edge computing. Towards the same, extensive survey of combination of three technologies namely IoT, edge computing and deep learning models has been done. Furthermore, a case study on IoT-based predictive edge analytics framework for diagnostic care of breast cancer patients has been proposed based on the framework developed. The methodology to be implemented in this framework is outlined.

This chapter has been organised into relevant sections. The second section gives an overview of various IoT frameworks. Deep learning models have inherent ability to learn and predict from large volumes of data collected. The mathematical background behind deep learning models is discussed in Section 4.3. This ability is exploited through deep learning-based IoT frameworks which has been detailed in Section 4.4. Subsequently, deep learning empowered by edge computing is discussed in Section 4.5. Section 4.6 brings out insights on deep learning-based IoT frameworks which employ edge computing techniques. Following this, Section 4.7 presents a case study on an IoT-based predictive edge analytics framework for diagnostic care of breast cancer patients. The case study details basic theme of the study, objectives and relevance of the work, proposed methodology, technical novelty and utility along with societal benefit through different sub sections. Finally, Sections 4.8 and 4.9 summarises the conclusions and future scope of the chapter. References section lists the references used in this chapter.

4.2 IoT FRAMEWORKS

Due to the emergence of concept of IoT, there would be significant change in the communication and networking scenario and this is bound to have an impact socially. There has been significant work on definition and design of different open architectures for IoT.

Veletic et al. [7] discussed the different available IoT architectural frameworks in terms of hardware and software. The paper also explored the various domains where the developed frameworks can be used such as in healthcare, agriculture as well as in nanoscale applications. The paper presented initial results with the usage of two protocols namely Message Queuing Telemetry Transport and Time Slotted Channel Hopping. The validation of the defined system was done in hardware frame of Open Mote and the operating systems used were Open Wireless Sensor Networks (WSN) as well as Contiki.

IoT involves millions of interconnected devices which are capable of communication amongst them in a dynamic fashion. The devices could range from very small devices to advanced processors with high processing ability. The major issue is with respect to managing all devices without any security related problems. Mohapatra et al. [8] discussed the challenges with respect to management, security and privacy in managing various aspects to IoT. The major work in this chapter was development of standard architecture based on cloud technology involving various aspects of IoT.

Though different architectures are available for particular applications, a unified solution is essential for IoT. Akasiadis et al. [9] described an IoT platform which supports different layers of communication protocols and is made of different open source frameworks. This platform developed offers a single approach for user access control and services that supports many communication protocols in application layer. The developed platform is being tested and found to be easily deployable in variety of infrastructure capable of computing. The different salient features of the platform were analysed and the outcomes of experiment with different communication protocol were being tested.

Sethi and Sarangi [10] proposed an original taxonomy for IoT technologies and discussed the importance of the technology. Furthermore, the paper also explored the applications where the technology could be made use of and specifically with respect to elderly and differently abled. The paper gave a comprehensive and complete coverage of all the concepts.

There has been a wide variety of applications where IoT has been deployed such as in smart homes, smart cities and health monitoring applications. Alharbi and Aspinall [11] illustrated the flaws that exist in home monitoring smart cameras and brought out the impact on security and privacy of users. The five major parts of the smart camera system was being covered in the framework with a set of specific test cases. The flaws or vulnerabilities were being found in the framework. The vulnerabilities found indicate that IoT devices were being utilised without regard to the user's security and privacy.

Proactive personal ehealth is a domain where IoT is widely used. Lake et al. [12] tried to address some major challenges that must be overcome with respect to privacy and security. This chapter also discussed the issues, use case scenarios and proposed a secure architecture framework. It further detailed the different standard organisations and the important role they play towards resulting in bloom of healthcare.

There has been a variety of applications where IoT has been applied. Different applications have varied requirements and different performance expectations. Because of the heterogeneity of the networks, different types of network architectures of IoT existed with different functionalities and components. However, a common framework having ability of cover different application domains acting as reference architecture could be of immense use as it could be modified according to application or requirement. Krčo et al. [13] presented a comprehensive overview of various activities in Europe towards defining a universal framework and how it can be utilised and the possible view of these endeavours.

There are different key enablers of IoT and there are some vital enablers along with main elements of IoT such as machine to machine communication, mobility of devices, discovery of devices and communication protocols which are appropriate for IoT background. Katole et al. [14] detailed the challenges in IoT and talks on the model for IoT-based solution for navigation and give a synopsis of projected architecture of framework of IoT.

There are different types of devices which are connected together through internet. Apart from the variety of devices, plethora of services can be availed through these devices. Due to the heterogeneous nature of these devices these could be different challenges which could arise. Bharati [15] discussed these issues related to coordination, control, management of data, interoperability and security in IoT. The paper discussed the details on the sensing devices which would keep track of recording and sending data towards analysis and making of decisions. The work also elucidated the key challenges and concerns in IoT.

Security is of prime importance and is a major issue of concern in IoT networks. Mahmoud et al. [16] provided a comprehensive survey on various important points of concern with respect to IoT security. Any IoT architecture has primarily three layers namely perception, network and application layer. Security is to be addressed and resolved in each of these layers by implementing countermeasures. This chapter gave a complete description of different security principles, challenges and future directions towards security in IoT networks.

Kim Abbas Ahmad et al. [17] described an original approach based on service for an IoT-based automated framework with an aim to mitigate all issues related to cost, coordination and scalability in terms of development of IoT devices and getting details on its design and implementation. The model developed is made of different kinds of testing which is suitable for IoT devices. To get into the details and provide a conceptual overview, the technical advancements are compared with traditional methods along with examples.

IoT brings together digital and physical objects on a same platform which calls in for security, authentication and data privacy. These characteristics can be strengthened by introduction of block chain in IoT. Bhushan et al. [18] detailed such a blockchain-based IoT (BIoT) with its architectural design along with security aspects and applications. The paper also described some centralised IoT challenges which have been addressed by BIoT. The future directions of BIoT have also been elucidated towards employing the technology in next-generation BIoT applications.

Among the different challenges in IoT-based systems, security is a major challenge. To overcome this challenge, block chain-based IoT offers a viable solution. Saxena et al. [19] discussed this new alternative on how improved security is accomplished by incorporation of block chain in IoT. Also, possible applications of BIoT were also being detailed.

Among the varied applications of IoT, the most significant is in healthcare industry. Goyal et al. [20] introduced the concept of Internet of Health Things in detail and how the interaction of various medical gadgets can be accomplished through IoT. The different attributes such as privacy and security was also being addressed in this work. Other than IoT, the scope of usage of other upcoming technologies such as augmented reality, big data and cloud computing in healthcare was also explored in this work. Finally, the work also presented open challenges in this field of study.

In parallel to the concept of IoT, WSN is another emerging technology which finds application in monitoring and controlling industrial equipment. Bhushan et al. [21] discussed the distinct feature of WSNs in factory automation along with industrial application of WSN. The reliability issue of industrial WSN was explored along with different types of security attack. Varied design considerations for Medium Access Control (MAC) protocols design were being detailed. The chapter concluded with brief on unresolved issues and open challenges during the design of the protocol.

Thus, millions of devices can be connected through such frameworks and huge volume of data can be collected from the various devices which are connected together. Such massive data could be of tremendous use where the data can be processed and subsequently analysis of the data could be done leading to prediction or some forecast based on the processed data. The main tool used towards building such predictive models is deep learning. Thus, the next upcoming sections details work on deep learning models and deep-learning-based IoT frameworks.

4.3 DEEP LEARNING MODELS

Machine learning has been successful in building models with intelligence and which tries to emulate human brain. There are different types of learning such as supervised learning, unsupervised learning, sequential learning, reinforcement learning, transfer learning and federated learning. Depending on the type of learning, there are various algorithms and network models which have been developed. Among the supervised models, artificial neural networks are powerful tool for classification, regression and building predictive models. There are different algorithms based on which neural network models are built. Among the different models deep learning models are gaining attention because of improved accuracy with different size of datasets. Convolutional neural networks (CNN) is special family of deep neural networks that contain convolution layers.

Deep learning models are based on CNN. These networks involve increased number of connections other than weights. The architecture of CNN is capable of regularisation as well as translation invariance. This model is capable of extracting the features which characterise the input. In a classical CNN, the initial analysis will be made of alternating convolution and sub-sampling operations. The final stage of the architecture is made of generic multilayer network. Ghosh et al. [22] described the basic architecture and mathematical model of CNN is derived based on back propagation algorithm and its model. The back propagation model consists of feed forward and backward pass. In forward pass, mean square error is being considered and being found out for *N* training samples.

$$Err^N = 1/2\sum_{a=1}^{N}\sum_{b=1}^{c}\left(Tar_b^a - yact_b^a\right)2 \tag{4.1}$$

where *t* is the target output and *y* is the actual output. When there are many classes then *c* comes into existence where *c* is the number of classes. The error parameter is being found for each training sample. If *u* be the input, for every current layer l, the output *x* is being calculated. Here the input *u* is being calculated using weights *w* and bias *b*. The input is subjected to an activation function to produce an output.

$$\text{op}^l = f\ (\text{inp}^l) \tag{4.2}$$

$$\text{With inp}^l = \text{We}^l\text{inp}^{l-1} + \text{bi}^l \tag{4.3}$$

Depending upon the input and output, the activation function is being chosen. There are linear and non-linear activation functions.

In the backward pass, the errors are propagated back into the network towards improved network performance. The backward propagation happens through the biases and the weights. The change in weights is found which will be a function of change in error with respect to weight as well as learning rate. The weights are subsequently appended or updated accordingly to aid in improvised performance.

$$\frac{\partial Err}{\partial bi} = \frac{\partial Err}{\partial u}\frac{\partial u}{\partial bi} = \delta \tag{4.4}$$

$$\delta^l = (We^{l+1})^T\ \delta^{l+1} * f'\ (inp^l) \tag{4.5}$$

$$\delta^L = f'(\text{inp}^L) * (\text{out}^n - \text{tar}^n) \tag{4.6}$$

$$\frac{\partial Err}{\partial We^l} = inp^{l-1}\ (\delta^l)^T \tag{4.7}$$

$$\Delta We^l = -\rho\frac{\partial Err}{\partial We^l} \tag{4.8}$$

Equations (4.4)–(4.8) describe the weight updating process. Weight updating factor is generated in Equation (4.8). The partial derivative of the error along with bias and weights is taken in as a contributing factor towards weight updating. ρ is the learning rate which also plays a role in weight updating. The above equations are forward and backward pass equations in back propagation algorithm which forms the basis for CNN.

These networks have convolution layer along with sub-sampling layer in order to reduce the time of computation. In the convolution layer, a learnable kernel is convolved along with the feature map of previous layers. This is subsequently passed through the activation function in order to result in output feature map. Every output map will combine the convolutions along with many input maps.

Equations (4.9)–(4.11) describe the updating of weights in convolution layer which takes along the learnable kernels also.

$$inp_d^l = f\left(\sum_{i \in M_d} inp_i^{l-1} * ker_{id}^l + bi_d^l\right) \tag{4.9}$$

$$\delta_d^l = \beta_d^{l+`}\left(f'\left(inp_d^l\right) * up\left(\delta_d^{l+1}\right)\right) \tag{4.10}$$

$$\frac{\partial Err}{\partial bi_d} = \sum_{p,q}\left(\delta_{d\,pq}^l\right) \tag{4.11}$$

The sub-sampling layer results in versions of input maps that are down sampled. Though the number of input and output maps will remain the same, the smaller output maps will be produced. This is shown by the following equation

$$inp_d^l = f\left\{\beta_d^l down\left(inp_d^{l-1}\right)\right\} + bi_d^l \tag{4.12}$$

It is desirous to produce an output map which involves the sum of many convolutions of different input maps as shown in the following equation

$$inp_d^l = f\left\{\sum_{g=1}^{N_{in}} \alpha_{pq}\left(inp_g^{l-1} * ker_g^l\right)\right\} \tag{4.13}$$

The various abbreviations and symbols used in the chapter are also detailed in Table 4.1.

In a network when there are sub-sampling as well as convolution layers there are many difficulties. The feed forward pass and back propagation alters where the sampling down of output maps of convolution layer has to happen in forward pass. Also up sampling happens during back propagation with higher sub-sampling layer output has to map with convolution layer output maps. Also, it is challenging to apply sigmoid and derivatives as activation function. These challenges are overcome in the modelling of systems with CNN. These networks are very versatile and are popularly used in building predictive models. The next section describes application of these deep learning models in IoT frameworks.

TABLE 4.1
List of Abbreviations and Symbols Used

	Abbreviations
IoT	Internet of Things
CNN	Convolutional Neural Networks
DL	Deep Learning
WSN	Wireless Sensor Networks
BIoT	Block chain-based Internet of Things
DeepIoT	Deep Learning-based Internet of Things
NBIoT	Narrow Band Internet of Things
AIoT	Autonomous Internet of Things
MAC	Medium Access Control
FEC	Flexible Edge Computing
FPGA	Field Programmable Gate Array
MQTT	Message Queuing Telemetry Transport
	Symbols
T	Target output
Y	Actual Output
U	Input
W	Weights
B	Bias
Err	Error
ΔWe^{l}	Weight updation
ρ	Learning Rate

4.4 DEEP LEARNING-BASED IoT FRAMEWORKS

Inherently, deep learning has many advantages which can be exploited in the development of emerging scenario of embedded IoT devices. Yao et al. [23] discussed the various challenges in embedding deep learning in mobile and IoT applications. Furthermore, the possibility of building robust deep learning models in an IoT environment is also demonstrated. A Deepsense framework has been developed which integrates CNN and Recurrent Neural Networks. The architecture enables learning of multisensory fusion tasks for classification as well as estimation oriented problems in noisy environment also. The developed model has been applied to human activity recognition and user identification system. The performance metrics of the developed method was compared with existing techniques to demonstrate the superiority in performance. A tailor made model for IoT applications called DeepIoT has also been presented which utilises novel compressor neural network model towards addressing resource constraint problems in IoT. Testing accuracy, execution time and energy consumption are the parameters based on which performances have been evaluated. Uncertainty estimation is demonstrated using a Multilayer perceptron model developed called RSense and again performance for IoT applications have been recorded.

The usage of deep learning for IoT big data analytics and IoT streaming data analytics was explored by Mohammadi et al. [24]. Initially the IoT data characteristics was analysed to bring out the need for analytics. A comprehensive survey of various deep learning models which can be utilised for various IoT applications was tabulated. The different deep learning frameworks such as H_2O, Tensorflow, Torch, Caffe was presented with their features. The applications of these frameworks in foundational services and applications such as smart city, smart home, health services, transport systems, agriculture, industry, government, sports and entertainment has been detailed.The challenges in terms of performance, complexity of implementation, pitfalls was discussed. Fog and cloud centric deep learning for IoT application and common datasets for deep learning in IoT was explained.

Mauldin et al. [25] discussed experimental study of Ensemble Deep learning method for analysis of time series data on IoT devices. This chapter presented a case study of ensemble techniques of Stacking and Adaboosting for a fall detection application in healthcare domain. This approach was found effective for small size dataset and outperformed conventional deep learning methods which also require large datasets for training.

Narrow band Internet of Things (NB-IoT) and the application of deep learning in NB-IoT for reliable communication was elaborated by Jiang et al. [26]. In this approach, optimisation approach for configuration selection is done through Q learning and deep learning-based Q learning methods. Compared to conventional tabular Q learning approaches, this method of deep-learning-based Q learning proved superior. The superiority in performance was established in terms of number of served IoT devices, training speed and configuration dimensions.

Security in IoT devices is a major challenge. Varied methods such as encryption, access control, authentication etc are not very effective with IoT-based devices. Hence, Al-Garadi et al. [27] discussed the usage of deep-learning-based algorithms in providing secure IoT-based ecosystem. Deep learning provided powerful tools to provide data exploration and handle security threats in IoT systems. The opportunities, challenges in applying deep learning to IoT systems were also detailed.

The integration of IoT with Automonous Control System (ACS) resulting in a new theory called Autonomous IoT (AIoT) was explored by Lei et al. [28]. The system status was recorded by the sensors which collects information. Based on the data collected, the intelligent agents in IoT devices and cloud centres take decisions on what are the actions to be taken by the actuators. Deep learning-based reinforcement learning methods was being used in the decision-making. The various deep learning-based algorithms that have been implemented in AIoT was discussed in detail, along with open end research challenges in the domain.

A unique narrow space deep-learning-based architecture for concentrating on low cost, time requirements and strict memory for IoT system with near sensor platform was discussed by Scheidegger et al. [29] This method was an emphirical demonstration with 3000 training models with improved precision, accuracy with limited memory and low latency.

Deep learning models are capable of processing large volume of sensor data in IoT and learning the underlaying features of the data which was detailed by Ma et al. [30] The usage of deep learning models in specifically four major domains of IoT namely smart home, smart healthcare, smart transportation and smart industry was exhaustively covered. Furthermore, the possibility of application of deep learning in less explored IoT domains such as disease analysis, traffic monitoring and predictions, autonomous driving, home robotics and indoor localisations was also brifed to give reasearchers scope to explore emerging fields.

These deep learning predictive models can be implemented nearer to the user instead of computation being done in the cloud server. This leads to the concept of deep learning implemented on the edge, i.e., deep learning empowered by edge computing which is being detailed in the next section.

4.5 DEEP LEARNING EMPOWERED BY EDGE COMPUTING

Smart devices and sensors are generating massive volumes of data. The ever increasing demands of the network and consumers are forcing computations and services to move from cloud on to edge. Consequently, the deployment of deep learning algorithms is also moving from cloud to edge leading to the emergence of edge intelligence as explained by Wang et al. [31]. This culminates in formulation of edge computing frameworks capable to performing intelligent operations as they are empowered by deep learning models in the background. The paper has explored this through the application scenarios and the practical implementations of the concepts discussed. Edge computing on its own has inherent ability of computing making it intelligent edge and when amalgamated with deep learning algorithms results in edge intelligence. This edge intelligence along with intelligence in computing is also bestowed with intelligent decision-making and predictive abilities.

IoT-based sensors and end devices are capable of generating volumes of data which can be processed by deep learning-based models. Edge computing involves operating the computing nodes near the edge device and is capable of performing high computation and has low latency requirements. This greatly enhances scalability, privacy and efficiency in terms of bandwidth. Chen and Ran [32] worked in providing insights on the intersection of edge computing and deep learning. The paper has provided comprehensive overview on how deep learning can be applied at the edge of the network along with methods that can be used for deep learning interference execution across different end devices, servers and cloud. Various popular deep learning models can be applied across many edge devices. The challenges involved in terms of performance of the system, different technologies of the network and privacy has also been discussed. The methodology towards increasing the deep learning interference and performing training distributing on edge devices was also detailed in the paper.

Wang et al. [33] provided a complete review of the latest techniques and contributions of how the advancements in deep-learning-empowered edge-computing-based applications. Specifically, four domains have taken into consideration namely smart industry, smart city, smart transportation and smart multimedia. The main research bottlenecks in the domain were being discussed along with the contributions in the specific applications.

The deployment of deep learning services to edge computing has resulted in edge intelligence. The integration of deep learning into frameworks of edge computing enables dynamic and adaptive maintenance in edge networks. Cui [34] discussed the application scenario under which edge networks and deep learning can be integrated. Furthermore, the practical implementation methods and technologies were also being explored in customised edge-computing-based frameworks. The key factors which play a role in the connecting of the DL and intelligent edge was also being detailed which results in a fused framework developed.

McClellan et al. [35] had given a description on how Machine Learning along with mobile edge computing will play a crucial role in 5G networks. The usage of mobile edge computing within 5G networks will bring the abilities of cloud computing, analysis and storage closer to the end user. These advances aid in different functioning of 5G networks such as adaptive resource allocation, modelling of mobility, improved security and energy efficiency in 5G networks.

These deep learning models empowered by edge computing can be amalgamated in IoT-based frameworks. This would greatly benefit and enhance the quality of the functioning of IoT networks. Hence, the next section details deep-learning-based IoT frameworks on the edge.

4.6 DEEP-LEARNING-BASED IoT FRAMEWORKS ON THE EDGE

The study so far has been analysing the usage of deep learning in edge computing. It is understood that the multilayer structure of deep learning enables its usage in edge environment. The proposed model developed in the work by Sureddy et al. [36] discussed the performance of IoT deep learning applications with edge computing abilities. The architecture was termed as flexible edge computing architecture as it has the potential to overcome the rigidity which is present in IoT-based edge computing systems. The major characteristics of the model were to have adaptability to changes in environment and ability to orient according to the user. The flexible edge computing (FEC) Architecture was explored with multiple agents.

Khelifi et al. [37] studied and discussed the various applications of combining deep learning models such as CNN, Recurrent Neural Network and Reinforcement learning along with IoT over the edge. The paper comprehensively gave a comparison of different machine learning models which could be used on IoT platform along with its advantages and disadvantages.

IoT along with edge computing can potentially be utilised in healthcare replacing the usage of cloud computing. Compared to cloud computing, edge computing better handles issues regarding battery power consumption, bandwidth, data safety and privacy. Swain [38] explored the possibility of integration of edge computing and machine learning algorithms in an IoT framework. The main objective of the paper was to get the necessary data which is of interest amongst the huge volume of data which has been generated from the sensor framework in IoT devices

Machine learning algorithms were extensively used only in high-performance servers initially. With the advancements in technology, embedded mobile processors are being used in varied applications. Improvements in characteristics of these devices in terms of processing power, energy storage and memory capacity has resulted in evolution of these devices as edge devices. The implementation of machine learning algorithms on these edge devices will greatly facilitate reduced processing abilities. Hence, Narayanan and Makler [39] explored the feasibility of training and validating machine learning algorithms on a Raspberry Pi device.

The different algorithms which were trained and tested include Random Forest, Support Vector Machine and Multilayer perceptron. Improved performance metric in terms of speed, accuracy and power consumption was observed.

Most of the IoT applications depend on Machine Learning algorithms. Traditionally most of these ML algorithms run on the cloud. With the advent of AI-powered microcontrollers embedded software and hardware and the increasing resource constraints on IoT devices, Deep Edge is a concept gaining momentum towards IoT virtualisation concept. Campolo et al. [40] described the design of AI-embedded IoT devices which have cognitive capabilities in order to cater to IoT applications. The two major outcomes of the proposed models would be mitigating the constraints on the existing devices and enabling interoperability amongst different AI platforms. The robustness of the proposed solution is demonstrated by proof of concept which was also detailed.

Ghosh and Grolinger [41] presented investigation of amalgamation of cloud and edge computing for IoT data analytics. The data reduction happens on the edge through deep learning approaches along with machine learning techniques applied on the cloud. The presence of auto encoder's encoder part on the edge is to reduce the dimensions of the data. This data is then subsequently sent to the cloud for machine learning techniques.

Azimi et al. [42] discussed the usage of edge-based deep learning model for healthcare IoT-based systems. The paper compared the results obtained using two types on deep learning models namely CNN and hierarchical computing architecture. The system developed benefited from the feature of both these models resulting in high accuracy and high-level availability. A real-time health monitoring system was used as a case study on ECG classification and the performance of the system was being evaluated for accuracy as well as response time.

There are varied applications where deep-learning-empowered edge-computing-based IoT frameworks can be utilised. Amongst the different application smart city is a promising domain where the requirements, architecture and security issues become important. Haque et al. [43] had given insights on holistic overview and discusses the different crucial issues in the implementation along with future research directions.

Deep learning can be applied across end devices which can be based on IoT, at edge servers as well as in cloud data centres. Based on the above discussions, it is evident that IoT frameworks which are empowered by deep learning have tremendous potential to have improved performance metrics and can be of use to wide range of applications. The subsequent section presents a case study of an IoT-based predictive edge analytics-based framework for diagnostic care of breast care patients.

4.7 CASE STUDY: AN IoT-BASED PREDICTIVE EDGE ANALYTICS FRAMEWORK FOR DIAGNOSTIC CARE OF BREAST CANCER PATIENTS

4.7.1 Basic Theme of the Study

Among different healthcare domains cancer diagnosis and treatment is a major challenge where cancer is major reason for death of millions of people every year as illustrated by Panda et al. [44]. There are different types of cancers and each has distinct characteristics and hence diagnosis and treatment changes according to the type of cancer and patient as discussed by Ibnouhsein et al. [45]. Breast cancer is a syndrome that causes huge numbers of casualty every year due to ineffectiveness of proper filtering and appropriate classification methods as detailed by Venkatesan et al. [46]. Breast cancer is not one of the homogeneous diseases that differ greatly among different categories of cancer sufferer and even within each individual tumour. The field of edge computing and predictive analytics is extremely powerful and is rapidly expanding. These technologies have started playing pivot role in the emerging healthcare industry. The existing databases can be converted into useful knowledge. Analytical models based on the useful information could aid in prediction of diseases or in improvisation of healthcare mechanisms as discussed by Sheoran [47]. Hence, this case study explores and discusses the application of predictive analytics towards building forecasting models to detect presence of breast cancer and also works towards providing solutions for treatment and care depending on specific cases using predictive models. The predictive models can also be moved to the edge on hand held devices using edge computing. Furthermore, towards treatment, and care of the patients, IoT, which has potent to connect millions of devices, can be involved and potentially deployed. Such a comprehensive framework built will greatly improve existing methods to predict breast cancer and greatly enhance the current treatment methods for patients.

4.7.2 Existing Work

The primary method of detection of breast cancer is through mammography. Once cancerous cells are detected it is imperative to find whether they are malignant or benign. Most of the existing work, depending on the shape of the mass being malignant or benign is found using image processing techniques. The finding out of malignant or benign is considered a predictive model and Support vector Machines or Back Propagation-based techniques is being used towards finding them out. The rate of performance, state of training and error histogram are the results depicted based on the predictions. Furthermore, the next stage to which cancer belongs is popularly found out using cloud as well big data-based techniques as detailed by Umesh and Ramachandra [48].

Towards increasing the classification accuracy and improving the performance, *K* nearest neighbor algorithms are also being employed. Big Data is used towards amalgamation of datasets, support vector machines and eigenvalue decomposition is used for building forecasting models as discussed by Sankari et al. [49]. These techniques are efficient methods resulting in a mathematical framework for incorporation of data fusion as well as handling non-linear classification issues. Data-mining-based algorithms have also been used for building predictive models. The three data-mining-based algorithms whose ability to predict is compared are Bayes model, Radial Bias function-based model and J48 as detailed by Gomathi and Sandhya [50]. There are other techniques available in data mining which are capable building predictive models. Amongst them three techniques namely Decision Tree Support Vector Machine (DT-SVM), instance-based learning and sequential minimal optimisation is being compared. It has been found that DT-SVM outperforms the other two methods in terms of prediction accuracy. Shailaja et al. [51] found that data mining algorithms are of tremendous help in prediction algorithms in the early stages of breast cancer.

4.7.3 Objectives and Relevance of the Work

The following are the main objectives of the proposed work:

To provide a sustainable solution to the inherent problems of diagnosis, treatment, care and cure by the introduction of advanced systems and means of processing a dataset with the state-of-the-art high-performance computing systems.

- To convert the data sets generated for possible breast cancer patients into useful knowledge in accordance to requirements.
- To apply predictive analytics (deep learning) processes on database and build predictive models for premature detection of breast cancer
- To develop edge-computing-based analytics frameworks resulting in forecasting done on handheld devices both on the patient and hospital end.
- To implement IoT-based framework for systematic treatment and cure of breast cancer patients.

Breast cancer is increasingly becoming a threat to women around the globe. About 1 in 8 women will develop invasive breast cancer over the course of her lifetime. In 2020, an estimated 276,480 new cases of invasive breast cancer are expected to be diagnosed in women in the United States as illustrated by Chaurasia et al. [52]. If predicted at an early stage, the cancer can be treated easily and patient can be cured of ailment without complicated surgery or chemotherapy or radiotherapy. This will greatly empower the women and help her get back with her life with minimum complications. Mammography is existing technique for breast cancer detection. This technique is not comfortable and often cumbersome as discussed by Sivakami [53]. Furthermore, there are some techniques developed by researchers based on image processing as explored in Mulatu and Gangarde [54]. However, accuracy and efficiency of these techniques are not very promising. Hence, under this context, a comprehensive framework based on predictive analytics, edge computing and IoT to be developed for diagnosis, treatment and cure of breast cancer patients would be beneficial and will have profound effect in improving the quality of healthcare offered for breast cancer as detailed by Shukla et al. [55].

4.7.4 Proposed Research Methodology

The different levels in the research methodology are as follows:

- Level 1: Processing of available data sets to convert to useful knowledge or required parameters
- Level 2: Building predictive models towards forecast of type of cancer (benign or malignant), stage of cancer and survivability model.

The predictive models are built towards forecasting certain key parameters related to improvement of performance. These predictive models are based on deep learning process.

- Level 3: Implementing Predictive Models using edge computing so that the models could be run on hand held devices as per requirements.
- Level 4: Building IoT-based framework for systematic treatment, cure and patient care
- Level 5: Integrating various levels and technologies to build a comprehensive framework
- Level 6: Validating results and deducing inferences based on the built framework.

4.7.5 Expected Impact of the Work

Integration of diagnosis, treatment, care and cure for breast cancer into a single framework is the major outcome. From the existing databases formulation of an information base based on classification of stage and intensity of cancer which will be useful for further research in the area is another result of proposed work,

- More accurate and efficient framework compared to the existing systems will be accomplished.
- Reduced Processing and implementation time compared to the existing systems will be achieved.
- Availability of customised treatment, care and cure depending on the outputs of the predictive models is another significant outcome.
- Classification of cancer as benign or malignant will be done.
- A predictive model capable of predicting stage of cancer along with survivability model will be built.
- Implementation of predictive models in hand held devices through edge computing will be accomplished.
- Comprehensive, wholesome solution towards diagnosis, treatment and care for breast cancer patients will be formulated.
- Major impact of the proposed work would be providing superior healthcare facility towards diagnosis and treatment for breast cancer patient resulting in reducing losses.
- The proposed and developed framework will pave way to formulation of information base and pathway sufficient to drive many similar research projects.

4.7.6 Technical Novelty and Utility

The amalgamation of various technologies such as IoT, predictive analytics and edge computing though a herculean task is the technical novelty in the comprehensive framework proposed which is an unexplored terrain amongst the existing work. Furthermore, existing work used image processing algorithms which were not very accurate and efficient. Development of more efficient, accurate and less time consuming framework would be another uniqueness of the work. In the proposed work, separate levels are implemented using different technology and then integrated. Customisation according to the need and type of patient is another distinctive feature of the proposed work. Such a wholesome solution will have tremendous utility for problems associated with breast cancer in terms of its prediction, treatment and subsequent cure. The software solution developed could be cost effective and integrated in hand held devices making the comprehensive framework compact and very useful.

4.7.7 Societal Benefit

The implementation of this case study is of huge societal relevance in developing countries like India where healthcare facilities are poor. Hence, this work will be extremely beneficial in prediction of breast cancer which has become common around the globe because of changing lifestyle and increased pollution. This will enable provide a complete solution for treatment and care of breast cancer patients and hence greatly aid towards improvement in the health of Indian women and in turn empower women which would indirectly help in improving the economic conditions of the country. After successful trial run of the proposed model the template or the built product can be commercialised and cost of the product could be optimised depending on the functionalities and requirements of the consumers of the implemented project. The developed product could be patented before commercialisation and could be potentially be of use to hospitals, cancer institute and research centres.

4.8 FUTURE RESEARCH DIRECTIONS

The future line of action could be on preparation of list of variety of applications where the proposed framework could be made use of. Furthermore, implementation of the framework can be done towards examining the contribution of the above approach through different verification and validation experiments. Also the proposed methods could be enhanced with real time and dynamic decision-making abilities and intelligence. Usage of multiple IoT-based edge servers could enhance the performance. The interaction of the edge servers amongst them and with end devices could also be explored towards improvements in the model. Future work could also be directioned in the line of pruning the deep learning algorithms and models in sync with recent research works. Energy consumption reduction or optimisation of energy in the proposed frameworks could also be the factor based on which further work can be carried. Security of data in the edge is also a matter of concern and there is huge scope of work in exploring security concerns and techniques of deep-learning-empowered edge-computing-based IoT frameworks. Managing and scheduling in edge computing resources along with migrating edge computing applications with an objective to reduce latency is a major challenge which can also be worked on.

4.9 CONCLUSIONS

In this chapter, we have introduced several approaches for deep-learning-based edge computing along with deep-learning-empowered IoT framework which can be applied to variety of applications. The main advantage of the proposed framework is to reduce latency in the case of applications which are time critical. The large volume of data produced by the IoT devices can be processed and predictive models can be built based on the obtained data through deep learning techniques. Furthermore, the predictive models developed are brought much nearer to the user by adopting edge computing. Hence, the chapter brings a comprehensive review of deep learning and edge-computing-based IoT frameworks. A specific case study on IoT-based Predictive Edge Analytics framework for diagnostic care of breast cancer patients is also presented to enforce the impact on the proposed framework in a specific application. The proposed amalgamation of technologies and framework can be applied in a variety of applications ranging from agriculture, healthcare, assistive living and many more.

REFERENCES

1. Shanzhi Chen, Hui Xu, Dake Liu, Bo Hu, and Hucheng Wang, 2014, A Vision of IoT: Applications: Challenges and Opportunities With China Perspective, IEEE Internet Of Things Journal, 1(4), 349–359.
2. Jonathan Tournier, François Lesueur, Frédéric Le Mouël, Laurent Guyon, and Hicham Ben-Hassine, 2020, A Survey of IoT Protocols and Their Security Issues through the Lens of a Generic IoT Stack, Internet of Things, 16(4), 24–31.
3. Dinh Dung and Van K. Nguyen, 2021, Deep ReLU Neural Networks in High-Dimensional Approximation, Neural Networks, 142, 619–635.
4. Muder Almiania, Alia AbuGhazlehb, Amer Al-Rahayfeha, Saleh Atiewia, and Abdul Razaque, 2020, Deep Recurrent Neural Network for IoT Intrusion Detection System, Simulation Modelling Practice and Theory, 101, 1–20.
5. Weisong Shi, George Pallis, and Zhiwei Xu, 2019, Edge Computing, Proceedings of The IEEE, 107(8), 1474–1481.
6. Mohamed A. Amanullaha, Riyaz A. A. Habeebb, Fariza H. Nasaruddinc, Abdullah Ganid, Ejaz Ahmede, Abdul S. M. Nainarf, Nazihah Md Akimb, and Muhammad Imran, 2020, Deep Learning and Big Data Technologies for IoT Security, Computer Communications, 151, 495–517.
7. Mladen Veletic, Igor Radusinovic, and Slavica Tomovic, 2017, The IoT Architectural Framework, Design Issues and Application Domains, Wireless Personal Communications, 92, 127–148.
8. Sukant K. Mohapatra, Jay N. Bhuyan, Pankaj Asundi, and Anand Singh, 2016, A Solution Framework For Managing Internet Of Things (IoT), International Journal of Computer Networks & Communications (IJCNC) 8(6), 73–87.
9. Charilaos Akasiadis, Vassilis Pitsilis, and Constantine D. Spyropoulos, 2019, A Multi-Protocol IoT Platform Based on Open-Source Frameworks, Sensors, 19(4217), 2–25.
10. Pallavi Sethi and Smruti R. Sarangi, 2017, Internet of Things: Architectures, Protocols, and Applications, Hindawi, Journal of Electrical and Computer Engineering, 9324035, 1–25.
11. R. Alharbi and D. Aspinall, 2018, An IoT Analysis Framework: An Investigation of IoT Smart Cameras' Vulnerabilities, Living in the Internet of Things: Cybersecurity of the IoT – 2018, 1–10.

12. David Lake, Rodolfo Milito, Monique Morrow and Rajesh Vargheese, 2013, Internet of Things: Architectural Framework for eHealth Security, Journal of ICT, 3(4), 301–328.
13. S. Krčo, B. Pokrić, and F. Carrez, 2014, Designing IoT Architecture(s): A European Perspective, IEEE World Forum on Internet of Things (WF-IoT), 79–84. doi:10.1109/WF-IoT.2014.6803124
14. Bhagyashri Katole, Manikanta Sivapala, and V. Suresh, 2013, Principle Elements and Framework of Internet of Things, International Journal Of Engineering And Science, 3(5), 24–29.
15. Taran Singh Bharati, 2019, Internet Of Things (Iot): A Critical Review, International Journal Of Scientific & Technology Research, 8(10), 227–232.
16. R. Mahmoud, T. Yousuf, F. Aloul, and I. Zualkernan, 2015, Internet of things (IoT) security: Current status, challenges and prospective measures, 10th International Conference for Internet Technology and Secured Transactions (ICITST), 336–341.
17. H. Kim Abbas Ahmad, Jaeyoung Hwang, Hamza Baqa, Franck Le Gall, Miguel Angel Reina Ortega, and Jaeseung Song, 2018, IoT-TaaS: Towards a Prospective IoT Testing Framework, IEEE Access, 6, 15480–15493.
18. B. Bhushan, C. Sahoo, and P. Sinha, 2021, Unification of Blockchain and Internet of Things (BIoT): Requirements, Working Model, Challenges and Future Directions. Wireless Networks, 27, 55–90.
19. Shivam Saxena, Bharat Bhushan, and Mohd Abdul Ahad, 2021, Blockchain Based Solutions to Secure IoT: Background, Integration Trends and a Way Forward, Journal of Network and Computer Applications, 181, 1–25.
20. Sukriti Goyal, Nikhil Sharma, Bharat Bhushan, Achyut Shankar, and Martin Sagayam, 2021, IoT Enabled Technology in Secured Healthcare: Applications, Challenges and Future Directions, Cognitive Internet of Medical Things for Smart Healthcare, 25–48.
21. B. Bhushan, and G. Sahoo, 2020, Requirements, Protocols, and Security Challenges in Wireless Sensor Networks: An Industrial Perspective. In: Gupta B., Perez G., Agrawal D., Gupta D. (eds) Handbook of Computer Networks and Cyber Security. Springer, Cham., 683–713.
22. AnirudhaGhosh, A. Sufian, FarhanaSultana, andAmlanChakrabarti, 2020, Fundamental Concepts of Convolutional Neural Network, Recent Trends and Advances in Artificial Intelligence and Internet of Things, 519–567. doi:10.1007/978-3-030-32644-9_36
23. Shuochao Yao, Yiran Zhao, Aston Zhang, Shaohan Hu, Huajie Shao, Chao Zhang, Lu Su, and Tarek Abdelzaher, 2018, Deep Learning for the Internet of Things, Computer, 51(5), 32–41.
24. M. Mohammadi, A. Al-Fuqaha, S. Sorour, and M. Guizani, 2017, Deep Learning for IoT Big Data and Streaming Analytics: A Survey, IEEE Communications Surveys & Tutorials, 20(4), 2923–2960.
25. Taylor Mauldin, Anne H. Ngu, Vangelis Metsis, Marc E. Canby, and Jelena Tesic, 2019, Experimentation and Analysis of Ensemble Deep Learning in IoT Applications, Open Journal of Internet of Things (OJIOT), 5(1), 133–149.
26. N. Jiang, Y. Deng, A. Nallanathan, and J. A. Chambers, 2019, Reinforcement Learning for Real-Time Optimization in NB-IoT Networks,IEEE Journal on Selected Areas in Communications, 37(6), 1424–1440.
27. M. A. Al-Garadi, A. Mohamed, A. K. Al-Ali, X. Du, I. Ali, and M. Guizani, 2020, A Survey of Machine and Deep Learning Methods for Internet of Things (IoT) Security, IEEE Communications Surveys & Tutorials, 22(3), 1646–1685.
28. L. Lei, Y. Tan, K. Zheng, S. Liu, K. Zhang, and X. Shen, 2020, Deep Reinforcement Learning for Autonomous Internet of Things: Model, Applications and Challenges, IEEE Communications Surveys & Tutorials, 22(3), 1722–1760.

29. Florian Scheidegger, Luca Benini, Costas Bekas, and Cristiano Malossi, 2019, Constrained Deep Neural Network Architecture Search for IoT Devices Accounting for Hardware Calibration, Advances in Neural Information Processing Systems, 32, 6054–6064.
30. Xiaoqiang Ma, Tai Yao, Menglan Hu, Yan Dong, Wei Liu, Fangxin Wang, and Jiangchuan Liu, 2019, A Survey on Deep Learning Empowered IoT Applications, IEEE Access, 7, 181721–181732.
31. Xiaofei Wang, Yiwen Han, Victor C. M. Leung, Dusit Niyato, Xueqiang Yan, and Xu Chen, 2020, Convergence of Edge Computing and Deep Learning: A Comprehensive Survey, IEEE Communications Surveys and Tutorials, 22(2), 869–904.
32. Jiasi Chen and Xukan Ran, 2019, Deep Learning With Edge Computing: A Review, Proceedings Of The IEEE 107(8), 1655–1675.
33. F. Wang, M. Zhang, X. Wang, X. Ma, and J. Liu, 2020, Deep Learning for Edge Computing Applications: A State-of-the-Art Survey, IEEE Access, 8, 58322–58336.
34. Wei Cui, 2020, Research and Application of Edge Computing Based on Deep Learning, Journal of Physics: Conference Series, 1646, 1–5.
35. Miranda McClellan, Cristina Cervelló-Pastor, and Sebastià Sallent, 2020, Deep Learning at the Mobile Edge: Opportunities for 5G Networks, Applied Science, 10, 4735.
36. Sneha Sureddy, K. Rashmi, R. Gayathri, and Archana S. Nadhan, 2018, Flexible Deep Learning in Edge Computing for IoT, International Journal of Pure and Applied Mathematics, 119 (10), 531–543.
37. Hakima Khelifi, Senlin Luo, Boubakr Nour, Akrem Sellami, Hassine Moungla, Syed H. Ahmed, and Mohsen Guizani, 2019, Bringing Deep Learning at the Edge of Information-Centric Internet of Things, IEEE Communications Letters, 23(1), 52–55.
38. Sangram K. Swain, 2017, Use Of Big Data Analytics In Lung Cancer Data, International Journal of Computational Engineering Research, 7(12), 2250–3005.
39. Ramaswamy Narayanan and Amy Makler, 2016, Big Data Analytics and Cancer, MOJ Proteomics & Bioinformatics, 4(2), 1–4.
40. C. Campolo, G. Genovese, A. Iera, and A. Molinaro, 2021, Virtualizing AI at the Distributed Edge towards Intelligent IoT Applications. Journal of Sensor, Actuator Networks, 10(13), 1–14.
41. A. M. Ghosh and K. Grolinger, 2019, Deep Learning: Edge-Cloud Data Analytics for IoT, IEEE Canadian Conference of Electrical and Computer Engineering (CCECE), 1–7.
42. I. Azimi, J. Takalo-Mattila, A. Anzanpour, A. M. Rahmani, J. Soininen, and P. Liljeberg, 2018, Empowering Healthcare IoT Systems with Hierarchical Edge-Based Deep Learning, IEEE/ACM International Conference on Connected Health: Applications, Systems and Engineering Technologies (CHASE), 63–68.
43. K. M. Bahalul Haque, Bharat Bhushan, and Gaurav Dhiman, 2021, Conceptualizing Smart City Applications: Requirements, Architecture, Security issues, and Emerging Trends, Expert Systems, https://doi.org/10.1111/exsy.12753.
44. Ritu P. Panda, Prakalpa P. Barik, and P. A. K. Prusty, 2016, A Review Paper on Big Data in Lung Cancer Big Data Analytics in Lung Cancer, International Journal of Trend in Research and Development, 3(5), 451–454.
45. Issam Ibnouhsein, Stéphane Jankowski, Karl Neuberger, and Carole Mathelin, 2018, The Big Data Revolution for Breast Cancer Patients, European Journal Of Breast Health, 14, 61–62.
46. Mithra Venkatesan, Anju V. Kulkarni, Radhika Menon, and V. Ramakrishnan, A Comprehensive Big Data Analytics Based Framework for Premature Recognition of Breast Cancer, International Journal of Innovative Technology and Exploring Engineering, 9(3), 2188–2192.

47. Savita K. Sheoran, 2018, Breast Cancer Classification Using Big Data Approach, Paripex – Indian Journal of Research, 7(1), 401–403.
48. D. R. Umesh, and B. Ramachandra, 2016, Big Data Analytics to Predict Breast Cancer Recurrence on SEER Dataset using MapReduce Approach, International Journal of Computer Applications, 150(7), 7–11.
49. L. Sankari, R. Rajbharath, and G. T. Arasu, 2017, Predicting Breast Cancer using Novel Approach in Data Analytics, International Journal of Engineering Research & Technology, 6(5), 72–76.
50. N. Gomathi, and P. Sandhya. 2017, Prognosis and Diagnosis of Breast Cancer Using Interactive Dashboard Through Big Data Analytics, Biotechnology: An Indian Journal, 13(1), 1–17.
51. K. Shailaja, B. Seetharamulu, and M. A. Jabbar, 2018, Prediction of Breast Cancer Using Big Data Analytics, International Journal of Engineering & Technology, 7 (4.6), 223–226.
52. Vikas Chaurasia, Saurabh Pal, and BB Tiwari, 2018, Prediction of Benign and Malignant Breast Cancer Using Data Mining Techniques, Journal of Algorithms & Computational Technology, 12(2) 119–126.
53. K. Sivakami, 2015, Mining Big Data: Breast Cancer Prediction using DT – SVM Hybrid Model, International Journal of Scientific Engineering and Applied Science, 1 (5), 418–429.
54. Desta Mulatu and Rupali R. Gangarde. 2017, Survey of Data Mining Techniques for Prediction of Breast Cancer Recurrence, International Journal of Computer Science and Information Technologies, 8(6), 599–601.
55. Rati Shukla, Vikash Yadav, Parashu Ram Pal, and Pankaj Pathak, 2019, Machine Learning Techniques for Detecting and Predicting Breast Cancer, International Journal of Innovative Technology and Exploring Engineering, 8(7), 2658–2662.

5 A Geo-Referenced Data Collection Microservice Based on IoT Protocols for Smart HazMat Transportation

Ghyzlane Cherradi and Azedine Boulmakoul
LIM Lab, FSTM, Hassan II University,
Casablanca, Morocco

Lamia Karim and Meriem Mandar
LISA Lab, ENSAB, Hassan First University,
Settat, Morocco

CONTENTS

DOI: 10.1201/9781003219620-5

5.1 INTRODUCTION

Recently, continuous real-time monitoring of the transport of hazardous materials (HazMat) has gained a great deal of attention. This is due to the ability to provide clear insights into dynamically moving vehicles through web mapping systems [1, 2]. However, the major challenge faced by such systems is designing an efficient architecture for collecting telemetry and real-time event data transmitted by the vehicle's embedded system, such as positioning data. With the appearance of the Internet of Things (IoT) [3], wireless sensor networks (WSNs) [4] play a significant role in the continuous monitoring of moving objects (such as HazMat vehicles). WSN is a set of several varieties of nodes that can detect and interact with other devices and platforms. These elements can detect events and notify them via wireless links. Extensive deployment of WSN-based applications for use in environmental monitoring, transportation, and building management shows that WSN networks are one of the most promising technologies for capturing environmental information and providing event services. In this sense, the publish/subscribe communication model [3, 5] has been adopted, in most cases, for the design of event services in WSN networks in order to deploy numerous applications in the aforementioned fields. Indeed, the publish/subscribe paradigm is considered an effective communication model for WSNs because it includes brokers and intermediary nodes, the role of which is to collect all emissions from sensor nodes (publishers), manipulate these information (filtering, aggregation, detection, etc.), and finally transmit the resulting data to the collection nodes (the subscribers). Brokers are the only nodes that connect data and their users. Indeed, the publish/subscribe paradigm is considered an effective communication model for WSNs because it includes brokers and intermediary nodes, the responsibility is to collect all emissions data from sensor nodes (the publishers), handle these data (filtering, aggregation, processing, etc.), and eventually send the resulting data to the acquisition nodes (the subscribers). Brokers are the single nodes that connect data and its users. Therefore, publishers and subscribers do not need to know their identity and location (decoupling in space) and do not need to be synchronized (decoupling in time). In other words, an event can be notified asynchronously to a subscriber while simultaneously performing an activity; during this period, publishers are not blocked from offering events. This allows more scalability and flexibility, also a more dynamic network topology, which is typical for WSN-based applications. In addition, the continuous innovation of hardware, software, and connectivity solutions in recent years has led to the expansion of the IoT, and the amount of connected objects continues to increase exponentially [6, 7]. With the sheer amount of data sensed by these devices, it is necessary to provide a suitable system design that can collect and process all the generated data. Even though the cloud-based paradigm is currently used for this end, but the new pattern of fog computing is designed to adjust and improve the IoT infrastructure. The fog computing paradigm can be defined as an infrastructure responsible for collecting, processing, and storing data arising from connected objects. It provides layered computing abilities with the benefit of edge server nodes [8–10]. In the area of data storage, processing, and IoT, fog computing creates an additional interface that can be situated between edge computing and cloud computing. And it should be considered an extension of cloud computing, essentially because it can offer local processing and analyses, which complete and aggregate the streams of data before sending it to the cloud. Therefore, the use of edge, fog, and cloud computing as complementary technologies provides the optimization of workflows and the cost minimization of the treatment of the large volume of data in geographically distributed data centers [9, 11–13].

The main object of this chapter is to present a smart geo-referenced data collection system for HazMat vehicles using IoT protocols such as Message Queuing Telemetry Transport (MQTT). The system uses a publish/subscribe middleware for scalable data acquisition, real-time data processing, and data sharing through mobile devices. The system is designed to be part of a larger IoT infrastructure, which means it should serve as a data provider to a larger IoT platform. Thus, the data collected can be used in other contexts, for example in application related to smart parking or fleet management. With the fact that sensors, platforms, and services operate in different places, and with the influence of distributed architectures and the fog computing paradigm, the proposed system is designed as event-driven microservices. This design gives the system great flexibility, scalability, and adaptability to different data applications. The remainder of this chapter is structured as follows: in Section 5.2, we describe the emergent technologies and architectures that have formed the IoT. In Section 5.3, we describe the proposed data collection microservice based on IoT protocols for smart HazMat transportation. Section 5.4 concludes the chapter.

5.2 IoT TECHNOLOGIES AND ARCHITECTURES

This section describes the IoT architectures and technologies that form the basis of IoT solutions. Therefore, we dive into an overview of the IoT's architectural foundation to provide a clear view of detection, collection, analytics, and network technologies.

5.2.1 IoT Technologies

Over the last decade, the IoT paradigm has received considerable interest not only in academic domains but also in industry fields. The main reason for this interest is the capabilities provided by the IoT [14]. It also ensures the establishment of a world where all smart objects and devices are connected to the internet and interact with each other with minimal human interference. These objects all have very different characteristics. Some are mobile (i.e., lightweight service robots), some are controlled by a human user, while others only interact with other connected objects (machine-to-machine or M2M communications). Some are just sensors that take a measurement regularly when others have the ability to act in their environment, such as a vehicle that stops after having received an instruction [15]. The IoT contains a huge number of technologies that can be grouped into four layers: sensing (or hardware), middleware, communication, and application layers.

5.2.1.1 Sensing Technologies

In the field of HazMat transport, IoT technologies are introduced mainly to increase the safety and efficiency of this type of transport. It has a certain practical advantage in decreasing the negative impact on society and the environment. HazMat transport management includes geolocation, real-time monitoring of speed, fuel, and cargo status to identify potential issues. Effectively, four essential elements are taken into consideration for smart HazMat transportation: sensing, information management, intelligent alerting, and decision-making. IoT sensing involves collecting data from connected objects and sending it back to a database, or a cloud platform. The sensing technologies can be divided into two classes. The first, seen as interne, allows detecting the HazMat vehicle state, such as mechanical components state, speed, and position. The second informs about the surrounding environment state, such as distance detection, and environmental condition sensing, among others. Indeed, geolocation technologies using systems such as RFID and GPS have greatly evolved [16, 17]. In addition, the power of radio detection and telemetry technology as well as integrated smart cameras allows for recognizing distances, directions, lanes, and moving objects. All of these technologies provide useful information that enables us to make decisions and act automatically and instantly.

5.2.1.2 IoT Middleware

In IoT ecosystems, processing, storage, and communication services are designed to be highly distributed. In addition, entities such as people, objects, and the surrounding environment are intended to form a common set of highly decentralized nodes, which need to be interconnected by a dynamic network of networks [6, 18, 19]. Thus, the smart integration of these heterogeneous entities leads to an effective IoT ecosystem. Recently, the proposed IoT architecture needs to address several factors like scalability, interoperability, reliability, QoS, and security. In this context, IoT middleware is considered an essential factor in producing IoT services that are highly solicited to be very ubiquitous and decentralized. Middleware can be defined as software that provides an abstraction between IoT devices and application. IoT middleware is required for two major reasons:

- Enables distributed processing and intercommunication of services between heterogeneous networks, devices, and applications;
- Provides portability and supports standard protocols to enable interoperability.

5.2.1.3 IoT Protocols

The IoT protocol ensures that information from one device or sensor is readable and understandable by another device, gateway, or service. Various standardized IoT protocols have been designed and optimized for different scenarios and applications [20, 21]. Given the variety of IoT devices available, it is important to use the right protocol in the right context. In our case, we chose MQTT because of its lightweight header, which consumes much less power and makes it one of the most suitable solutions within tight environments. A great advantage given by MQTT is the capacity to store some messages for new subscribers by setting a keep flag into published messages. By using the TCP protocol, MQTT can be critical for strained devices. Thus, a solution has been proposed for sensor networks (MQTT-SN) using the UDP protocol and supporting the indexing of topic names. The MQTT-SN was specially designed for sensor networks [22] and is considered an improved version of MQTT. Moreover, the MQTT protocol cannot specifically integrate spatiotemporal data flows. To meet this, Herle et al. [23] propose the GeoMQTT (Geospatial Message Queuing Telemetry Transport) extension, which addresses this problem by adding the functionality of marking published data with a timestamp and location and allowing subscriptions to some areas and/or time intervals. It preserves the efficiency of MQTT resources and therefore constitutes a standard well adapted to the integration of spatiotemporal data flows of IoT devices.

5.2.2 IoT Architectures

Several architectures have been proposed to manage and exploit exchanged data in IoT. Many approaches are proposed to overcome the lack of resources on connected objects while allowing scaling and low latency calculations.

5.2.2.1 Fog-Cloud Architecture

The fog-cloud paradigm uses a node-oriented approach that is especially useful for integrating IoT devices. It includes three elements that are considered the principal criteria to take into account to decide whether a fog-cloud architecture is required by an IoT application (see Figure 5.1).

- **Resource nodes:** Include the organization of distributed compute nodes that represent message broker, data connection, IoT device connector, and data processing. Geographically, adjacent computing nodes deployed on the edge, fog, and cloud are typically connected through a large number of communication networks.
- **Analytical nodes:** Include most excellent practical solutions for performing analytical functions, which are crucial to meeting the demand of IoT applications.
- **Storage nodes:** Include efficient distributed storage solution deployed to manage the large quantity of data generated by massive IoT devices.

For better management of HazMat transportation, it is necessary to monitor this type of transport in space (destination, roads, stops, etc.) and in time (cargo status, schedule, speed, etc.). In this way, the data collection system using IoT devices is expected to have the following functions:

- **Scalability:** HazMat transportation management applications will not require extensive analytics all the time. Thus, the collection, processing, and analysis of raw data close to IoT devices can produce temporary relevant content, which can be utilized as a foundation for intelligent decision-making.
- **Mobility and geo-distribution:** In most transport management systems, mobility and geo-distribution are essential for supporting real-time analytics.
- **Low latency:** For many decision-making systems, the fog analytics can identify actions in near real time and avoid delays among the sensor-generated event and the response to that event.

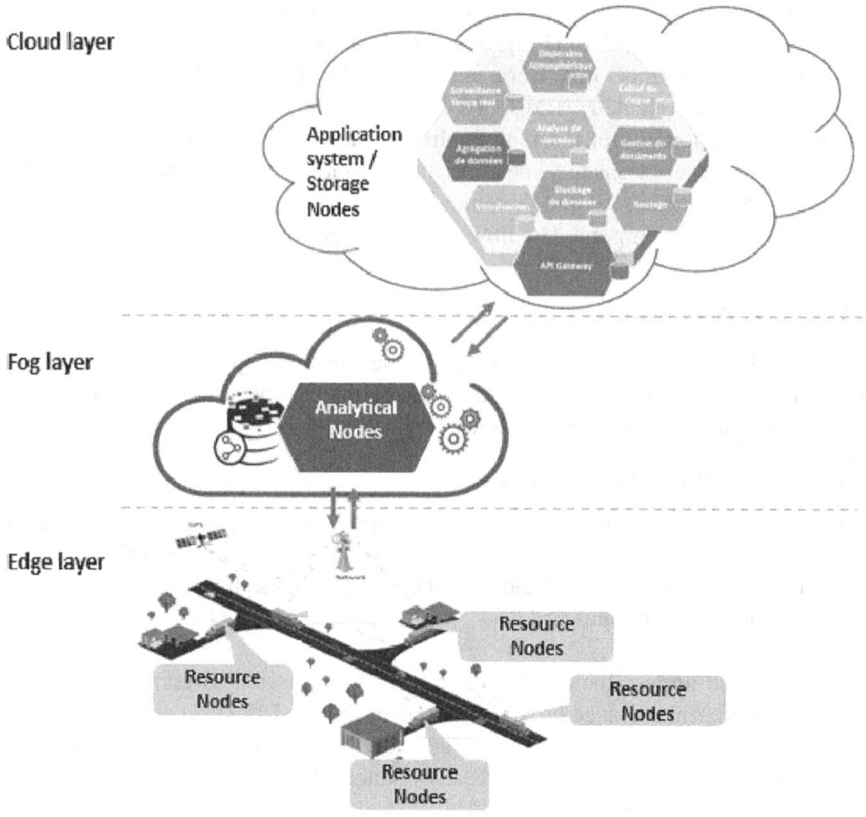

FIGURE 5.1 Edge-fog-cloud architecture.

The fog-cloud architecture might be the appropriate solution to address these challenges. This highly distributed infrastructure is needed from our perspective to support large parallel IoT data streams. The fog-cloud architecture offers various advantages [24]:

- **Network load reducing:** The fog-cloud offers data processing near IoT generators, thus reducing the circulation of a large amount of data in the network.
- **Providing context:** Fog-cloud devices also produce contextual knowledge on data generated by sensors using location or application contexts.
- **No single point of failure:** As computation in the fog-cloud paradigm is fully distributed, the pattern has no single point of failure. Moreover, several instances of an application can be deployed on the cloud for increased reliability.

5.3 THE PROPOSED GEO-REFERENCED DATA COLLECTION MICROSERVICE

In this section, we present a geo-referenced data collection system based on fog-cloud architecture, exploiting its advantages in low-level devices to improve the HazMat transport while taking into account mobility constraints.

5.3.1 System modeling

The collection system was designed to be part of a larger IoT infrastructure, which means it should serve as a data provider to a larger IoT platform. In this way, the data collected can be used in other contexts, for example related to smart parking or fleet management. This was inspired by the fact that sensors, platforms, and services operate in different locations and are provided by different entities. With the influence of distributed architectures and the fog computing paradigm, the proposed system is designed as event-driven microservices (see Figure 5.2). This design gives the system great flexibility in scalability and adaptability to various data applications [2, 25]. The system needs to integrate dynamic data streams to create predefined tasks in near real time. Several streams of spatiotemporal data are sent regularly from the onboard system via the MQTT protocol. To enable the integration and management of these streams, we have created a cluster of scalable local brokers to which the onboard systems send their data.

This cluster is realized with a flexible number of nodes supporting the GeoMQTT protocol. Each broker processes the storage and transfers the data. It represents the main bus that entities and applications can subscribe to and access data.

The main entity in the system is an event detector that encapsulates one or more microservices. In fact, the proposed system uses an event detector to evaluate the

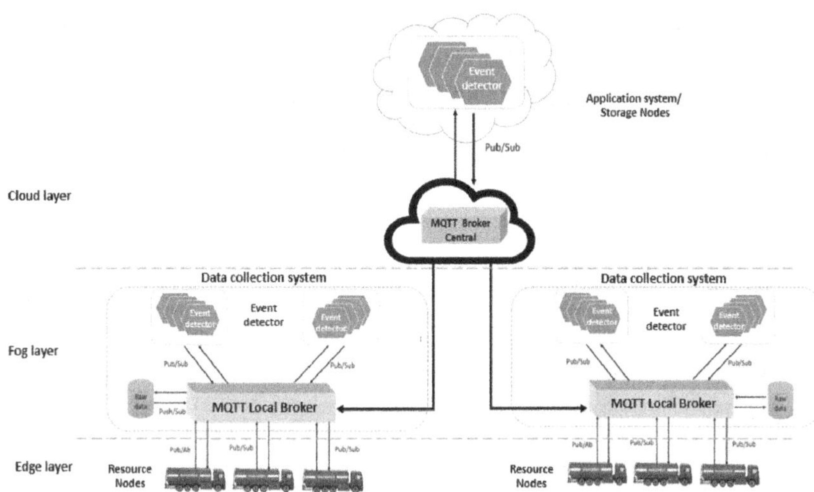

FIGURE 5.2 An overview of the data collection architecture.

data it receives; this may expose more detailed data to the application system. This detailed data is sent to the ITS application via the central broker within a limited time frame. The central broker acts as middleware to ensure communication between these two systems. With its powerful computational resources and overall knowledge of roads and networks, the ITS application system makes final decisions based on further analysis of the data. The proposed system is deployed at the fog level to support real-time analysis of local data. In fact, cloud fog allows you to process data near the source instead of sending it to the cloud or a remote data center. This allows for low latency communication and fast response to data [26]. In addition, fog-cloud applications are useful in situations where the network is geographically very large and many units are connected to the network. You can also enhance your privacy and security by not having to send all your data to the cloud platform. It also reduces bandwidth, allowing you to optimize your data collection and event detection processes. In what follows, we will describe in detail the elements of the proposed architecture.

5.3.2 Cluster of GeoMQTT Brokers

Publish/subscribe middleware typically provides distributed, asynchronous, loosely coupled communication between multiple message generators (publishers) and message consumers (subscribers). The data producer publishes a message about the subject observed by the middleware. This subscription mechanism facilitates the integration of shared data. Taking a scenario where a particular service is designed to work only with data in a particular area. For example, it detects prohibited vehicle intrusions (geofences) into sensitive areas. This service subscribes to buffers around the area of interest and specifically consolidates the data of interest. This use case is shown in Figure 5.3. The line represents the course of the vehicle. Consolidate all vehicle passpoints into the red zone by specifying the GeoMQTT [27] spatial filter as shown. In this work, we used a cluster of local GeoMQTT brokers as middleware to provide interoperability and abstraction between the physical (embedded system) and the cloud layers (ITS application system).

A GeoMQTT broker cluster is a distributed system that represents a logical GeoMQTT broker. It consists of several GeoMQTT broker nodes, usually deployed

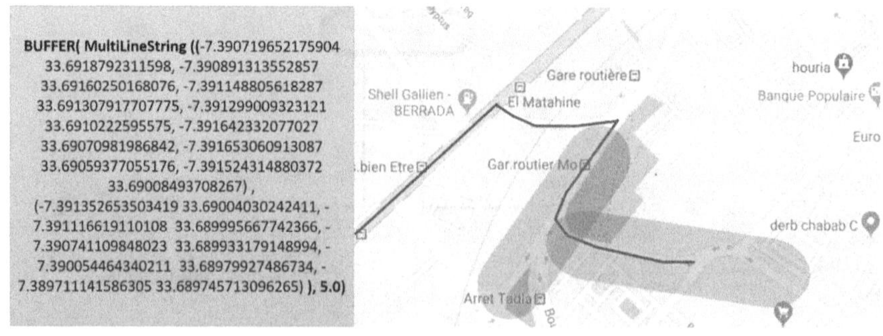

FIGURE 5.3 Geofencing use cases.

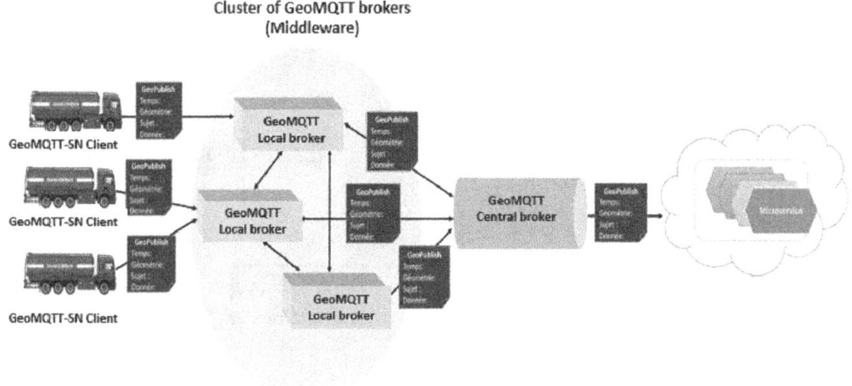

FIGURE 5.4 A cluster of GeoMQTT brokers.

on distinct physical machines and connected over the network. In the perspective of the GeoMQTT client, the broker cluster behaves like a single GeoMQTT broker (see Figure 5.4).

The advantages of using a group (cluster) of brokers are as follows:

- **Elimination of single points of failure:** Due to its architecture, MQTT communication usually has a single point of failure (broker). MQTT communication is not possible if this central component is not available. With a broker cluster, various brokers (deployed on separate machines) operate as a single unit, and even if one of these brokers is unavailable, the entire logical cluster is still available, so this single point of failure is eliminated.
- **Clients can resume sessions:** Since a broker cluster is a logical broker, the node must ensure that the publish and subscribe mechanism works the same on all nodes and messages are distributed to the cluster nodes as needed. This means that the customer can receive the message in the queue, resume the QoS flow, and maintain the subscription. This is also true if the node to which the client originally connected is no longer available.
- **Resilient and fault-tolerant:** Broker clusters are designed for use in vulnerable ecosystems and are sufficiently resilient and fault-tolerant in the event of infrastructure problems. It means that if any number of nodes fails, the cluster will continue to operate.

5.3.3 Event Detector

Incident detection and prevention is one of the key activities of hazardous substance transport systems. In fact, the event detection process relies on a real-time extraction of the collected data stream. However, due to vehicle mobility and wireless communication bandwidth limitations, events generated by the onboard system can be delayed and sent to remote application systems or lost. In this system, high-level event detection tasks are divided into various low-level detection tasks (that is,

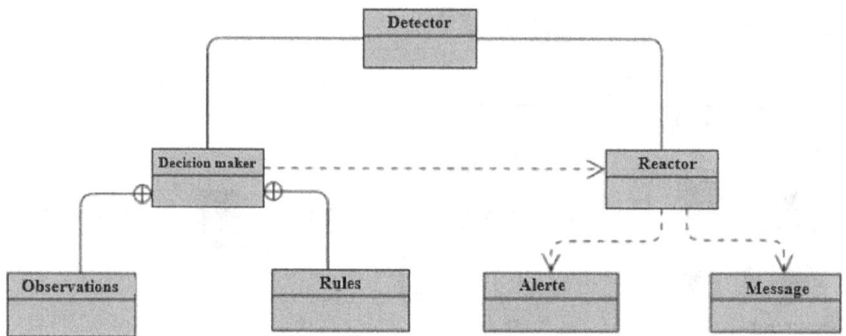

FIGURE 5.5 Event detector design.

secondary detectors), depending on specific application logic and device characteristics of data flow. The event detection process is modeled as an event tree topology with routes as the source of the data flow (that is, the onboard system), each leaf as the detection result and action, the corresponding value, and all other vertices as the secondary detector. These detectors are programs encapsulated in lightweight, containerized microservices for a high degree of automation, deployment, resilience, and reconfiguration and provide the available computing resources for these nodes. With this in mind, it will be distributed and deployed across different fog nodes, as CPU, memory, storage, and network bandwidth. Communication between these microservices is via standardized messages called topics published in the middleware mentioned above. The event detector design is shown in Figure 5.5.

Each event detector manages two main components: the decision-maker and the reactor. Decision-makers subscribe to one or more topics to observe the various parameters and other secondary detectors involved in event generation. These observations apply to predefined rules (e.g., comparison with standard values). Reactors are used to perform actions triggered by decision-makers. The main function of the reactor is to forward event messages and trigger alerts.

5.3.4 Internal System

As mentioned earlier, the need for a new architecture in the TMD management ecosystem is clear and depends not only on how the system is designed but also on how the system interacts with other entities. The system described in this section uses several independent microservices. Each microservice is organized around features in a limited context, with presentation, business domain, and persistence layers. It is responsible for end-to-end development of functions, data models, persistence, user interfaces, service access contracts using Application Programming Interfaces (APIs), and more. In fact, well-designed microservices should not expose their internal data model directly to the outside world. Instead, APIs should be used to avoid corruption of the business layer, which includes functional logic and represents the added value of the application. This is the part that is most likely to be retained if the technology changes. The API is intended to act as an "adapter" to the business layer. For example, the REST

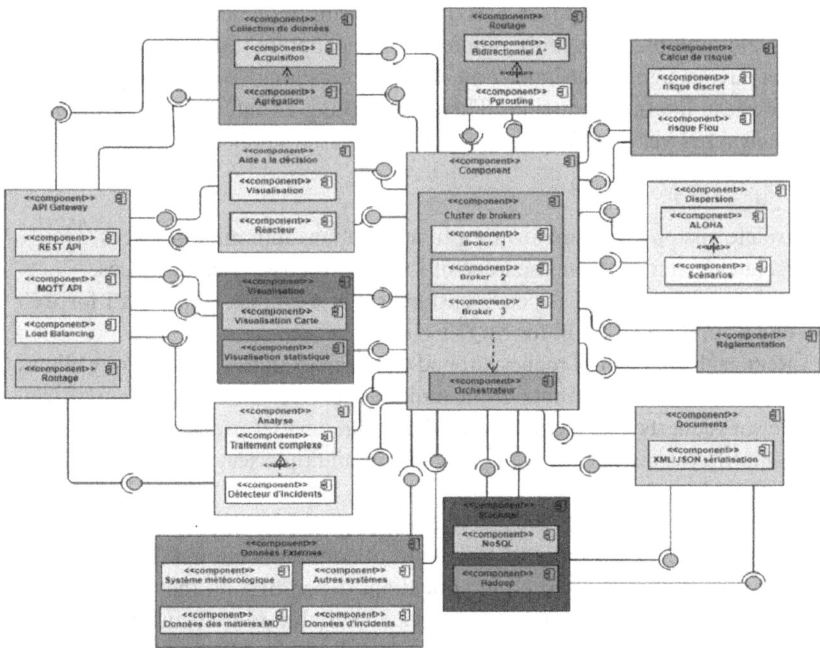

FIGURE 5.6 Component diagram of internal system.

API is used to accept requests from external users or other components of microservices. Internally, call the main business layer to perform this operation and generate a response. Similarly, the database adapter (data access layer) is used by the business layer to interact with external databases, as shown in Figure 5.6. Microservices architectures often require multiple microservices to interact with each other in order to perform a particular task. These dialogues are provided through communication services based on a non-blocking remote call model that uses some form of asynchronous communication mechanism. To completely abstract the complex message flow between microservices, we utilize the API gateway to offer a single-entry point into the set of microservices that make up the internal ecosystem. The promising features of our ecosystem and the microservices that implement them are discussed in the next subsection.

5.3.4.1 API Gateway

An API gateway is a service or system that precedes one or more APIs and performances as a single-entry point into the system. The API gateway acts as a reverse proxy server by pulling data from the back end server and writing its content to the client or application. Installing API gateway in front of the system as a single-entry point improves the isolation level of all internal system microservices. This allows functions such as authentication and monitoring to be centralized within the API gateway. Other features such as caching and load balancing can also improve system performance. In fact, in our ecosystem, customers typically need to use the capabilities of multiple microservices. If this consumption occurs directly, the client must handle multiple calls to the microservices endpoint. Future ecosystem evolutions can have a significant impact on client applications, as client applications are bound to these internal endpoints. Therefore, having an intermediate-level indirection (gateway) is very useful for microservices-based applications and has the following advantages:

- Isolate customers from how to split applications into microservices.
- Isolate customers from the issue of locating microservice instances.
- Provide the best API for each customer. For example, REST API for web clients and MQTT API for MQTT clients.
- Reduce the number of requests/round trips. For example, API gateway allows customers to extract data from multiple services in a single round trip. Fewer requests also mean less overhead and a better user experience.
- Simplify the client by moving logic to call multiple services from the client to the API gateway.

The proposed ecosystem supports two types of communication, HTTP-based communication and MQTT-based communication, depending on the various application protocols adopted. To make the service available to our customers, the API gateway provided includes two types of APIs, the Rest API and the MQTT API. The REST API enables hypertext exchange between the customer's application and the internal ecosystem. This type of communication allows customers to register, log in, or log out before further operation. To support communication with onboard devices and systems, use the MQTT API to communicate with these devices via the MQTT protocol. This API allows the end user to send commands to the device and the device to report the status to the system. In addition, the microservices that make up the ecosystem are deployed in the cloud and are typically dynamically allocated. In addition, all instances of the service are dynamically modified by autoscaling and upgrades. Therefore, API gateway must use the service discovery mechanism of the client-side system, just like any other service client in the system. API gateway availability is essential to ensure application availability. To be able to use API gateway, you need a load balancer that can adapt to rapid changes in microservices, such as version control and dynamic resizing. In addition, because API gateway is exposed to the external network, it must be able to provide secure transport and authentication, as well as a variety of access policies for external and internal customers.

5.3.4.2 Communication Service

The communication service uses a message broker at its center to send messages between microservices. The message broker provides content and subject-based routing using the publish/subscribe model, which makes the sender and recipient independent of each other. All viewing microservices subscribe to one or more topics that they can use to receive messages and also connect to topics that they can post to other microservices. All interactions carried out through message brokers are asynchronous in nature and do not block the sender. It helps grow publishers and subscribers independently. All requests generated by users or microservices are posted to a topic to which one or more microservices can subscribe and receive the message to be processed. The generated results can also be written on a topic that can be later selected by another microservice, which can either return the response to the application or keep it in a database. If a subscriber fails, the broker can reread the message. Likewise, if all the subscribers are busy, the broker can accumulate messages until they are processed by the subscribers.

5.3.4.3 Data Aggregation Service

Data acquisition takes place primarily at the fog layer. In fact, all sensing devices (such as onboard systems and sensor data from smart phones) are part of fog nodes based on their physical location. Most of the raw data is therefore collected at the fog level. This data will be periodically moved to the data aggregation service in the cloud, which will allow the integration of all the data collected and kept for traceability. Note that the data in the fog layer can be used immediately (real-time data). In addition, additional data (such as meteorological data, HazMat data, incident history data, or data from third-party applications) can optionally be integrated via this aggregation service. The data aggregation service integrates and aggregates the data collected and correlates them with the continuous transmission of data from other providers' designs of the global data ecosystem. In fact, the use of such a service, in the context of data management, is a means of performing any treatment for the combination, cleaning, the elimination of redundancy, and the heterogeneity or storage of the data. This can be obtained by the application of some basic operations, such as the sum, multiplication, or grouping. Many other data aggregation techniques could be easily applied in this architecture. However, in this work, we use a basic program for data aggregation. A detailed study of the theory and methods of aggregation exceeds the framework of this work. The data aggregation service retrieves the data generated by the data collection system in real time and generates a database integrating data from all data providers. Thus, importing all this data into a new large database could consolidate data concerning the road, the states of vehicles, and substances transported with their locations.

5.4 CONCLUSION

This work presents the design and overall description of a geo-referenced data collection microservice based on IoT protocols for smart HazMat transportation. The HazMat transport management systems respond to a concept that is increasingly important in the way data collection and analysis are carried out: add the data computation process at the edge as close as possible to the data source devices. This presents general advantages that would benefit almost any IoT application. It makes possible to receive fast feedback by the analysis of the data. Thus, we used edge-fog-cloud architecture to offer a high level of insulation, autonomy, and responsiveness applications in sensitive areas such as the transport of dangerous goods. In addition, the proposed ecosystem can be reused and easily readapted for other application fields.

ACKNOWLEDGMENT

This work was partially funded by the CNRST project in the priority areas of scientific research and technological development "Spatio-temporal data warehouse and strategic transport of dangerous goods".

REFERENCES

1. Boulmakoul, A., Laurini, R., & Zeitouni, K. (2001). Spatial monitoring and routing system for the transportation of hazardous materials. In Environmental Information Systems in Industry and Public Administration (pp. 227–236). IGI Global.
2. Cherradi, G., El Bouziri, A., Boulmakoul, A., & Zeitouni, K. (2017). Real-time HazMat environmental information system: a micro-service-based architecture. Procedia Computer Science, 109, 982–987.
3. Atzori, L., Iera, A., & Morabito, G. (2010). The internet of things: a survey. Computer Networks, 54(15), 2787–2805.
4. Akyildiz, I. F., Su, W., Sankarasubramaniam, Y., & Cayirci, E. (2002). Wireless sensor networks: a survey. Computer Networks, 38(4), 393–422.
5. Antonić, A., Marjanović, M., Pripužić, K., & Žarko, I. P. (2016). A mobile crowd sensing ecosystem enabled by CUPUS: cloud-based publish/subscribe middleware for the Internet of Things. Future Generation Computer Systems, 56, 607–622.
6. Bansal, S., & Kumar, D. (2020). IoT ecosystem: a survey on devices, gateways, operating systems, middleware and communication. International Journal of Wireless Information Networks, 27, 340–364. https://doi.org/10.1007/s10776-020-00483-7.
7. Borgia, E. (2014). The internet of things vision: key features, applications and open issues. Computer Communications, 54, 1–31.
8. Bakshi, K. (2016). Big data analytics approach for network core and edge applications. In 2016 IEEE Aerospace Conference (pp. 1–10). IEEE.
9. De Brito, M. S., Hoque, S., Steinke, R., & Willner, A. (2016, September). Towards programmable fog nodes in smart factories. In 2016 IEEE 1st International Workshops on Foundations and Applications of Self* Systems (FAS* W) (pp. 236–241). IEEE. Augsburg, Germany.
10. Rahmani, A. M., Liljeberg, P., Preden, J. S., & Jantsch, A. (Eds.). (2017). Fog Computing in the Internet of Things: Intelligence at the Edge. Springer, Cham.
11. Giang, N. K., Leung, V. C., & Lea, R. (2016, November). On developing smart transportation applications in fog computing paradigm. In Proceedings of the 6th ACM symposium on development and analysis of intelligent vehicular networks and applications (pp. 91–98).
12. Saxena, S., Bhushan, B., & Ahad, M. A. (2021). Blockchain based solutions to secure IoT: background, integration trends and a way forward. Journal of Network and Computer Applications, 181, 103050.
13. Tang, B., Chen, Z., Hefferman, G., Wei, T., He, H., & Yang, Q. (2015). A hierarchical distributed fog computing architecture for big data analysis in smart cities. In Proceedings of the ASE Big Data & Social Informatics 2015 (pp. 1–6).
14. Kumar, S., Tiwari, P., & Zymbler, M. (2019). Internet of Things is a revolutionary approach for future technology enhancement: a review. Journal of Big Data, 6(1), 1–21.
15. Zanella, A., Bui, N., Castellani, A., Vangelista, L., & Zorzi, M. (2014). Internet of things for smart cities. IEEE Internet of Things Journal, 1(1), 22–32.
16. Djuknic, G. M., & Richton, R. E. (2001). Geolocation and assisted GPS. Computer, 34(2), 123–125.

17. Jaselskis, E. J., Anderson, M. R., Jahren, C. T., Rodriguez, Y., & Njos, S. (1995). Radio-frequency identification applications in construction industry. Journal of Construction Engineering and Management, 121(2), 189–196.
18. Agarwal, P., & Alam, M. (2020). Investigating IoT middleware platforms for smart application development. In Smart Cities—Opportunities and Challenges (pp. 231–244). Springer, Singapore.
19. Ngu, A. H., Gutierrez, M., Metsis, V., Nepal, S., & Sheng, Q. Z. (2016). IoT middleware: a survey on issues and enabling technologies. IEEE Internet of Things Journal, 4(1), 1–20.
20. Al-Fuqaha, A., Khreishah, A., Guizani, M., Rayes, A., & Mohammadi, M. (2015). Toward better horizontal integration among IoT services. IEEE Communications Magazine, 53(9), 72–79.
21. Sachs, K., Appel, S., Kounev, S., & Buchmann, A. (2010). Benchmarking publish/subscribe-based messaging systems. In International Conference on Database Systems for Advanced Applications (pp. 203–214). Springer, Berlin, Heidelberg.
22. Stanford-Clark, A., & Truong, H. L. (2013). MQTT for sensor networks (MQTT-SN) protocol specification. International Business Machines (IBM) Corporation Version, vol. 1, no 2.
23. Herle, S., & Blankenbach, J. (2016). GeoPipes using GeoMQTT. In Geospatial Data in a Changing World (pp. 383–398). Springer, Cham.
24. Mohan, N., & Kangasharju, J. (2016, November). Edge-Fog cloud: a distributed cloud for Internet of Things computations. In 2016 Cloudification of the Internet of Things (CIoT) (pp. 1–6). IEEE, Paris, France.
25. Cherradi, G., El Bouziri, A., & Boulmakoul, A. (2017). Smart data collection based on IoT protocols. In JDSI'16.
26. Sarkar, S., Chatterjee, S., & Misra, S. (2018). Assessment of the suitability of fog computing in the context of internet of things. IEEE Transactions on Cloud Computing, 6(1), 46–59.
27. Hillar, G. C. (2017). MQTT Essentials—A Lightweight IoT Protocol. Packt Publishing Ltd, Birmingham.

6 Impact of Dimensionality Reduction on Performance of IoT Intrusion Detection System

Susanto and M. Agus Syamsul A.
University of Sriwijaya, Palembang, Indonesia
University of Bina Insan, Lubuklinggau, Indonesia

Deris Stiawan
University of Sriwijaya, Palembang, Indonesia

Mohd. Yazid Idris
University of Teknologi Malaysia, Johor, Malaysia

Rahmat Budiarto
University of AlBaha, Al Bahah, Saudi Arabia

CONTENTS

DOI: 10.1201/9781003219620-6

6.1 INTRODUCTION

The Internet of Things (IoT) enables merging of digital and physical objects using viable communication technologies and introduces a vision of a future where computing systems, users, and objects work together for convenience and economic benefits (Bhushan et al. 2021). This fact has led to an increase in the use of IoT technology. This technology still has many security vulnerabilities and privacy risks ranging from attacks on data to attacks on devices (Sharma et al. 2021). Current IoT devices are insecure and unable to sustain themselves mainly due to their limited resources, poor interoperability, and lack of security design in software development as well as device design (Saxena, Bhushan, and Ahad 2021). With these conditions, it is not easy to create a security intrusion detection system (IDS); thus, the biggest challenge is to provide efficient security techniques (Bhushan and Sahoo 2018), with low computational and storage costs (Kumar et al. 2021).

Dimensional reduction is suitable to be applied to IDS IoT and it is very beneficial because it will reduce computational resource and memory usage while preserving the accuracy detection. The dimensionality reduction (DR) technique is implemented during data preprocessing as well as data analysis (Yan et al. 2006). The way the DR technique works is by mapping high-dimensional data into respective low-dimensional data, which then works on the mapped data to solve the problems in the high-dimensional data, in this case, the IoT network traffic data.

This chapter contributes toward revealing the impacts of the use of DR technique on IoT IDS. This chapter also provides IoT network traffics data classification using six classifier algorithms, i.e. k-Nearest Neighbor (k-NN), decision tree (DT), random forest (RF), adaboost (AD), gradient boosting (GB), and Naïve Bayes (NB), for IoT network traffic characterization purpose. Finally, this work measures the performances of the classifiers in terms of processing time, accuracy, and false positive rate (FPR).

The rest of this chapter is structured into six sections as follows. Section 6.2 presents an overview of DR. Section 6.3 describes the data source, technical design implementation of DR, and analysis tool. Section 6.4 presents the analysis of the experimental results and comparison with other works. Section 6.5 presents the visualization of IoT IDS. Section 6.6 provides the analysis and discussion. Lastly, Section 6.7 presents the conclusions.

6.2 GENERAL OVERVIEW OF DIMENSIONALITY REDUCTION

This section begins with a brief overview of DR techniques along with their functionalities and ends with the works that implement the DR in IoT IDS.

6.2.1 Dimensionality Reduction Techniques

DR is data projection process from high-dimensional data to lower dimension data by converting data from n-dimension to k-dimension, where k < n. DR has been proven as important part of data analysis for machine learning (ML), especially in preprocessing data. Dimensional reduction implementation in preprocessing data can increase efficiency and effectivity of the learning process. It eliminates irrelevant or exceeding data, increases learning accuracy, and produces comprehension result. In literatures, DR techniques are categorized into two groups, i.e. linear dimension reduction and nonlinear dimension reduction.

- **Linear dimension reduction**. It uses basic linear function to convert high-dimension data to lower dimension data. DR techniques, which are grouped in this category, include Principal Component Analysis (PCA), Singular Value Decomposition (SVD), Latent Semantic Analysis (LSA), Locality Preserving Projections (LPP), Independent Component Analysis (ICA), Linear Discriminant Analysis (LDA), Projection pursuit, Incremental PCA (I-PCA), Probabilistic PCA (P-PCA), Sparse PCA (S-PCA), Factor Analysis (FA), Fast ICA, Truncated SVD, Large-Margin Nearest Neighbor (LMNN) Metric Learning, Linear Local Tangent Space Alignment, Nonnegative Matrix Factorization (NMF), and Random Projection (RP).
- **Nonlinear dimension reduction**. It uses basic nonlinear function to change high-dimensional data to lower dimension data. DR techniques that are included in this category are as follows: Kernel Principal Component Analysis (KPCA), Multidimensional Scaling (MDS), Isomap, Locally Linear Embedding (LLE), Learning Vector Quantization (LVQ), Self-Organizing map, T-Stochastic Neighbor Embedding (T-SNE), Landmark Isomap, Variants Hessian LLE, Modified LLE, Local Tangent Space Alignment (LTSA), Laplacian Eigenmaps, Diffusion Maps, Manifold Charting, Local Linear Coordination (LLC), Local Affine Multidimensional projections (LAMP), Projection by Clustering (PBC), Interactive Document Maps (IDMAP), Maximally Collapsing Metric Learning (MCML), Uniform Manifold Approximation and Projection (UMAP), Nonmetric Multidimensional Scaling (N-MDS), Landmark MDS, Gaussian Process Latent Variable Model (GPLVM), Sparse Random Projection (S-RP), Maximum Variance Unfolding (MVU), Fast MVU, Landmark MVU, Generalized Discriminant Analysis (GDA), Kernel LDA, Least Square Projection (LSP), Fastmap, Piecewise Least Square Projection (PLSP), Autoencoder, Stochastic Proximity Embedding (SPE).

6.2.2 Function of DR

DR is considered a part of ML and it is a dynamic algorithm. DR is able to take decision because of its functions, which are as follows:

- Increases classification accuracy,
- Requires low data storage because data are converted to lower dimension,
- Increases execution speed,
- Reduces data duplication,
- Optimizes computation resource,
- Reduces computation complexity,
- Solves problem of "curse of dimensionality",
- Reduces data redundancy,
- Removes irrelevant feature,
- Extracts feature,
- Selects feature,
- Visualizes data.

6.2.3 Dimensionality Reduction on IoT IDS

In general, there is an implementation of IDS on IoT network, and the DR technique is used during the data preprocessing stage. The technique is used for feature extraction and selection process (Khalid, Khalil, and Nasreen 2014; Velliangiri, Alagumuthukrishnan, and Iwin Thankumar Joseph 2019), while in the data analysis stage, it is used for data visualization.

6.2.3.1 Feature Extraction

Feature extraction is a transformation process from the original feature set to other feature sets, which contains the best relevant and significant features. The extraction is done by projecting original data with high dimensionality into space with lower dimensionality using algebra transformation (Yan et al. 2006). The feature extraction has to be able to combine the obtained new information while prheserving the previously obtained knowledge without accessing the trained data in the previous process (Diaz-Chito, Ferri, and Hernández-Sabaté 2018). Besides, data complexity is also being reduced when the attributes of data are decreased (Baaqeel and Saqib 2020). Feature extraction in IoT IDS is used to extract raw data to make the data able to be analyzed. Implementation of dimensional reduction technique in feature extraction process is through reducing the size of feature space from raw data without limiting the original information. So it reduces data complexity and gives simple data representation. Selecting dimensional reduction technique used in feature extraction process is based on data type. The feature extraction process of dimensional reduction technique in IoT IDS is illustrated in Figure 6.1.

Dataset, which is still in the data raw form or .pcap format, is extracted using DR technique without reducing the information in the data itself. Then it is followed by

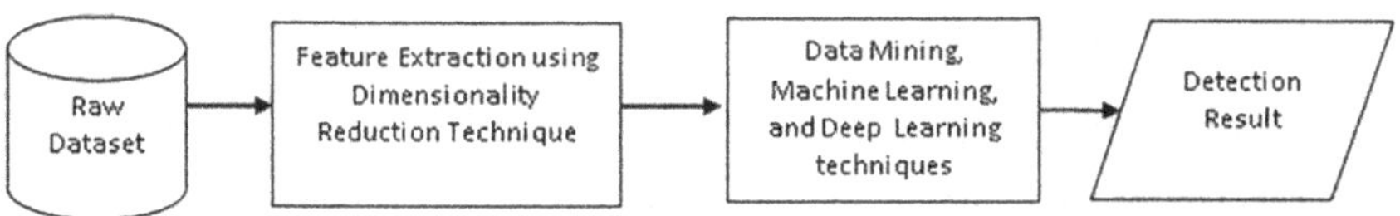

FIGURE 6.1 Feature extraction process using dimensionality reduction technique.

the detection process of whether using data mining technique, ML, or deep learning (DL) that provide optimal detection results.

6.2.3.2 Feature Selection

Feature selection is a process to select the best features from all relevant features based on certain clusters to differentiate classes of features. In feature selection process, available subset of feature data is chosen for learning process. The best subset is the subset with the lowest dimensionality and contributes toward the increase of accuracy during learning process (Khalid, Khalil, and Nasreen 2014). The selection of feature subset works by deleting irrelevant and redundant features; a subset of features that was selected will provide the best performance according to several main objectives (Wang et al. 2020). This approach is slightly different with DR technique used as feature selection because it firstly compresses the data with high features into lower features data. Implementation of DR technique as feature selection in IoT IDS will be beneficial; the rationale is because removing features that are irrelevant will reduce search space and detection time. Feature selection process in IDS using DR technique is illustrated in Figure 6.2.

During the intrusion detection experiments, dataset in CSV format is processed using feature selection methods, while DR technique functions to reduce/eliminate features. Then the detection process is performed either using data mining technique, ML, or DL to provide optimum detection results.

6.2.3.3 Visualization

One of the techniques that can be used to analyze data is visualization. Visualization is important step in ML. Visualization can be implemented on big or small dataset. Advantage of visualization is to explore a group of data that are unknown and have unclear views about dataset. IDS visualization can be used to analyze attack and normal traffic patterns by visualizing traffic data in two-dimensional or three-dimensional view.

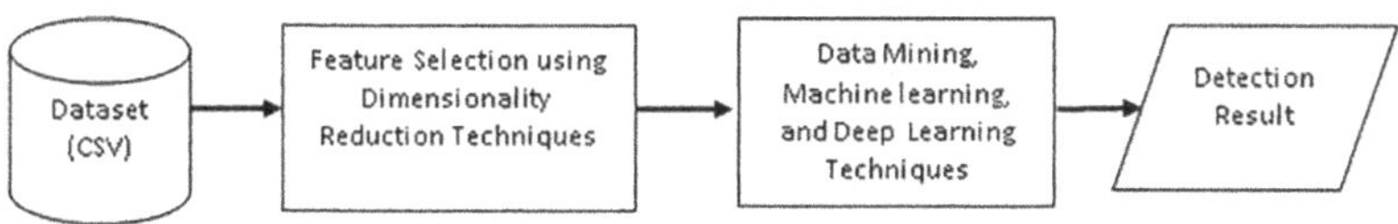

FIGURE 6.2 Feature selection process using dimensionality reduction technique.

6.3 METHODOLOGY

6.3.1 Data Source

In this chapter, the DR techniques are implemented on the IoT IDS and experimented on N-BaloT dataset (Meidan et al. 2018). The IoT network testbed for creating the dataset consists of nine devices, i.e. two doorbells, thermostats, baby monitors, four camera security, and webcam. An example of entries of the N-BaIoT dataset, which is captured from one of the device, i.e. Danmini Doorbell, is shown in Table 6.1. This dataset is chosen because it is collected from a big organization network that is estimated to increase in a number of IoT devices. N-BaIoT dataset has a total of 115 independent features on each file, with additional class label that will decrease from each file name, so this dataset has a very high dimension. The data dimension is reduced with randomly selected 2 features, 3 features, 6 features, and 10 features. We called these 2-, 3-, 6-, and 10-feature schemes.

6.3.2 Technical Design Implementation of Dimensionality Reduction

Due to the limited resources in IoT systems, high-dimension data may cause computation problem in implementation of its IDS. This section presents some techniques of DR, i.e. PCA, I-PCA, S-PCA, and FA. The techniques are then combined with classification algorithms/methods, i.e. k-NN, DT, RF, AD, GB, and NB. Performance metrics that will be evaluated are execution time, accuracy, and FPR value. The overall design of the proposed DR implementation has several steps as presented in Figure 6.3.

TABLE 6.1
An Example of Entries of the N-BaloT Dataset from Danmini Doorbell Device

No	Device	Malware Type	Attack Type	Total Data
1	Danmini Doorbell	Benign	Benign	49548
			Combo	59718
			Junk	29068
		Bashlite	Scan	29849
			Tcp	92141
			Udp	105874
			Ack	102195
			Scan	107685
		Mirai	Syn	122573
			Udp	237665
			Udpplain	81982
Total				1018298

Source: Data from Meidan et al. (2018).

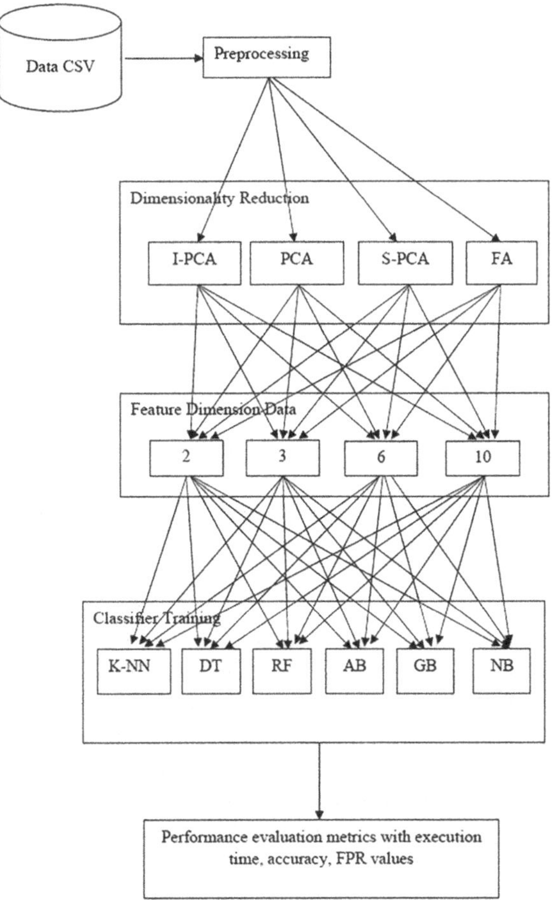

FIGURE 6.3 Technical design implementation of dimensionality reduction.

6.3.3 Analysis Tool

The DR simulation on IoT IDS is implemented on computer with the following specifications: Intel core i7 processor 9th gen, 16-GB RAM, and 512-GB SSD storage. Running operating system is Windows 10. The DR analysis uses Scikit-learn libraries from Python.

6.4 RESULT

6.4.1 Principal Component Analysis (PCA)

PCA is one of the statistics techniques, which is widely used in the data mining research field for reducing dimensionality of original data and identifying a data point with the highest variants (Ikram and Cherukuri 2016). PCA is a dimension reduction technique without basic control and learning about projection matrix;

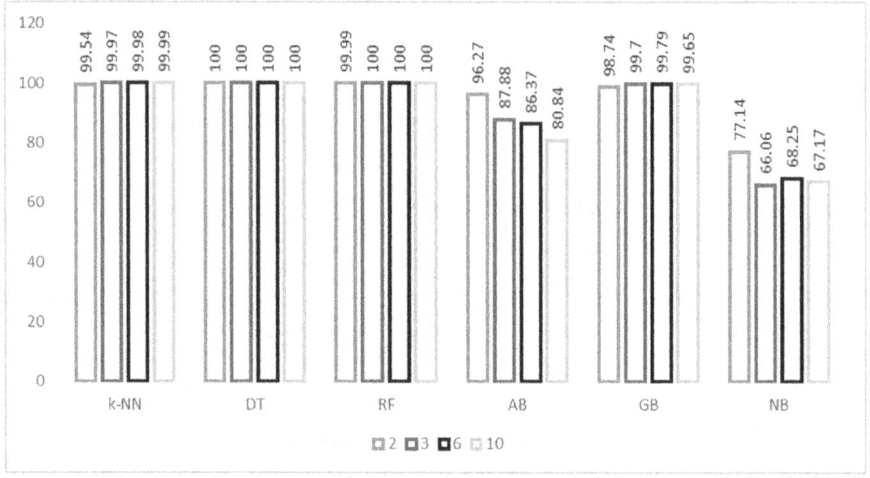

FIGURE 6.4 Accuracy level of PCA technique combined with k-NN, DT, RF, AD, GB, and NB classification methods.

then data variants that have low dimension are maximized (Zhao, Wang, and Nie 2019). It reduces feature space and keeps maximum variants from original data (Ahmadkhani and Adibi 2016). PCA is used to simplify data by linear transformation and create new coordinates with maximum variants using a group of vectors from eigenvalues (Erwin et al. 2019). To understand how the PCA reduces data dimension and its impact on the detection performances, we present here some experimental results. DR implementation will definitely reduce execution time in attacks/malware/anomaly detection.

Experiments on IoT botnet malware detection using DR PCA technique combined with classification methods, k-NN, DT, RF, AD, GB, and NB, were carried out. Figure 6.4 exhibits the accuracy level comparison. Overall, the accuracy level is not significantly affected by dimension reduction. We observed that only detection accuracy using RF with 2-feature scheme is slightly decreased. Meanwhile, detection accuracy using k-NN as well as GB with any features showed small decrement. In contrast, detection accuracy using NB and AB relatively increased. The lesser the number of features used, the higher the accuracy yielded.

For the FPR value, DT and RF have the lowest value of 0 for any number of features used. For k-NN, AB, and GB, there was increment of FPR value when the number of feature used is less. This condition is contradicted with NB that has lower FPR value when the number of feature used is less. Figure 6.5 shows the results.

In terms of execution time, the use of 3-feature scheme, combinations of DR techniques with all classification methods provide fastest execution time. NB is fastest with 9.12 seconds. Figure 6.6 shows the execution time results.

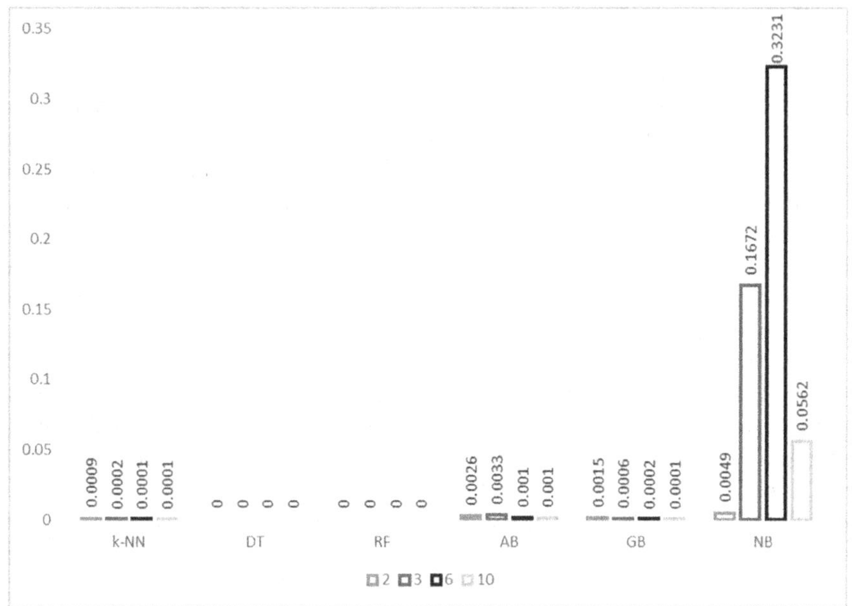

FIGURE 6.5 FPR values of PCA technique combined with k-NN, DT, RF, AD, GB, and NB classification methods.

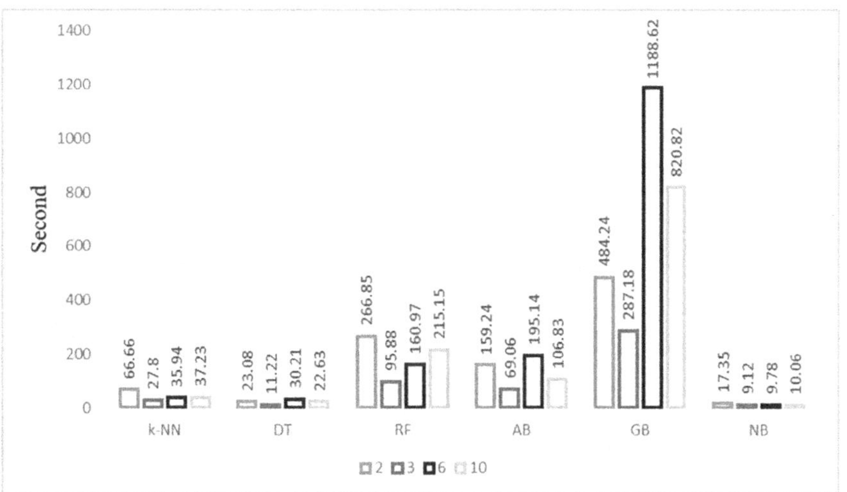

FIGURE 6.6 Execution time of PCA technique combined with k-NN, DT, RF, AD, GB, and NB classification methods.

6.4.2 Incremental Principal Component Analysis (I-PCA)

I-PCA is an extension version of PCA. I-PCA algorithm calculates vector eigen and eigenvalue for covariant sample matrices that differentiate from known matrix data, splitting the eigenvalue system problem. I-PCA makes data matrix available before solving the problem (batch method) and renews the eigenvector when there is upcoming data (Dagher 2010). Then, I-PCA renews the principal components by using new sample data without indirectly calculating the covariant matrices (Park and Lee 2020). Experimental results show that I-PCA is able to reduce big-scale data dimension into significant lower dimension. DR implementation boosts up execution time in botnet malware detection.

Figure 6.7 shows accuracy level of I-PCA combined with classification methods k-NN, DT, RF, AD, GB, and NB in detecting botnet malware. The results in the figure showed that the detection accuracy is not significantly affected by dimension reduction. Accuracy of RF with 2-feature scheme is slightly decreased, while k-NN and GB have slightly different decrement of accuracy for any feature schemes used. For AB, the less used features increase accuracy. While for NB, the accuracy varies.

A comparison of FPR values is shown in Figure 6.8. DT and RF have the lowest FPR value of 0 for any feature schemes. For On k-NN, AB, and GB, the FPR values are slightly increased when the less features are used. In contrast, NB has various FPR values.

Observation on detection time, k-NN, DT, RF, AD, and NB, provides fluctuations in execution times for any features used. For GB, the use of lesser number of features gives faster execution time. Figure 6.9 shows the experimental results.

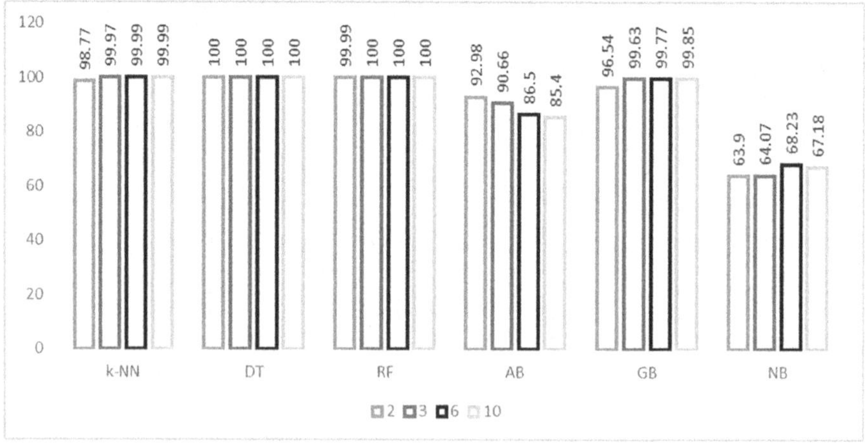

FIGURE 6.7 Accuracy level of I-PCA technique combined with k-NN, DT, RF, AD, GB, and NB classification methods.

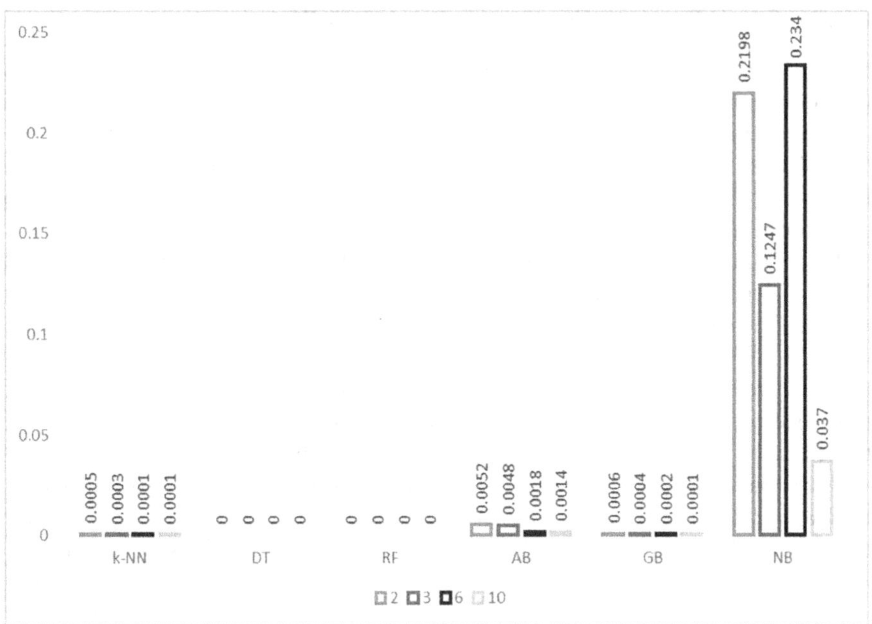

FIGURE 6.8 FPR value of I-PCA technique combined with k-NN, DT, RF, AD, GB, and NB classification methods.

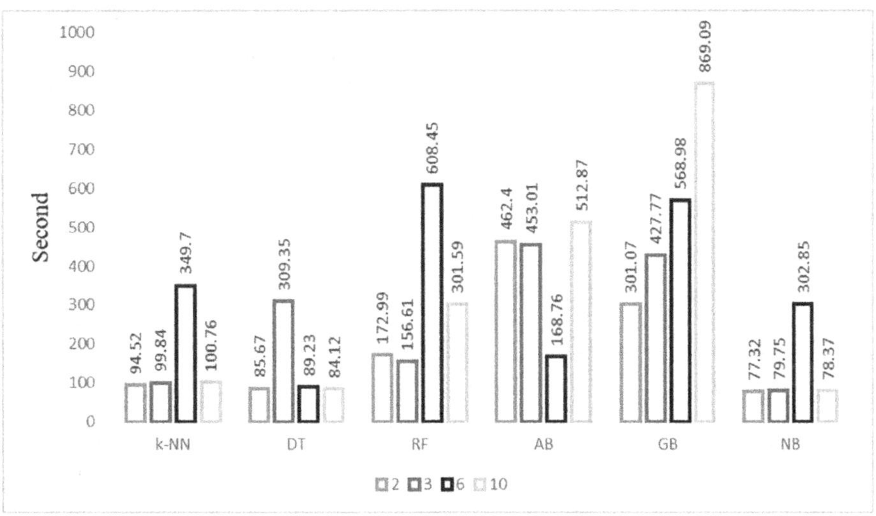

FIGURE 6.9 Execution time of I-PCA technique combined with k-NN, DT, RF, AD, GB, and NB classification methods.

6.4.3 Sparse Principal Component Analysis (S-PCA)

S-PCA is popular approaching factorization matrix based on PCA. It combines maximization and variants' diffusion with final aim to increase data interpretation (Camacho et al. 2020). S-PCA uses lasso (elastic net) in producing principle components that are then modified with scarce weights (Zou, Hastie, and Tibshirani 2006). Experimental results on S-PCA implementation show that the data DR speeds up execution time in botnet detection.

Figure 6.10 shows accuracy level of S-PCA combined with classification methods k-NN, DT, RF, AD, GB, and NB in detecting botnet malware. Accuracy detection using DT technique is not impacted by dimensional reduction. Accuracy detection using RF method with 2-feature scheme is slightly decreased, while k-NN and GB methods are slightly decreased for any features used. The decrement of number of features used in the detection increases the accuracy of AB and NB methods. Overall, the lesser the features used, the higher the accuracy achieved.

A comparison of FPR values is shown in Figure 6.11. DT and RF have the lowest FPR value of 0 for any feature schemes. There is slight increment on FPR values for k-NN method when using less number of features. Then FPR values for AB and NB are fluctuative, while GB has high FPR values when using higher number of features.

Execution (detection) times of k-NN, DT, RF, AD, and NB by using less number of features are relatively fast. Figure 6.12 shows execution time on each classification methods.

6.4.4 Factor Analysis (FA)

FA is used to find attributes observed from a set of data with a lot of initial attributes (variables). In a large dataset, there are a lot of attributes and some of them are not related to the analysis goal. FA is superlative approach for big data analysis (Usman Ali et al. 2017). There are two types of FA: explorative FA and confirmation FA

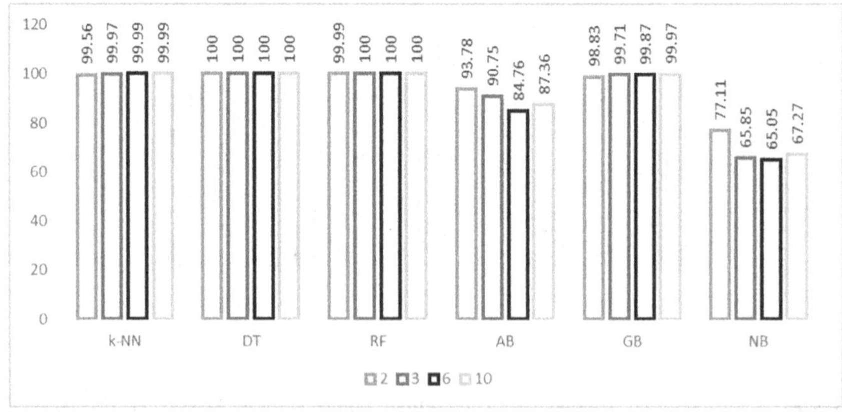

FIGURE 6.10 Accuracy level of S-PCA technique combined with k-NN, DT, RF, AD, GB, and NB classification methods.

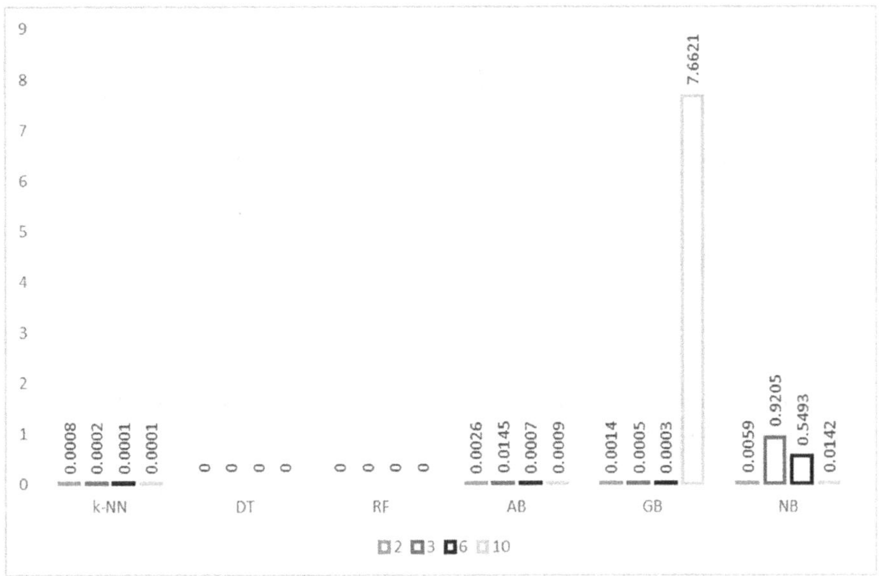

FIGURE 6.11 FPR value of S-PCA technique combined with k-NN, DT, RF, AD, GB, and NB classification methods.

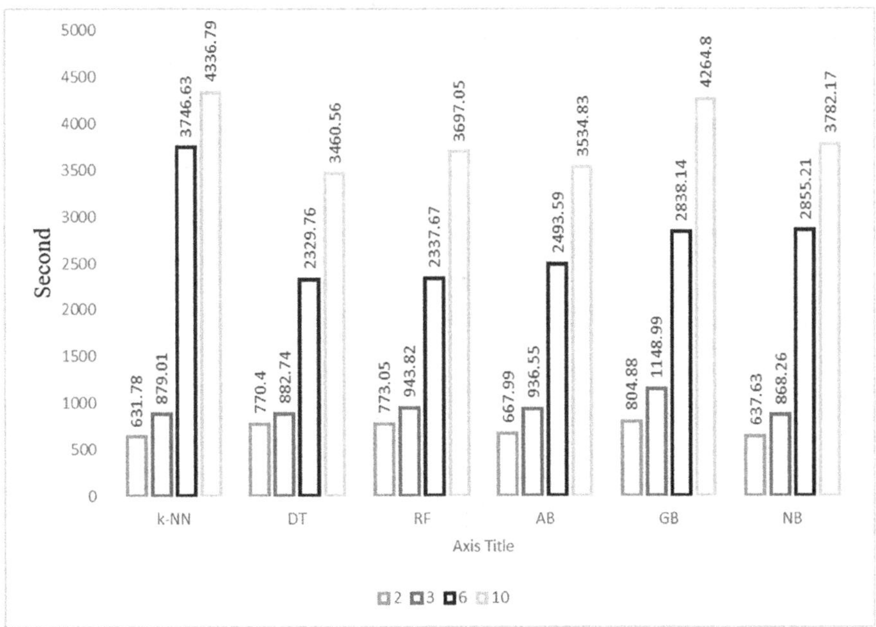

FIGURE 6.12 Execution time of S-PCA technique combined with k-NN, DT, RF, AD, GB, and NB classification methods.

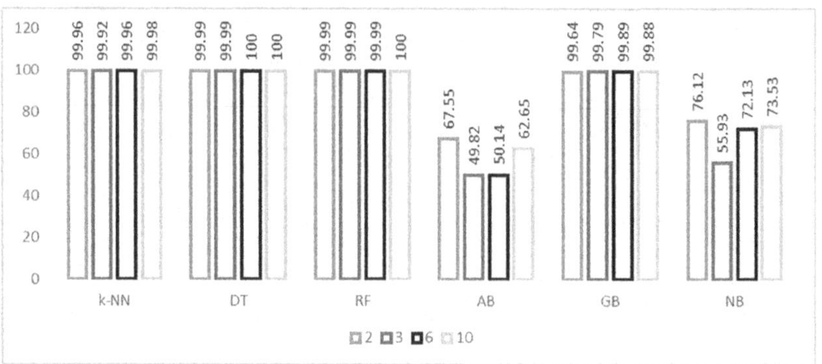

FIGURE 6.13 Accuracy level of FA technique combined with k-NN, DT, RF, AD, GB, and NB classification methods.

(You and Hung 2021). Experimental results show that FA was well functioned in reducing the data dimensionality.

Figure 6.13 shows accuracy level of FA combined with classification methods k-NN, DT, RF, AD, GB, and NB in detecting botnet malware. Comparison result shows that there is slightly decrement of accuracy on DT and RF method. The accuracy decrement on DT method happens when the number of features used is less than 3, while accuracy decrement on RF happens when the number of features used is less than 6. This fact is different with accuracy of k-NN and GB, where it is slightly different accuracy decrement for any number of features used. The reduction of number of features increases accuracy on AB and NB and the accuracy will become higher when 2-feature scheme is used.

A comparison of FPR values is shown in Figure 6.14. DT has the lowest FPR value of 0 for any feature number used. k-NN and GB have slightly increased FPR

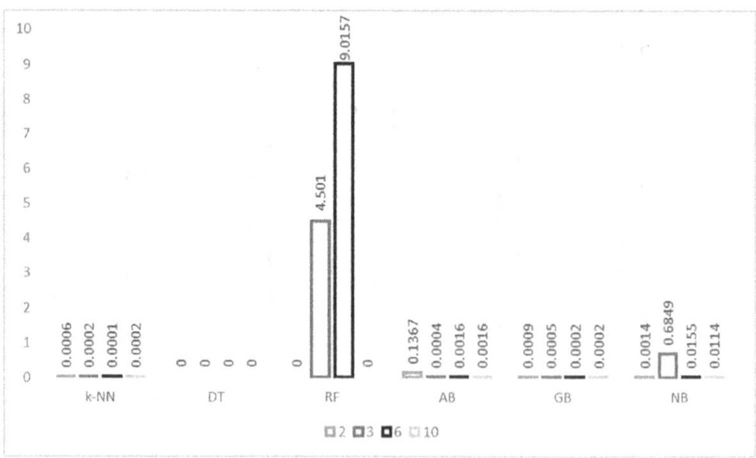

FIGURE 6.14 FPR value of FA technique combined with k-NN, DT, RF, AD, GB, and NB classification methods.

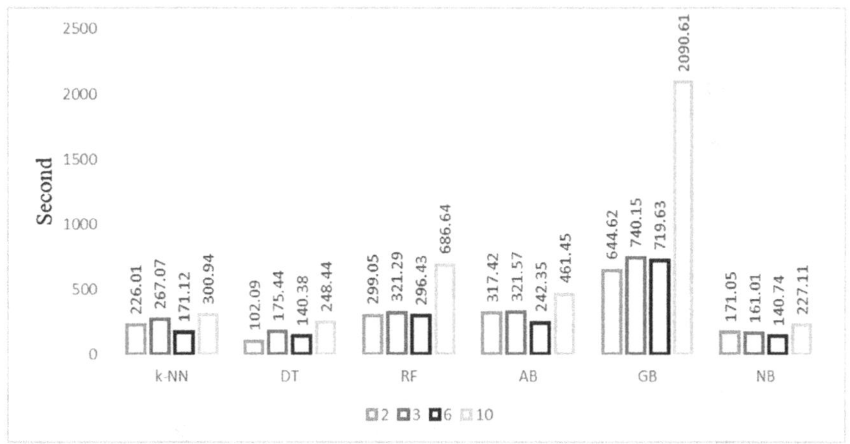

FIGURE 6.15 Execution time of FA technique combined with k-NN, DT, RF, AD, GB, and NB classification methods.

values when less number of features is used. This fact is different with RF, which has a high FPR value when a 6-feature scheme is used, but a very low FPR value of 0 when used 2- and 10-feature schemes.

From time execution perspective, k-NN, DT, RF, AD, and NB have fast execution times when less number of features is used. Comparison of execution time for all selection methods is shown in Figure 6.15.

6.5 VISUALIZATION OF DIMENSIONALITY REDUCTION ON IDS IoT

The challenge in exploring high-dimensional dataset is that dataset has huge size of data. The use of visualization facilitates the understanding of the observed data, because the visualization offers intuitive meanings to obtain insights of the processed data (Zong, Chow, and Susilo 2019), better data interpretation (Roshni Mol and Immaculate Mary 2020), and easier in learning and identifying data (Alsadi and Hadi 2017).

Experiments were carried out to visualize IoT IDS dataset in two dimensions using Seaborn library and three dimensions using mpl_toolkits.mplot3d of Python. This visualization helps in identifying botnet and benign packet traffics in IoT network. Thus, correlation and better understanding of IoT botnet traffic patterns can be achieved.

6.5.1 Visualization on Factor Analysis Dimensionality Reduction

Figure 6.16 shows visualization result of botnet detection on IoT network using FA DR technique with 2-feature scheme.

Figure 6.17 shows visualization result of botnet detection on IoT network using FA with 3-feature scheme

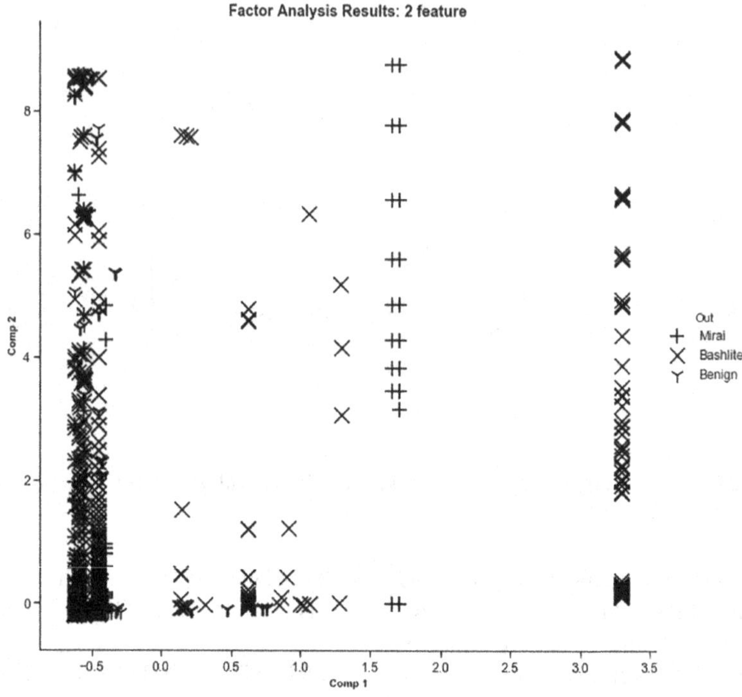

FIGURE 6.16 IoT Botnet visualization using Factor Analysis method with 2-feature scheme.

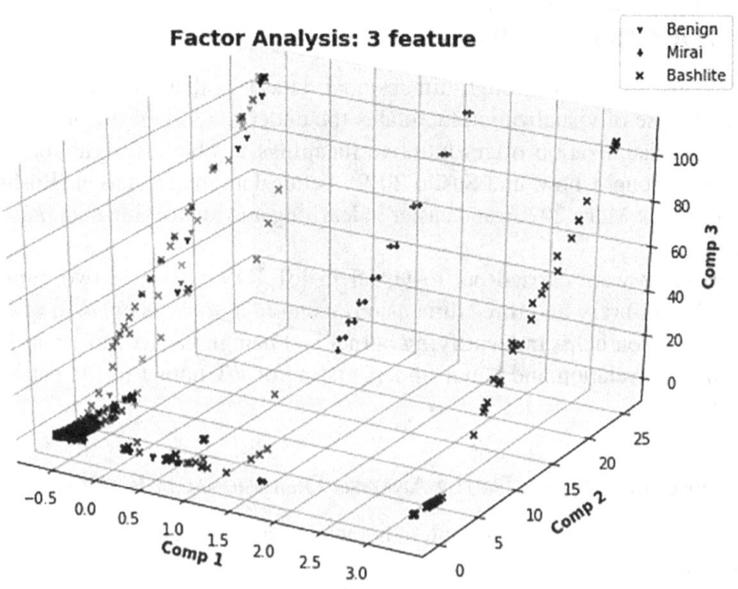

FIGURE 6.17 IoT Botnet visualization using Factor Analysis method with 3-feature scheme.

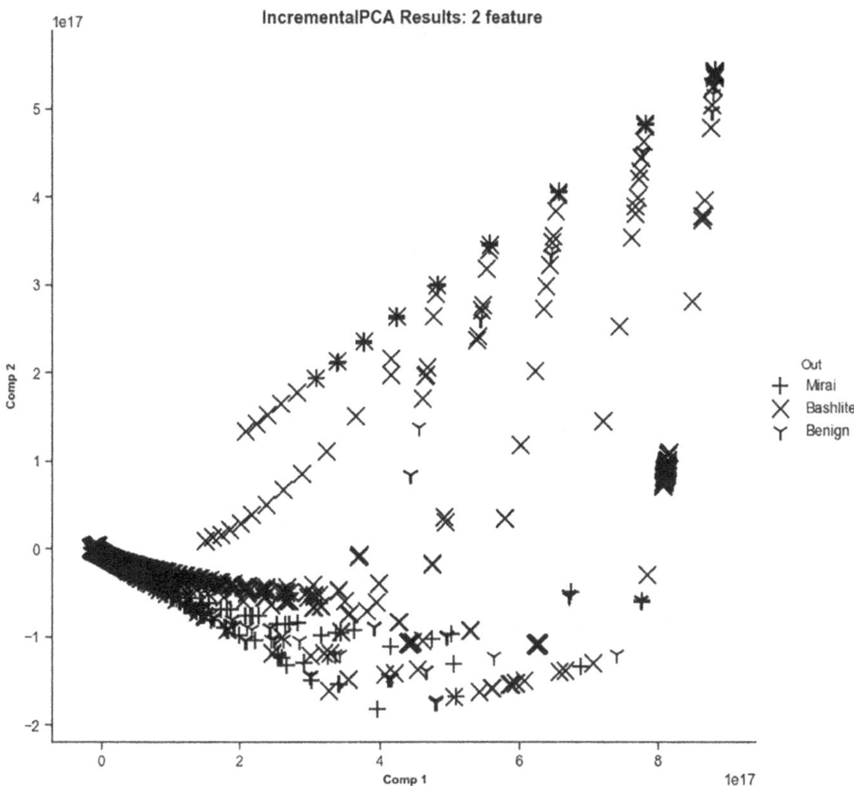

FIGURE 6.18 IoT Botnet visualization using I-PCA method with 2-feature scheme.

6.5.2 I-PCA Dimensionality Reduction Visualization

Figure 6.18 shows visualization result of botnet detection on IoT network using I-PCA DR technique with 2-feature scheme.

Figure 6.19 shows botnet IoT visualization using I-PCA method with 3-feature scheme.

6.5.3 Sparse PCA Dimensionality Reduction Visualization

Figure 6.20 shows visualization result of botnet detection on IoT network using S-PCA DR technique with 2-feature scheme.

Figure 6.21 shows botnet IoT visualization using S-PCA method with 3-feature scheme.

6.5.4 PCA Dimensionality Reduction Visualization

Figure 6.22 shows botnet IoT IDS visualization using PCA technique with 2-feature scheme.

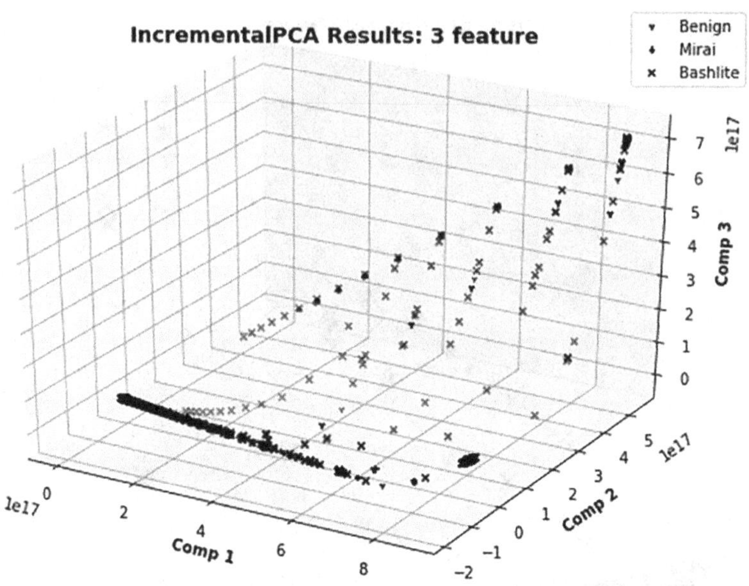

FIGURE 6.19 IoT Botnet visualization using I-PCA method with 3-feature scheme.

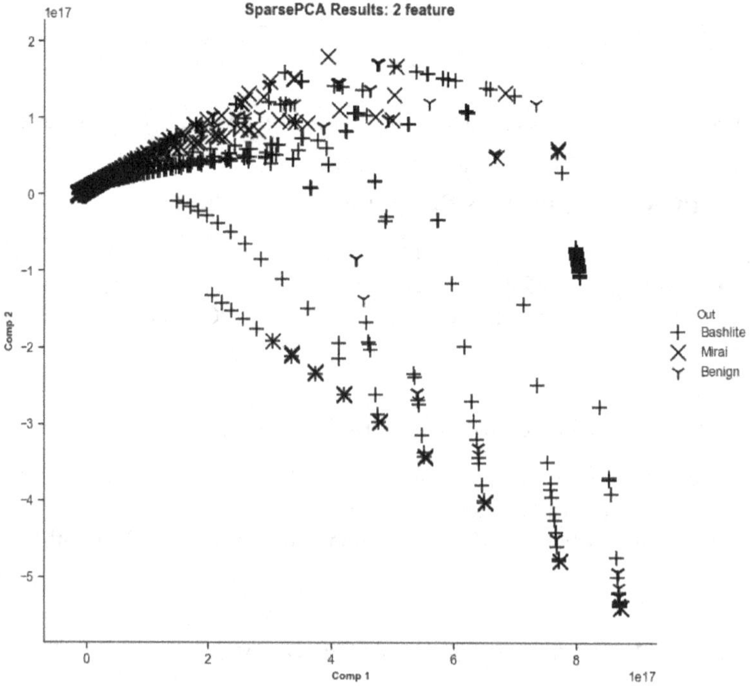

FIGURE 6.20 IoT Botnet visualization using S-PCA method with 2-feature scheme.

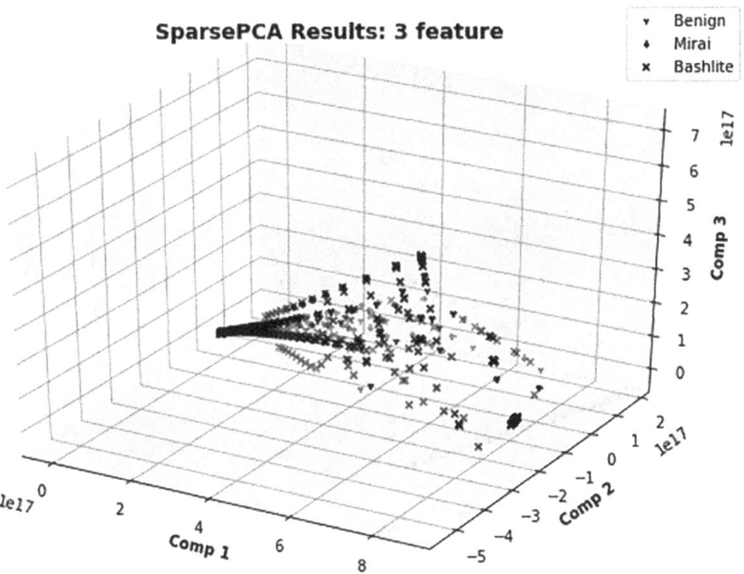

FIGURE 6.21 IoT Botnet visualization using S-PCA method with 3-feature scheme.

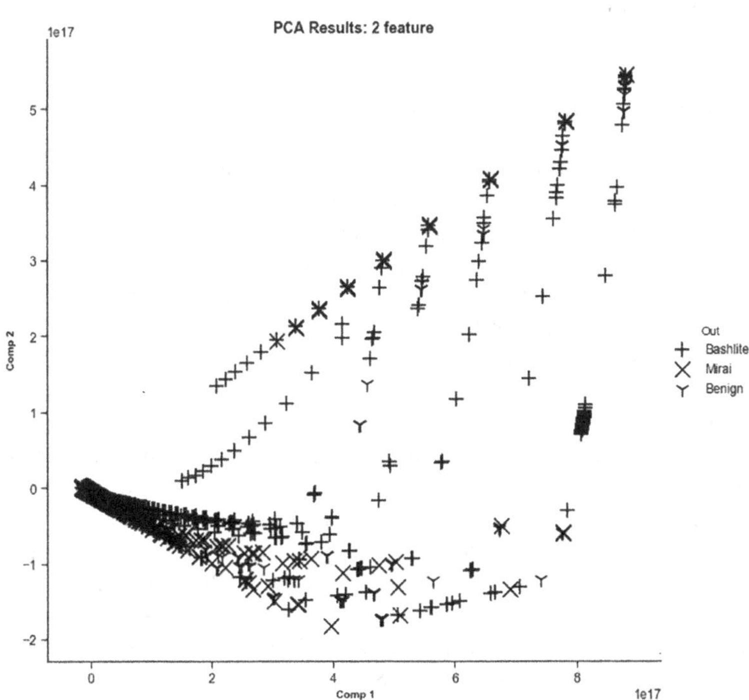

FIGURE 6.22 IoT Botnet visualization using PCA method with 2-feature scheme.

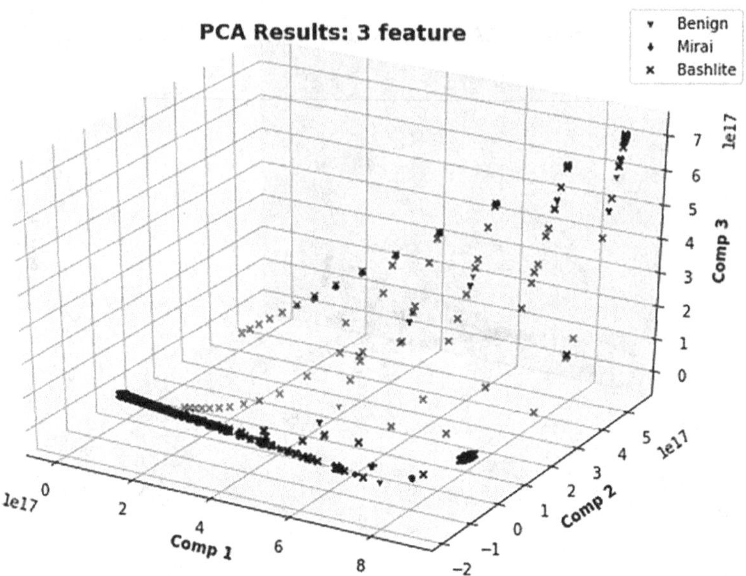

FIGURE 6.23 IoT Botnet visualization using PCA method with 3-feature scheme.

Figure 6.23 shows botnet IoT visualization using PCA method with 3-feature scheme.

6.6 ANALYSIS AND DISCUSSION

Experiments have been carried out using 2-, 3-, 6-, and 10-feature schemes. Implementation of DR technique on IoT IDS is very efficient and effective that can be seen from classification accuracy, FPR values, and execution time resulted from the experiments. The best achievement was obtained by PCA, I-PCA, S-PCA, and FA techniques that combined with DT and RF classification methods that provide high accuracy, ranging from 99% to 100%, followed by the combination with k-NN and GB with accuracy level between 98% and 99%, while combination with AB and NB gives the lowest accuracy level. Observation shows that with lesser number features used, the accuracy increases.

Experiments on measuring FPR values for combination of dimensional reduction techniques with DT and with RF have the lowest FPR values of 0 for all schemes. The combination with FA and RF also has FPR values of 0 for 2-feature and 10-feature schemes but increases up to 9.0157 when using 6-feature scheme. While the combination of the dimensional reduction techniques with k-NN, AB, and GB give FPR values that are relatively low, but there is slight increment when using less feature schemes. Nevertheless, FPR value on GB method increases higher when 10-feature scheme is used, which combined with S-PCA technique, which reaches 7.6621. While FPR values on NB are relatively higher, all combinations with the dimensional reduction techniques are used.

On the other hand, from the implementation of all dimensional reduction techniques combined with each classification methods, it is observed that the execution time is quite fluctuative. We observed that the PCA has the fastest execution time when using 3-feature scheme. For I-PCA technique, only combination with GB gives faster execution time where lesser feature scheme is used. This fact is different when using S-PCA technique where all combinations with less feature scheme provide faster execution time. While FA's execution time varies when using 2-, 3-, and 6- features of each classification method, whereas when using the 10- feature the execution time is getting longer.

6.7 CONCLUSION

This research work has implemented DR techniques PCA, I-PCA, S-PCA, and FA that combined with classification methods k-NN, DT, RF, AB, GB, and NB for IoT botnet detection. Dimensional reduction implementation on IoT IDS gives significant impact, especially in the detection performance in terms of accuracy. Besides, it reduces the storage for keeping the data during the detection process. As we can observe from the use of N-BaIoT dataset with 115 original numbers of features, which reduced into lower dimensions, yet it only affects a bit on the accuracy decrement. Even, the accuracy detection using combination of DT with PCA, S-PCA, and I-PCA was not affected at all with the reduction of the data dimension. Accuracy detection using combination of FA with all variants of feature selection methods was slightly affected by the use of dimensional reduction. In contrast, the combinations of NB with all variant of data dimensional reduction techniques except I-PCA increase the accuracy. Meanwhile, the combinations of NB with all variant of data dimensional reduction techniques significantly decrease the accuracy.

On the other hand, the FPR values of DT and RF are stable and the lowest among other techniques, while RF and FA have high fluctuating FPR values. Other feature classification methods have relatively low increment in their FPR values. We may conclude that the impact of the implementation of dimensional reduction on IDS makes the detection performance more effective and efficient.

REFERENCES

Ahmadkhani, Somaye, and Peyman Adibi. 2016. "Face Recognition Using Supervised Probabilistic Principal Component Analysis Mixture Model in Dimensionality Reduction without Loss Framework." *IET Computer Vision* 10(3): 193–201. https://ur.booksc.eu/book/50392163/d39471.

Alsadi, Muayyad Saleh, and Ali Hussein Hadi. 2017. "Visualizing Clustered Botnet Traffic Using T-SNE on Aggregated NetFlows." Proceedings – 2017 International Conference on New Trends in Computing Sciences, *ICTCS 2017* 2018–Janua: 179–84. https://ieeexplore.ieee.org/document/8250285.

Baaqeel, Hind and Nazar Saqib. 2020. "Using Dimensionality Reduction Technique for Valuable IoT Botnet Data Extraction." *International Journal of Advanced Computational Engineering and Networking* 8(8): 1–8. http://iraj.in/journal/IJACEN/paper_detail.php?paper_id=17414&name=Using_Dimensionality_Reduction_Technique_for_Valuable_IoT_Botnet_Data_Extraction.

Bhushan, Bharat, Chinmayee Sahoo, Preeti Sinha, and Aditya Khamparia. 2021. "Unification of Blockchain and Internet of Things (BIoT): Requirements, Working Model, Challenges and Future Directions." *Wireless Networks* 27: 55–90. https://doi.org/10.1007/s11276-020-02445-6.

Bhushan, Bharat, and Gadadhar Sahoo. 2018. "Recent Advances in Attacks, Technical Challenges, Vulnerabilities and Their Countermeasures in Wireless Sensor Networks." *Wireless Personal Communications* 98(2): 2037–77. https://link.springer.com/article/10.1007/s11276-020-02445-6.

Camacho, J., A. K. Smilde, E. Saccenti, and J. A. Westerhuis. 2020. "All Sparse PCA Models Are Wrong, but Some Are Useful. Part I: Computation of Scores, Residuals and Explained Variance." Chemometrics and Intelligent Laboratory Systems 196 (November 2019): 103907. https://www.sciencedirect.com/science/article/abs/pii/S0169743919303636.

Dagher, Issam. 2010. "Incremental PCA-LDA Algorithm." *International Journal of Biometrics and Bioinformatics* 4(2): 86–99. https://www.cscjournals.org/library/manuscriptinfo.php?mc=IJBB-58.

Diaz-Chito, Katerine, Francesc J. Ferri, and Aura Hernández-Sabaté. 2018. "An Overview of Incremental Feature Extraction Methods Based on Linear Subspaces." Knowledge-Based Systems 145: 1–14. https://www.sciencedirect.com/science/article/abs/pii/S0950705118300285.

Erwin, M. Azriansyah, N. Hartuti, Muhammad Fachrurrozi, and Bayu Adhi Tama. 2019. "A Study about Principle Component Analysis and Eigenface for Facial Extraction." *Journal of Physics: Conference Series* 1196 (1). https://iopscience.iop.org/article/10.1088/1742-6596/1196/1/012010/pdf.

Ikram, Sumaiya Thaseen, and Aswani Kumar Cherukuri. 2016. "Improving Accuracy of Intrusion Detection Model Using PCA and Optimized SVM." *Journal of Computing and Information Technology* 24(2): 133–48. http://cit.fer.hr/index.php/CIT/article/view/2701.

Khalid, Samina, Tehmina Khalil, and Shamila Nasreen. 2014. "A Survey of Feature Selection and Feature Extraction Techniques in Machine Learning." *Proceedings of 2014 Science and Information Conference, SAI 2014*, 372–78. https://ieeexplore.ieee.org/document/6918213.

Kumar, Ajay, Kumar Abhishek, Bharat Bhushan, and Chinmay Chakraborty. 2021. "Secure Access Control for Manufacturing Sector with Application of Ethereum Blockchain." Peer-to-Peer Networking and Applications. https://link.springer.com/article/10.1007/s12083-021-01108-3.

Meidan, Yair, Michael Bohadana, Yael Mathov, Yisroel Mirsky, Asaf Shabtai, Dominik Breitenbacher, and Yuval Elovici. 2018. "N-BaIoT-Network-Based Detection of IoT Botnet Attacks Using Deep Autoencoders." *IEEE Pervasive Computing* 17(3): 12–22. https://ieeexplore.ieee.org/document/8490192.

Park, Cheong Hee, and Gyeong Hoon Lee. 2020. "Comparison of Incremental Linear Dimension Reduction Methods for Streaming Data." *Pattern Recognition Letters* 135: 15–21. https://www.sciencedirect.com/science/article/abs/pii/S0167865520301124.

Roshni Mol, P., and C. Immaculate Mary. 2020. "Intrusion Detection System from Machine Learning Perspective." *International Journal of Engineering Research and Technology (IJERT)* 8(3): 1–3. https://www.ijert.org/intrusion-detection-system-from-machine-learning-perspective.

Saxena, Shivam, Bharat Bhushan, and Mohd Abdul Ahad. 2021. "Blockchain Based Solutions to Secure IoT: Background, Integration Trends and a Way Forward." *Journal of Network and Computer Applications* 181 (September 2020): 103050. https://www.sciencedirect.com/science/article/abs/pii/S1084804521000758.

Sharma, Nikhil, Ila Kaushik, Vikash Kumar Agarwal, Bharat Bhushan, and Aditya Khamparia. 2021. "Attacks and Security Measures in Wireless Sensor Network." In Intelligent Data Analytics for Terror Threat Prediction, 237–68. https://onlinelibrary.wiley.com/doi/10.1002/9781119711629.ch12.

Usman Ali, M., Shahzad Ahmed, Javed Ferzund, Atif Mehmood, and Abbas Rehman. 2017. "Using PCA and Factor Analysis for Dimensionality Reduction of Bio-Informatics Data." *ArXiv* 8(5): 415–26. https://arxiv.org/abs/1707.07189.

Velliangiri, S., S. Alagumuthukrishnan, and S. Iwin Thankumar Joseph. 2019. "A Review of Dimensionality Reduction Techniques for Efficient Computation." *Procedia Computer Science* 165: 104–11. https://www.sciencedirect.com/science/article/pii/S1877050920300879.

Wang, Wenjuan, Xuehui Du, Dibin Shan, Ruoxi Qin, and Na Wang. 2020. "An Intrusion Detection Method Based on Stacked Autoencoder and Support Vector Machine." *IEEE Transactions on Cloud Computing* 1453(1): 1–14. https://iopscience.iop.org/article/10.1088/1742-6596/1453/1/012010.

Yan, Jun, Benyu Zhang, Ning Liu, Shuicheng Yan, Qiansheng Cheng, Weiguo Fan, Qiang Yang, Wensi Xi, and Zheng Chen. 2006. "Effective and Efficient Dimensionality Reduction for Large-Scale and Streaming Data Preprocessing." *IEEE Transactions on Knowledge and Data Engineering* 18(3): 320–32. https://ieeexplore.ieee.org/document/1583582.

You, Shingchern D., and Ming Jen Hung. 2021. "Comparative Study of Dimensionality Reduction Techniques for Spectral–temporal Data." *Information (Switzerland)* 12 (1): 1–12. https://www.mdpi.com/2078-2489/12/1/1.

Zhao, Haifeng, Zheng Wang, and Feiping Nie. 2019. "A New Formulation of Linear Discriminant Analysis for Robust Dimensionality Reduction." *IEEE Transactions on Knowledge and Data Engineering* 31(4): 629–40. https://ieeexplore.ieee.org/document/8369159.

Zong, Wei, Yang Wai Chow, and Willy Susilo. 2019. "Dimensionality Reduction and Visualization of Network Intrusion Detection Data." Lecture Notes in Computer Science (Including Subseries Lecture Notes in Artificial Intelligence and Lecture Notes in Bioinformatics) *11547 LNCS* (July): 441–55. https://link.springer.com/chapter/10.1007/978-3-030-21548-4_24

Zou, Hui, Trevor Hastie, and Robert Tibshirani. 2006. "Sparse Principal Component Analysis." *Journal of Computational and Graphical Statistics* 15 (2): 265–86. https://web.stanford.edu/~hastie/Papers/spc_jcgs.pdf.

7 IoT-Based Resources Management and Monitoring for a Smart City

Prakash Panigrahi, Yashwardhan Kumar, Shivendra Pratap Singh, Sandipan Mallik, Kunjabihari Swain, and Murthy Cherukuri
NIST (Autonomous), Berhampur, Odisha, India

CONTENTS

DOI: 10.1201/9781003219620-7

7.1 INTRODUCTION

The Internet of Things (IoT) is growing pervasively around us, making a huge change in human life. It is estimated that more than 60–70% of the population are moving toward rural to urban cities seeking improved quality of life. With the rapid growth of the population density in metropolitan cities, smart infrastructure, digital devices, and advanced facilities are required to provide the necessities of the residents, leading to the idea of evolving a metro city into a Smart city [1]. The efficient management of these services has a significant impact on the quality of life of the residents. The objective of a Smart city is to improve the livability of cities without affecting the global resources [2].

The IoT is a global infrastructure consisting of physical devices, modern vehicles, home automation, healthcare via mobile, smart education, and easy access. It also interacts with a wide multitude of devices such as home appliances, surveillance cameras, actuators, sensors, monitoring, displaying system, and transportation controlling system [3]. IoT can self-organize and communicate with other devices without human intervention. IoT aims to create a smart world consisting of smart devices, smart cars, smartphones, smart homes, and smart cities [4]. The idea of the Smart city isn't simply restricted to the use of technologies in cities. The creation of smart cities is a characteristic technique to lessen the issues arising out of fast urbanization and public development. Despite the expenses related to them, smart cities, once executed, can diminish energy utilization, water utilization, fossil fuel byproducts, transportation necessities, and city waste [5, 6].

Nonetheless, this heterogeneous field of utilization identifies solutions for fulfilling the prerequisites of all conceivable applications difficult. This difficulty has provoked the development of different and opposite suggestions for the sensible affirmation of IoT frameworks. Thus, from a framework perspective, the zenith of an IoT organization, and the required back-end organization, devices miss the mark to set up the best practice because of their curiosity and unpredictability. Despite the particular inconveniences, this IoT paradigm's determination is upset by the shortfall of a generally recognized and sensible game plan that can attract ventures to improve the arrangement of action of these advances [7].

This chapter aims to discuss a general reference framework for the design and implementation of a Smart city. We describe the features of a Smart city, the concept of IoT in a Smart city, various technologies and communication protocols used in a Smart city. Along with these, we have also presented a practical application of a Smart city with its components. On the technical side, we describe the parts and devices used in a Smart city for observing climate, getting wind speed, humidity and city temperature, water and waste management using smart and effective techniques. The Smart city is incomplete without harvesting and monitoring renewable energy resources. Thus, this chapter discusses creating and monitoring sustainable energy like solar, wind, and biomass. We present the results obtained from the earlier devices for a small-scale area, which can be further expanded for an entire city, making the city smart.

The rest of the book chapter is organized as follows. Section 7.2 presents the features of a Smart city, highlighting its characteristics, pillars, and architecture. Section 7.3 gives a brief idea about the concept of a Smart city. Section 7.4 reviews the various IoT technologies and communication protocols used in this work. Section 7.5 elaborates the practical applications of the Smart city model proposed in this work, including weather monitoring, water management, waste management, renewable and sustainable energy harvesting necessary in a Smart city. Finally, conclusions of the proposed work are presented in Section 7.6.

7.2 FEATURES OF A SMART CITY

A Smart city generally contains attributes, themes, and architecture. Attributes are also called characteristics of a Smart city. The themes are the pillars of a Smart city, and architecture is a vital feature of any Smart city [8]. It is a global and paradigm shift that has revolutionized the entire world with the support of sensor-enabled IoT devices [9]. This section explains the features mentioned earlier, considering a generic Smart city model. Figure 7.1 shows the features of a Smart city.

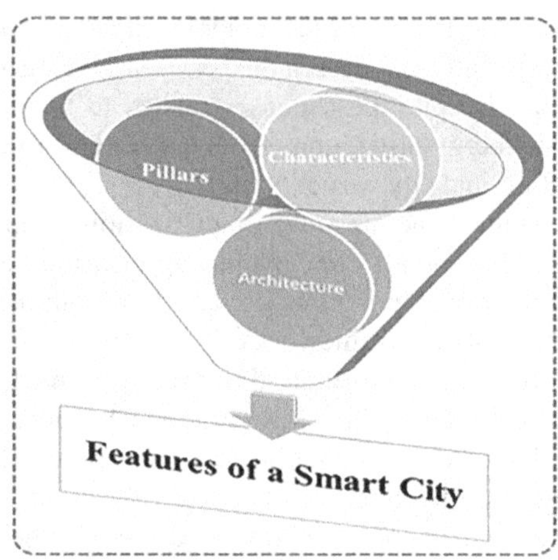

FIGURE 7.1 Features of a Smart city.

7.2.1 Characteristics of a Smart City

A Smart city generally includes smart buildings, transportation, smart energy management, smart healthcare, and smart factories [10]. Smartness is defined as the yearning to improve the city's social, economic, and environmental standards and its people [11]. Figures 7.2 and 7.3 show various characteristics of a Smart city.

7.2.1.1 Smart Building

Smart building is the elementary building block for smart cities, and the creation of smart cities is a major factor for rapid global urbanization [12]. It can reduce the consumption of redundant energy and improve the well-being of the residents.

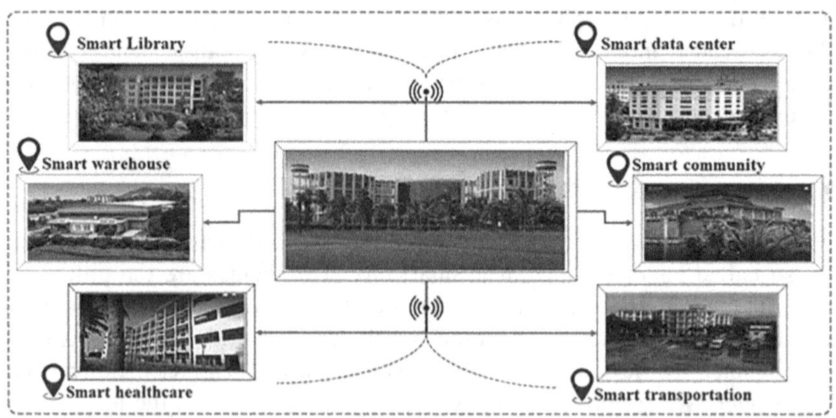

FIGURE 7.2 Characteristics/attributes of a smart city.

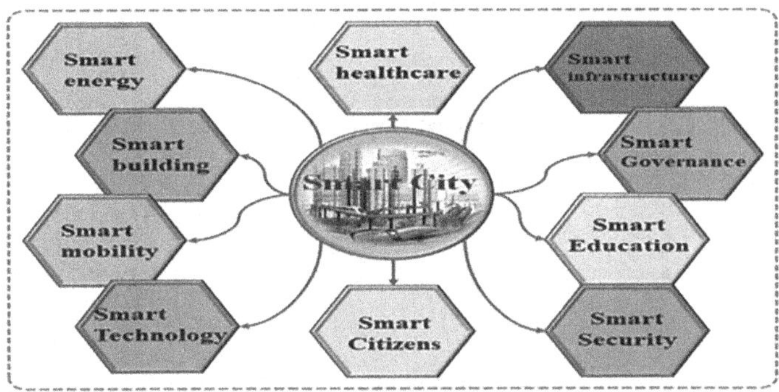

FIGURE 7.3 Various characteristics of a Smart city.

7.2.1.2 Smart Transportation

Smart transportation is considered a fundamental aspect of smart cities. The efficiency of a Smart city depends on the proficiency of the transport systems, that is, the effectiveness with which goods and people are moved throughout the city [13].

7.2.1.3 Smart Healthcare

The emergence of the IoT has revolutionized traditional healthcare and provided an opportunity to replace in-hospital medical systems with Internet-connected standalone systems (like wearable devices, smart bracelets, sugar and blood pressure detection systems), thus bringing the dawn of a new era of smart healthcare [14]. Doctors can collect real-time raw data using these IoT devices, reducing the in situ medical cost and enhanced experience.

7.2.1.4 Smart Factory

A smart factory empowers the constant collection, distribution, and access of industrial information anytime and anywhere. It represents a real-time manufacturing atmosphere that can handle instabilities in manufacturing using distributed information and communication structures for optimal manufacturing process management [15].

7.2.1.5 Smart Energy Management

A typical smart energy management system aims to make appropriate set points for all energy sources and storage to maintain economically optimized power to fulfill certain load demands [16]. It includes managing smart lighting, smart metering, smart grids, improving energy efficiency, and saving energy costs.

7.2.2 Pillars of a Smart City

There are four pillars/themes of a Smart city. These are institutional, social, economic, and physical infrastructure [5, 11]. Figure 7.4 shows the four pillars upholding the Smart city.

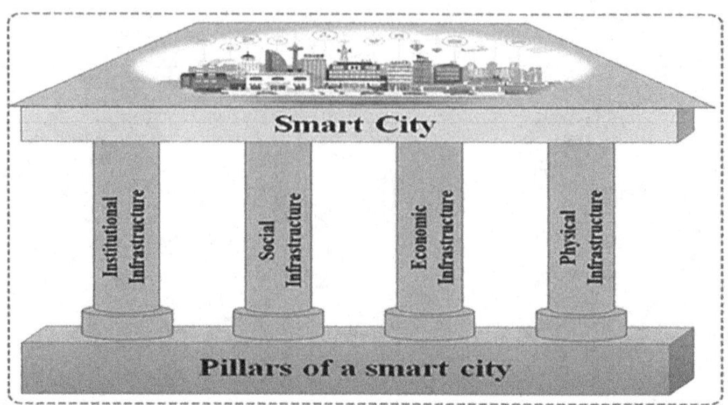

FIGURE 7.4 Four pillars of a Smart city.

7.2.2.1 Institutional Infrastructure

Governance of smart cities, associating with decision-making, social and public services, transport, and politically aware strategies [17], sensitive and careful consideration of political views makes the management of a city a lot easier. The institutional infrastructure acts as a bridge connecting central government and regional governments to increase the benefits of the Smart city [8]. It integrates private, public, civil, and national organizations when necessary. This alliance of different administration bodies serves citizens more efficiently, effectively, and reliably.

7.2.2.2 Social Infrastructure

The social infrastructure of a Smart city includes human capital, intellectual capital, and quality of life. A combination of culture/arts, education/training, and business/commerce with hybrid cultural, social, and economic enterprises make a Smart city sustainable. Even though smart cities utilize advanced technologies and are equipped with sophisticated equipment, the city cannot achieve guaranteed sustainability without social awareness [6]. Hence, social infrastructure becomes crucial for the development of a Smart city.

7.2.2.3 Economic Infrastructure

Economic infrastructure indicates the ability of the city to show stable jobs, and economic growth is also referred to as a smart economy [6]. The employment of best practices and applications of e-business and e-commerce to intensify the city productivity is known as the smart economy. It also comprises novel innovations in information and communication technology (ICT), manufacturing, and integration of advanced technologies that uplift the performance and reliability of economic management [18].

7.2.2.4 Physical Infrastructure

The physical infrastructure is the structural entity of the Smart city, including roads, buildings, railway tracks, water supply systems, and power supply lines [19]. It also consists of geographically dispersed devices (e.g., sensors and gateways), networking essentials (switches and routers), and large data centers [19]. The physical infrastructure is characteristically the non-smart component of smart cities. This pillar ensures the sustainability of resources to stay operational in the present and future [5].

7.2.3 Architecture of a Smart City

The architecture includes four layers: sensing layer, transmission layer, data management layer, and application layer [20, 21]. Figure 7.5 shows the interconnection between the four layers. Data protection is an important concern for any Smart city. Thus, security modules are added to each layer [21].

7.2.3.1 Sensing Layer

The work of the sensing layer is to collect data from physical devices. Data assortment is viewed as the most significant and challenging task because of enormous heterogeneity among data [20, 21]. A Smart city comprises divergent data from enhanced city activities, i.e., machine in a smart home to load balancing in a smart grid, waste management in communities to disaster management, individual well-being checking to pandemic illness of the executives, etc. In this layer, the sensor networks can automatically detect the change in environmental factors and exchange data among devices [21, 22].

7.2.3.2 Transmission Layer

The transmission layer transfers data to the upper layers using various communication technologies. This layer is also called the network layer, which connects all the IoT devices. It is used to aggregate data from existing infrastructures and transmit them to other layers, such as the sensing layer, application layer, etc. [21]. This layer consists of various types of wired, wireless, and satellite technologies [23]. There is a great need for information confidentiality and human privacy security as sensitive information is carried out through this layer [24, 25].

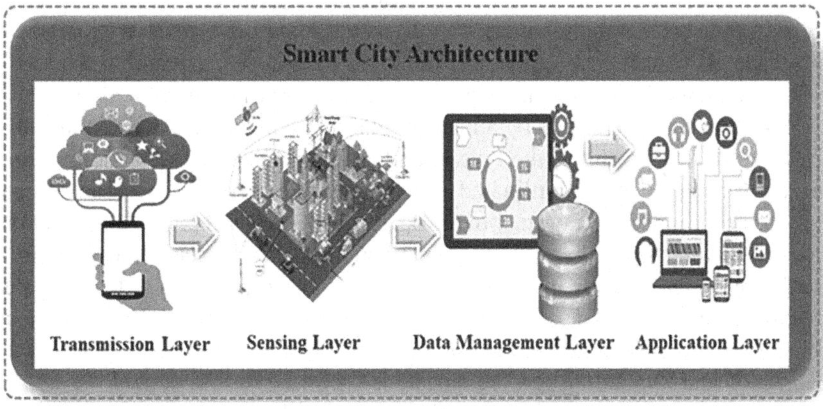

FIGURE 7.5 Smart city architecture showing the different layers.

7.2.3.3 Data Management Layer

A unified framework for data management is profoundly vital for a Smart city and every one of its applications. The data management layer processes and stores valuable information that is useful for various applications. This layer comprises data procurement, preparation, and distribution [24]. Data procurement and handling guarantee the measurements and nature of data are sufficient for the layer's powerful working [24]. Aside from these, different data access patterns and analysis instruments are needed to screen and improve the presentation of uses in a Smart city [26].

7.2.3.4 Application Layer

The application layer is an essential layer of Smart city architecture. The information attained from the sensing layer, managed by the data management layer via transmission layer, is received here to be used in various Smart city sectors [22]. These are referred to as applications, like smart homes, smart power system monitoring, integration of renewable energy generation, coordination of distributed power storage, etc. [24, 27]. The performance of this layer highly impacts user satisfaction, as the citizens are not aware of the intermediate data management layer [27].

7.3 CONCEPTUALIZATION

The term "Smart city" comes from a long course of metropolitan advancements which attempts to rethink the city and inject it with another tenacity and vision [28]. "Smart Cities" are characterized as an urban improvement vision pointed toward adapting various ICT arrangements to control the city's assets. These assets include libraries, water supply organizations, information frameworks, local area administrations, and law implementation. A Smart city is generally perceived as cities with smart people, smart transportation, smart living, smart governance, smart economy, and a smart environment [29].

The usage of the IoT paradigm in a metropolitan setting is explicitly important. It reacts to the requests of various public governments to accept the ICT path of action in general issues. Even though there isn't yet a legitimate and extensively recognized importance of "Smart city," it means using public resources, expanding the idea of the organizations offered to the occupants while decreasing the operational costs. This objective can be pursued by a metropolitan IoT context, i.e., a communication infrastructure that gives together, clear, and admittance to many public organizations, delivering potential coordinated efforts and growing simplicity to the residents [30]. A metropolitan IoT may smooth the operation of standard public organizations, transportation, lighting, observation, public regions, social heritage, the salubrity of crisis facilities, and schools. Furthermore, the local government can utilize the openness of different sorts of data assembled by metropolitan IoT to propel the government's actions toward the residents, increasing the active interest of the inhabitants in the organization and stimulating new organizations. This way, the use of the IoT paradigm to the Smart city is particularly interesting to the local associations that may transform on a broad scale [7].

For the vision of a Smart city, the scientific analysis of urban spatial data is critical to utilize the advancement of big data innovations to understand and redesign our urban spaces under real-time emerging patterns and dynamics [30]. A practical Smart city requires a realistic methodology for digitalization and advancements. The context of demand and supply, the state of affairs, and technology selection will have a basic impact in utilizing a few structural blocks for contextualizing the Smart city [28, 29].

7.4 IoT TECHNOLOGIES AND COMMUNICATION PROTOCOLS

Numerous devices deployed in various Smart city locations are connected with the help of different IoT technologies and communication protocols. These techniques and protocols make the reception and analysis of data much easier and secure [31]. These technologies and protocols are discussed below [22].

7.4.1 IoT Technologies

The concept of smart cities is possible to realize due to modern-day technological interventions. Many new technologies support developing a network of devices and units of a Smart city [32]. The principle idea of the IoT is the widespread presence of articles that can be measured, perceived and that can change the environment. It comprises smart gadgets like cell phones, machines, monuments, landmarks, and so on [33]. The appliances in a Smart city are embedded with sensors that are self-sustainable and computerized [32]. Figure 7.6 represents the IoT technologies and communication protocols used in a Smart city.

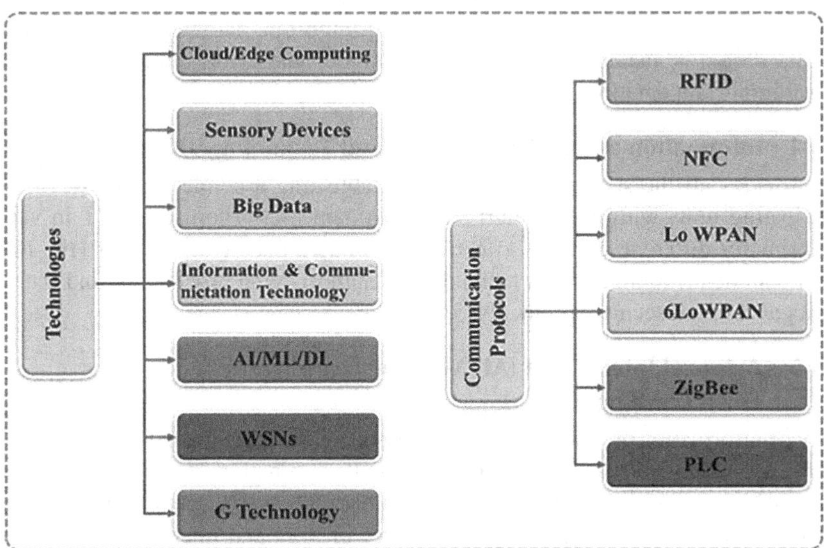

FIGURE 7.6 IoT technologies and communication protocols.

7.4.1.1 Cloud/Edge Computing

Cloud computing acknowledges and advances the conveyance of hardware and software over the Internet. Cloud services promise to deliver high availability, increased reliability, security, and improved quality of service [34]. The inhabitants are not needed to spend a gigantic measure of assets in setting up a solid foundation. Maybe, these can profit from assistance from any cloud specialist organizations like Amazon Web Services, Microsoft Cloud, Google Cloud Services, and so forth [31]. Edge computing makes the reaction time a lot quicker as it allows the client to perform fast and lighter calculations directly on the edge of a network instead of moving the entire information to the cloud for handling [32, 33, 35].

7.4.1.2 Sensory Devices and IoT

Sensory devices could be considered as the center of a Smart city. These are implanted gadgets interconnected to one another and with the Internet. Their essential point is to gather information (like humidity, wind speed, stress, temperature, and so on) from the general climate. These sensors can detect the adjustment of the encompassing conditions and pass this data to the data management layer for additional handling. These interconnected sensors that can communicate with each other, as well as the external surroundings, are termed the "IoT" [32, 36–38].

7.4.1.3 Big Data

Big data refers to collecting large and complex data sets that would be difficult to process using traditional database management tools or old-style data processing applications [5]. The data generated in smart cities can be big data due to its heterogeneity [27]. Big data has numerous difficulties, including visualization, mining, examination, capture, storage, search, and sharing. Storing a lot of information is possible because of the easily accessible and affordable storage devices. The stored data would then be able to be mined to recover important data when required. The big data should be handled with advanced analytics and algorithmic techniques to recover important data [5].

7.4.1.4 Information and Communication and Technology (ICT)

ICT provides all the services required for establishing a connection between the participating units within the Smart city ecosystem [32]. The use of ICT in various forms for different activities effectively smoothed the city operations [19]. ICT includes authentication, network technologies, authorization procedures, and access privileges through security protocols [5, 19].

7.4.1.5 Artificial Intelligence (AI)/Machine Learning (ML)/ Deep Learning (DL)

The huge information produced by a Smart city is useless if it is unanalyzed. Artificial intelligence (AI) works with the preparation and examination of the data produced from machine-to-machine communication in a Smart city. Perceptive and preventive decision-making, all-encompassing experiences of intra and interlayer settings are conceivable with the assistance of machine learning (ML) and deep learning (DL) advancements [32].

7.4.1.6 Wireless Sensor Networks (WSN)

A wireless sensor network (WSN) comprises remote sensor hubs, including a radio interface, numerous sensors, an analog-to-digital converter (ADC), memory, and a power supply [26]. WSNs make different information accessible and are applied in multiple utilizations like medical care, organization, and natural administrations. Also, WSNs are joined with additional sensors to get a few targets, such as information identified with objects, development, temperatures, etc.

7.4.1.7 G Technology

The availability of 3G and LTE made the development and expansion of wireless communication possible everywhere. This technology is generally applied for WANs, which require longer distance ranges [27]. 5G networks offer more prominent adaptability and empower the data stream to be greatly improved, advancing metro cooperation in a Smart city. This innovation gives upgraded availability, assigns more information to be overseen, working with networks to strengthen the shared experience and establish smart conditions [32].

7.4.2 Communication Protocols

The multitude of IoT applications makes it harder to monitor vital information from the surroundings. IoT-based protocols make monitoring and tracking easy and secure with features and attributes for device-to-device communication [31, 39]. Some of these protocols are discussed in this section.

7.4.2.1 Radio-Frequency Identification (RFID)

Radio-frequency identification (RFID) is a programmable and contactless innovation, giving an interface to the labeled items through wireless mediums to recover important data. It is conveyed in many uses, for example, fabricating, production management, stock control, and e-identifications. The fundamental components in an RFID framework are RFID labels, a reader, and a back-end server [40].

7.4.2.2 Near-Field Communication (NFC)

Near-field communication (NFC) is adopted for transferring the data in the infrastructure. They are applied on top of sensors that communicate wirelessly or with the network infrastructure [41]. NFC supports the existing RFID substructure and enables a reader device to communicate with an RFID tag operational at 13.56 MHz [42].

7.4.2.3 Low-Rate Wireless Personal Area Network (LoWPAN)

The low-rate wireless personal area network (LoWPAN) works according to IEEE 802.15.4 and is useful in many applications, including industrial monitoring and control, pressure monitoring, smart tags, and precision agriculture. However, one of the largest applications for LoWPAN is home automation and networking [43].

7.4.2.4 6LoWPAN

IPv4 was the main addressing innovation upheld by the Internet but has been supplanted by IPv6 because of the exhaustion of its address blocks and the inability to independently address billions of nodes, an attribute of IoT networks. The 6LoWPAN standard is indicated to adjust IPv6 communication [7].

7.4.2.5 ZigBee

ZigBee is an 802.15.4-based protocol used for IoT devices and applications. ZigBee is a high-level protocol used for mostly WSN applications. It provides services like encryption, authentication, association, and routing protocol [44].

7.4.2.6 Power-Line Communication (PLC)

Power-line communication (PLC) uses power line infrastructure in indoor and outdoor communication and networking. The main purpose of using PLC is to reduce the cost by utilizing already established wired networks. It plans to send information packets over the branches and trunks of the power lines. The geography of the power line network and the comfort of its power sockets as potential passageways help in networking the nifty gadgets dispersed everywhere in the city [45].

7.5 DEVELOPMENT OF SMALL-SCALE SMART CITY MODEL

In order to establish an actual "Smart city," a small-scale model of the city is realized and analyzed in this section. The various parts of the Smart city are discussed in detail.

7.5.1 Cloud-Based Weather Station in the City

This section discusses the weather station established in the small-scale model. The weather station is connected with Google Cloud Services for data storage and analysis and is incorporated with various sensors collecting data from the environment. An IoT-based weather monitoring system's main aim is low cost, low power, less maintenance, and minimal manual interference. It consists of Raspberry Pi, humidity and temperature sensor, barometric pressure sensor, and rain sensor. The system uses the inbuilt Wi-Fi in the Raspberry Pi as the communication medium for transferring data to the data center.

Humidity and temperature are monitored using a low-cost DHT22 sensor. It uses a thermistor and capacitive humidity sensor to measure the surrounding temperature and humidity, respectively. The hardware specifications are mentioned in Table 7.1. Atmospheric pressure is monitored using a BME280 sensor, developed specifically for weather/environmental sensing. This sensor features an extremely fast response time which supports performance requirements for emerging applications. Technical specifications are mentioned in Table 7.1. Rainfall intensity monitoring is achieved by using a rain sensor. It consists of two modules, a rain board that detects the rain and a control module that measures rainfall intensity. Figure 7.7 shows the block diagram of the weather station installed in the small-scale model.

The data collected from these sensors is transmitted over Wi-Fi to the data center for analysis. This information is additionally used to prepare new AI models for precise climate prediction. The models are conveyed in the cloud to anticipate the impact and study different climate examples and patterns. Residents can get to ongoing climate information and experiences distantly through an online interface.

TABLE 7.1
The Architecture of a Smart City

Sensing Layer	Transmission Layer	DM Layer	Application Layer
Sensors and transducers are the essential components of this layer.	Data Transceivers make up this layer.	Cloud-based data monitoring and management maintain this layer.	The services and devices provided to the residents are a part of this layer.
The huge amount of data generated from the sensors is called big data [9].	The big data is transmitted to the cloud, and useful information is received.	The cloud analyzes the data and makes decisions as required.	The information is delivered to the users as a service.
The data collected is encrypted and delivered to cloud servers via the transmission layer.	This layer acts as a bridge between all the layers, connecting them and transceiving data across them.	The data received is decrypted and analyzed.	This layer is an interface between the residents and the service providers/government.
This layer consists of components like a Temperature sensor, pressure sensor, rain sensor, etc.	This layer consists of components/ technology like, ZigBee, Wi-Fi, LTE, etc.	This layer consists of services like Cloud/ Edge computing, AWS, GCP, etc.	This layer consists of services like Web portals, weather forecasts, etc.

7.5.2 Water Management in the City

Water is a fundamental asset forever, and tragically, huge amounts of water are being wasted every day. The level of freshwater asset accessibility on the planet is lessening each year. Observing water use can control water utilization, and technological advancements can be a valuable part of an effective water management system.

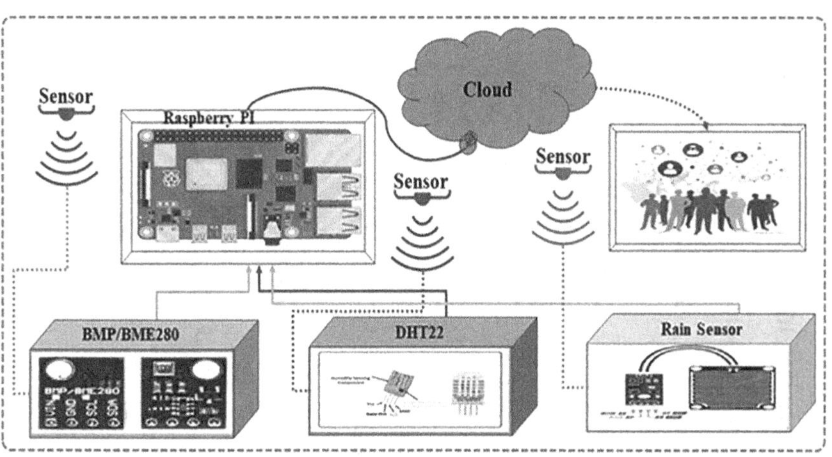

FIGURE 7.7 Block diagram of the weather station in a Smart city.

TABLE 7.2
Technical Specification of All the Sensors and Devices Used

Application	Humidity and Temperature	Barometric Pressure	Rain Sensor	Water flow Sensor	Temperature Sensor	pH Sensor	Ultrasonic Sensor	Load Cell	Wi-Fi Module	ZigBee	Raspberry Pi
Module name	DHT22	BME280	–	YF-S201	DS18B20	SEN0161	HC-SR04	Shear load cell	ESP8266	ZigBee	Raspberry Pi 4 Model B+
Supply voltage	3.3 V–6 V DC	1.71 V–3.6 V DC	3.3 V–6 V DC	5 V–18 V DC	3 V–5 V DC	5 V DC	5 V DC	5 V DC	2.5 V–3.6 V	1.8 V–3.8 V	5 V DC
Output signal	Digital signal	–	Digital switching output, and analog voltage output	Digital switching output, and analog voltage output	Analog voltage output	Analog voltage output	TTL output signal	Analog voltage output	Protocol: 802.11 b/g/n	Protocol: 802.15.4	Protocol: 802.11ac wireless, Bluetooth 5.0, BLE
Communication interface	–	I2C and SPI	–	–	–	–	TTL signal	–	UART/SDIO/SPI/I2C/GPIO/I2C/I2S	SPI	UART/SDIO/SPI/I2C/GPIO/I2C/I2S
Operating range	Temperature: –40°C–80°C Humidity: 0–100%RH	Temperature: –40°C–85°C Pressure: 300–1100 hPa Humidity: 0–100%RH	–	Temperature: –25°C to +80°C Flow rate: 1–30 L/min	Temperature: –55°C to +125°C	pH: 0–14	2 cm–4 cm at 15° angle	0 kg–24 kg	Frequency: 2.4 GHz–2.5 GHz Temperature: –40°C–125°C	Frequency: 2.4 GHz–2.5 GHz Temperature: –40°C–125°C	Frequency: 2.4 GHz–5 GHz Temperature: 0°C–50°C

(Continued)

TABLE 7.2 *(continued)*
Technical Specification of All the Sensors and Devices Used

Application	Humidity and Temperature	Barometric Pressure	Rain Sensor	Water flow Sensor	Temperature Sensor	pH Sensor	Ultrasonic Sensor	Load Cell	Wi-Fi Module	ZigBee	Raspberry Pi
Accuracy	Temperature: −0.5°C to +0.5°C Humidity: ±2% RH	Temperature: −0.5°C to +0.5°C Humidity: ±3% RH	–	±10%	±0.5°C	±0.1 pH	–	±10 g	–	–	–
Sensitivity	Temperature: 0.1°C Humidity: 0.1% RH	–	Adjustable	Pulses per Liter: 450	–	0.01 pH	–	Temperature: 1 g/°C Output: 1 mV/V	–	–	–
Use	Weather station	Weather station	Weather station	Water management system	Water management system	Water management system	Water management system, waste bin	Waste bin	Communication medium	Communication medium	Base station

Different advances are being applied to smart water management systems to screen and detect water spillages, theft, and dispersion.

This section discusses an IoT-based architecture for an underground water distribution system. It takes care of consumer utilization, water quality control, water theft detection, fault localization. By analyzing the water consumption data, daily water demand is determined and supplied to the residents. Nodes placed all over the city collect data related to water quality and supply and transmit, using the ZigBee technology, to the base station situated in the data center for control and analysis. Each node has a pH sensor, water flow sensor, water level sensor, and water temperature sensor. Hardware specifications of all the sensors are mentioned in Table 7.1. A pH sensor is used to determine the H^+ ions concentration in the water, checking the drinking water quality. The circuit of this sensor is highly compact and can be interfaced with any development board. It communicates with the node through UART communication protocols. The water level sensor indicated the quantity of water supplied through the pipelines. The sensor emits high-frequency radiation, generally in the 20 kHz–200 kHz range, and then listens for the echo. If the echo received is less than a certain predetermined value, it is assumed that there is some water leakage between the two nodes. Maintenance can be done when such information is received. A water flow sensor is a cost-effective hall-effect sensor that allows the nodes to determine the quantity of water (in liters) consumed daily to monitor the water consumption rate effectively. The nodes collect data from these sensors via wired and wireless mediums and are transmitted to the base station. Figure 7.8 shows the nodes connected to the cloud services and applications.

The base station consists of a Raspberry Pi connected with Google Cloud Services. Figure 7.8 shows the working of the base station. Data received over ZigBee from various nodes is decoded and stored in the cloud database. The analyzed data from the base station is monitored by the water control board and ensured that every node is working properly and the water supply is not hindered. The base station triggers the maintenance indicator in the water control board to indicate the leakage between the nodes during water leakage.

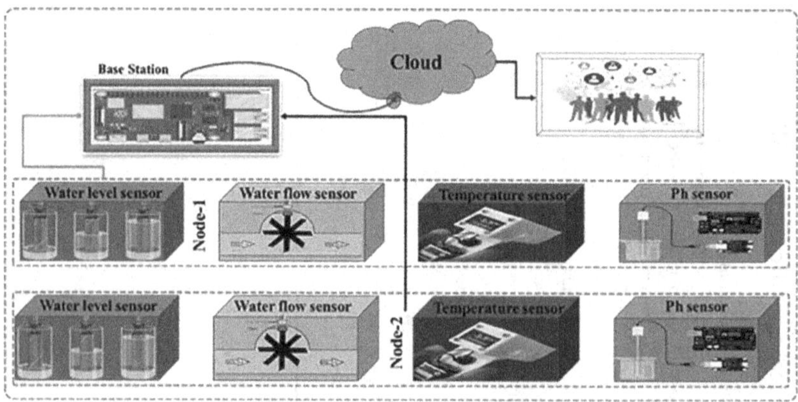

FIGURE 7.8 Nodes all over the city connected to the base station.

7.5.3 Waste Management in the City

This section focuses on realizing an IoT architecture to optimize waste management in the Smart city model. With the ever-increasing urbanization, migration issues, population, and change in lifestyle, solid waste generation levels are increasing significantly. The systems focus on sustainability issues in keeping with global trends, mainly through incorporating 3R (Reduce, Recycle, and Reuse) technologies. A multidimensional integrated approach for developing a sustainable waste management system is achieved in the small-scale Smart city. This advanced concept and designed framework would be fruitful across a variety of country-specific scenarios.

Waste management directly affects the lifestyle, environment, healthcare, recycling and disposal, and several other industries. It refers to handling the entire waste generated by the city. To achieve effective waste management, smart waste bins are placed all over the city. A smart waste bin consists of an ultrasonic sensor, load cell, and a Wi-Fi module for communication with nearby nodes.

An ultrasonic sensor generates high-frequency ultrasonic waves. When this wave hits the object, it reflects as an echo which is sensed by the receiver. It works on RADAR and SONAR system principles, used to determine the distance of objects. The distance is calculated by measuring the time needed for the echo to be received by the receiver. A load cell is utilized to gauge loads either directly or indirectly. Strain gauge load cells are being used in the waste bins. It is a transducer that converts a force or pressure, or strain into an electrical signal. Through a mechanical arrangement, the pressure disfigures a strain gauge component. The pressure misshapes the strain gauge elements resulting in a change in the effective electrical resistance of the wire.

The ultrasonic sensor connected to the hood of the bin monitors the amount of waste collected in the bin and informs the node about it. The load cell monitors the bin's weight and informs the node about the need for waste grouping. A Wi-Fi module is used to communicate with the base station. It will routinely send the data of waste weight and bin limit. When the waste bin is filled, it will send a brief notification for a laborer to pick the waste. Likewise, a versatile application is made for valuable waste management to help workers pick and deal with the waste bin. So, it can make the handle of a filled waste bin quicker and successful. All information from all waste bins is observed from a web portal. It will show the realistic every day, week after week, month to month, and yearly about the waste around there. Figure 7.9 depicts the general idea of the waste management procedure. The entire waste management involves four operations, as shown in Figure 7.10.

7.5.3.1 Generation

Solid waste is generated from homes, schools, stores, offices, hospitals, etc., collected and disposed of by the municipality. This waste generally comprises paper, food waste, plastics, leather, textiles, glass, metal, etc. Burning diminishes the volume of the trash, even though it isn't burnt totally, and open dumps regularly fill in as the favorite place for rodents and flies. Sanitary landfills were utilized as the substitute for open burning dumps. In a landfill, wastes are unloaded in a channel after compaction and covered with earth. Landfills are not an answer since wastes

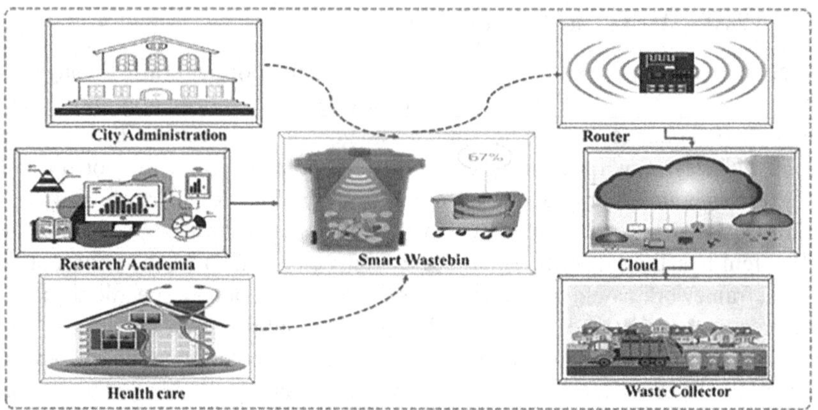

FIGURE 7.9 Waste management procedure.

have reached out to a lot that these are getting filled. In addition, there is a risk of spillage of chemical compounds from these landfills contaminating the underground water assets. Hospitals and medical facilities make poisonous materials that contain sanitizers and other unsafe substances and pathogenic microorganisms. Such waste also requires careful treatment and removal. Damaged electronic products are known as electronic wastes(e-waste) and are additionally delivered in huge sums in this modern world.

7.5.3.2 Collection and Separation

The assortment of waste is an important stage of waste management. A reasonable and proficient method of gathering waste prompts better well-being, cleanliness, and disposal. Waste collection in the small-scale Smart city model is achieved by getting an indication from the cloud servers/base station to the waste control board about the status of various waste bins. The filled waste bins get collected, and the trash is transferred for treatment and sorting.

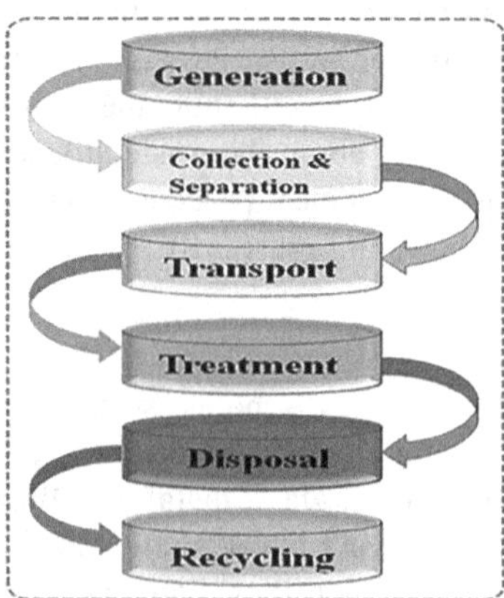

FIGURE 7.10 Flowchart of the waste management process.

7.5.3.3 Treatment

Various methods can treat hazardous waste. Chemical methods include oxidation and reduction, ion exchange, precipitation, and neutralization. Thermal methods can detoxify certain organic wastes and destroy them. Air pollution is a major drawback of hazardous-waste incineration. A technique used for the biological treatment of toxic waste is called landfarming. In this method, the waste is mixed with soil rich with microbes that can decompose waste. The chemical, thermal, and biological treatment techniques change the subatomic type of waste. Then again, physical treatment decreases the volume of trash. Biological cycles incorporate dissipation, flotation, sedimentation, and filtration.

7.5.3.4 Recycling and Disposal

Better recycling and disposal techniques are subject to each type of recyclable waste and its amount. With smart waste bins, separate for each sort of waste, the collectors will see through the cloud and examine what kind of waste is coming up and in what size. Along these lines, better plans are made, and effective recycling techniques are embraced progressively. Pertinent partners can be drawn in and offered freedoms to relieve waste disposal and different issues. All the generated waste is categorized into three types – (a) biodegradable, (b) recyclable, and (c) nonbiodegradable. The biodegradable materials are disposed into deep pits and left for natural breakdown. Nonbiodegradable waste is recycled as per their further use. Plastic wrappers and polythene can be recycled in various useful products like toys, etc. Hospital waste requires careful treatment and disposal. E-wastes are mostly incinerated or buried in landfills. Recycling in a protected environment is the best solution for the treatment of e-waste.

7.5.4 Renewable Resources Management in the City

The system's uniqueness lies in its monitor-cum-control functionality that enables the users to take control action based on monitored data remotely. It utilizes state-of-the-art clean technologies, telecom and information technologies, the IoT, and mobile technologies to deliver cleaner, greener, and sustainable development services as shown in Figure 7.11.

7.5.4.1 Solar and Wind Energy

It is one of the best and commonly used non-exhaustible energy sources and can be used in remote areas and villages easily and most efficiently. Figure 7.12(a) shows a 10-kW wind-PV hybrid setup generating almost 10,000 units of power at a rate of 50 units a day. Figure 7.12(b) shows a solar concentrator that produces heat to convert water into steam and facilitate steam-based cooking for about 800–1000 meals a day. It produces approximately 60 kWh energy per day and nearly 9000 kWh energy per year, equivalent to 51 LPG cylinders.

7.5.4.2 Biogas Plant

Biogas is a mix of gases delivered by microbial movement and fuel. Figure 7.13 shows the blockdiagram of a typical biogas plant. Various components of a typical biogas plant is depicted in Figures 7.14(a),(b), and (c).

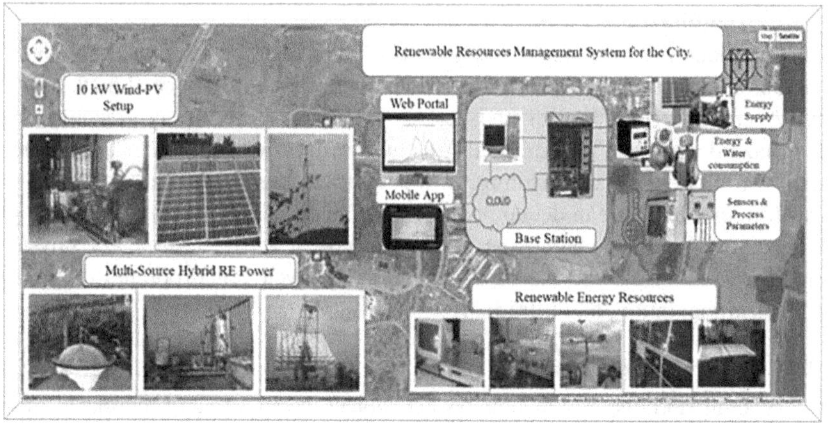

FIGURE 7.11 Smart renewable resources management system in the model Smart city.

FIGURE 7.12 (a) 10-kW WIND-PV Hybrid setup and (b) solar concentrator.

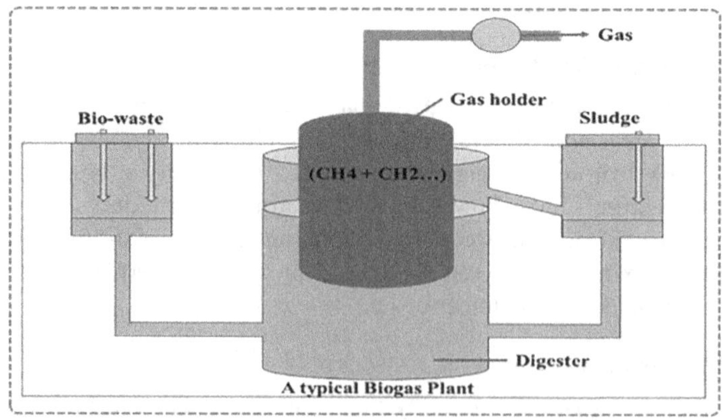

FIGURE 7.13 Block diagram of a typical biogas plant.

FIGURE 7.14 (a) Biogas fired 10-MT cold storage, (b) 15 kVA Biomass gasifier, and (c) kitchen waste biogas plant.

7.6 CONCLUSION

The Smart city is a necessity of today's world. So, Smart cities are growing at a fast pace in developing and developed countries. It consists of smart basic amenities with proper planning and plotting supported by IoT. The concept of smart cities has grown, evolved, and revolutionized with recent developments in ICT that combine wireless sensors and digital networks. Although it is very difficult to cover every aspect of the Smart city, we aim to focus on the most important areas of the Smart city proposed in this chapter. The small-scale model established here, working as a prototype, can be realized into an actual city. Various communication protocols and IoT technologies discussed here can be used in realizing a Smart city. Harnessing the power of the cloud in the city can greatly reduce the need for physical infrastructure while providing highly advanced applications or services. The city can accommodate more people and offer an improved quality of life while utilizing its resources to their highest potential. Using renewable resources as the primary energy source to power the entire city will reduce the strain on the environment due to the advancements in technologies and fast urbanization.

ACKNOWLEDGMENTS

This work was supported by the National Institute of Science and Technology, Berhampur, India.

REFERENCES

1. S. I. Saleem, S. R. M. Zeebaree, D. Q. Zeebaree, and A. M. Abdulazeez, "Building smart cities applications based on IoT technologies: A review," *Technol. Rep. Kansai Univ.*, vol. 62, no. 3, pp. 1093–1092, 2020.
2. P. Marques, "An IoT-based smart cities infrastructure architecture applied to a waste management scenario," *Ad Hoc Networks*, vol. 87, pp. 200–208, 2019.

3. A. J. Jara, Y. Sun, H. Song, R. Bie, D. Genooud, and Y. Bocchi, "Internet of Things for cultural heritage of smart cities and smart regions," in Proc. – IEEE 29th Int. Conf. Adv. Inf. Netw. *Appl.* Work. *WAINA 2015*, pp. 668–675, 2015.
4. H. Rajab and T. Cinkelr, "IoT based smart cities," *2018 Int.* Symp. Networks, Comput. Commun. ISNCC 2018, pp. 1–4, 2018.
5. S. P. Mohanty, U. Choppali, and E. Kougianos, "Everything you wanted to know about smart cities: The internet of things is the backbone," *IEEE Consum. Electron. Mag.*, vol. 5, no. 3, pp. 60–70, 2016.
6. V. Albino, U. Berardi, and R. M. Dangelico, "Smart cities: Definitions, dimensions, performance, and initiatives," *J. Urban Technol.*, vol. 22, no. 1, pp. 3–21, 2015.
7. A. Zanella, N. Bui, A. Castellani, L. Vangelista, and M. Zorzi, "Internet of things for smart cities," *IEEE Internet Things J.*, vol. 1, no. 1, pp. 22–32, 2014.
8. B. N. Silva, M. Khan, and K. Han, "Towards sustainable smart cities: A review of trends, architectures, components, and open challenges in smart cities," *Sustain. Cities Soc.*, vol. 38, no. August 2017, pp. 697–713, 2018.
9. A. H. Sodhro, S. Pirbhulal, Z. Luo, and V. H. C. de Albuquerque, "Towards an optimal resource management for IoT based Green and sustainable smart cities," *J. Clean. Prod.*, vol. 220, pp. 1167–1179, 2019.
10. L. Zhao, J. Wang, J. Liu, and N. Kato, "Optimal edge resource allocation in IoT-based smart cities," *IEEE Network*, vol. 33, no. 2, pp. 30–35, 2019.
11. B. Bhushan, A. Khamparia, K. M. Sagayam, S. K. Sharma, M. A. Ahad, and N. C. Debnath, "Blockchain for smart cities: A review of architectures, integration trends and future research directions," *Sustain. Cities Soc.*, vol. 61, p. 102360, 2020.
12. T. K. L. Hui, R. S. Sherratt, and D. D. Sánchez, "Major requirements for building Smart Homes in Smart Cities based on Internet of Things technologies," *Future Gener. Comput. Syst.*, vol. 76, pp. 358–369, 2017.
13. D. H. Mrityunjaya, N. Kumar, S. Ali, and H. M. Kelagadi, "Smart transportation," in *2017 International Conference on I-SMAC (IoT in Social, Mobile, Analytics and Cloud)(I-SMAC)*, pp. 1–5, 2017.
14. H. Yin, A. O. Akmandor, A. Mosenia, and N. K. Jha, "Smart healthcare," Foundations and Trends in Electronic Design Automation, now publishers, pp. 401–466, 2018.
15. D. Lucke, C. Constantinescu, and E. Westkämper, "Smart factory-a step towards the next generation of manufacturing," in *Manufacturing Systems and Technologies for the New Frontier*, Springer, London, pp. 115–118, 2008.
16. C. Chen, S. Duan, T. Cai, B. Liu, and G. Hu, "Smart energy management system for optimal microgrid economic operation," *IET Renew. Power Gener.*, vol. 5, no. 3, pp. 258–267, 2011.
17. P. Sharma and S. Rajput, "Perspectives of smart cities: Introduction and overview," in *Sustainable Smart Cities in India*, Springer, India, pp. 1–13, 2017.
18. T. M. V. Kumar and B. Dahiya, "Smart economy in smart cities," in Smart *Economy* in *Smart Cit*ies, Advances in 21st Century Human Settlements, Springer, Singapore, pp. 3–76, 2017.
19. H. Chourabi et al., "Understanding smart cities: An integrative framework," in *2012 45th Hawaii International Conference on System Sciences*, pp. 2289–2297, 2012.
20. M. Wu, T.-J. Lu, F.-Y. Ling, J. Sun, and H.-Y. Du, "Research on the architecture of Internet of Things," in 2010 3rd international conference on advanced computer theory and engineering (ICACTE), vol. 5, pp. V5–484, 2010.
21. P. Wang, S. Chaudhry, L. Li, S. Li, T. Tryfonas, and H. Li, "The Internet of Things: a security point of view," *Internet Res.*, vol. 26, no. 2, pp. 337–359, 2016.
22. B. Bhushan, C. Sahoo, P. Sinha, and A. Khamparia, "Unification of Blockchain and Internet of Things (BIoT): Requirements, working model, challenges and future directions," *Wireless Networks*, vol. 27, no. 1, pp. 55–90, 2021.

23. V. C. Gungor et al., "A survey on smart grid potential applications and communication requirements," *IEEE Trans. Ind. Inf.*, vol. 9, no. 1, pp. 28–42, 2012.
24. L. Nastase, "Security in the internet of things: A survey on application layer protocols," in *2017 21st international conference on control systems and computer science (CSCS)*, pp. 659–666, 2017.
25. H. Zhang, M. Babar, M. U. Tariq, M. A. Jan, V. G. Menon, and X. Li, "SafeCity: Toward safe and secured data management design for IoT-enabled Smart city planning," *IEEE Access*, vol. 8, pp. 145256–145267, 2020.
26. C. Jing, M. Du, S. Li, and S. Liu, "Geospatial dashboards for monitoring Smart city performance," *Sustainability*, vol. 11, no. 20, p. 5648, 2019.
27. S. Talari, M. Shafie-Khah, P. Siano, V. Loia, A. Tommasetti, and J. P. S. Catalão, "A review of smart cities based on the internet of things concept," *Energies*, vol. 10, no. 4, p. 421, 2017.
28. A. Ogra, "Conceptualization of Smart city: A methodological framework for smart infrastructure, smart solutions and smart governance," in *Data-Driven Multivalence in the Built Environment*, Springer, Switzerland, pp. 57–72, 2020.
29. A. K. M. B. Haque, B. Bhushan, and G. Dhiman, "Conceptualizing Smart city applications: Requirements, architecture, security issues, and emerging trends," Expert Syst., 2021. DOI: 10.1111/exsy.12753.
30. K. M. Nahiduzzaman, M. Holland, S. K. Sikder, P. Shaw, K. Hewage, and R. Sadiq, "Urban transformation toward a Smart city: An E-commerce–induced path-dependent analysis," *J. Urban Plann. Dev.*, vol. 147, no. 1, p. 4020060, 2021.
31. A. Kondoro, I. Ben Dhaou, H. Tenhunen, and N. Mvungi, "Real time performance analysis of secure IoT protocols for microgrid communication," *Future Gener. Comput. Syst.*, vol. 116, pp. 1–12, 2021.
32. M. A. Ahad, S. Paiva, G. Tripathi, and N. Feroz, "Enabling technologies and sustainable smart cities," *Sustain. Cities Soc.*, vol. 61, no. May, p. 102301, 2020.
33. H. Arasteh et al., "IoT-based smart cities: A survey," EEEIC 2016 - Int. Conf. Environ. Electr. Eng., pp. 2–7, 2016.
34. G. Suciu, A. Vulpe, S. Halunga, O. Fratu, G. Todoran, and V. Suciu, "Smart cities built on resilient cloud computing and secure internet of things," in *2013 19th International Conference on Control Systems and Computer Science*, pp. 513–518, 2013.
35. Y. C. Hu, M. Patel, D. Sabella, N. Sprecher, and V. Young, "Mobile edge computing—A key technology towards 5G," *ETSI White Pap.*, vol. 11, no. 11, pp. 1–16, 2015.
36. K. Ashton, "That 'internet of things' thing," *RFID J.*, vol. 22, no. 7, pp. 97–114, 2009.
37. B. Hammi, R. Khatoun, S. Zeadally, A. Fayad, and L. Khoukhi, "IoT technologies for smart cities," *IET Networks*, vol. 7, no. 1, pp. 1–13, 2018.
38. E. Park, A. P. del Pobil, and S. J. Kwon, "The role of Internet of Things (IoT) in smart cities: Technology roadmap-oriented approaches," *Sustainability*, vol. 10, no. 5, pp. 1–13, 2018.
39. S. Saxena, B. Bhushan, and M. A. Ahad, "Blockchain based solutions to secure IoT: background, integration trends and a way forward," *J. Network Comput. Appl.*, p. 103050, 2021.
40. I. Erguler, "A potential weakness in RFID-based Internet-of-things systems," *Pervasive Mob. Comput.*, vol. 20, pp. 115–126, 2015.
41. T. Anagnostopoulos et al., "Challenges and opportunities of waste management in IoT-enabled smart cities: A survey," *IEEE Trans. Sustain. Comput.*, vol. 2, no. 3, pp. 275–289, 2017.
42. R. Want, "Near field communication," *IEEE Pervasive Comput.*, vol. 10, no. 3, pp. 4–7, 2011.
43. E. Callaway et al., "Home networking with IEEE 802.15. 4: A developing standard for low-rate wireless personal area networks," *IEEE Commun. Mag.*, vol. 40, no. 8, pp. 70–77, 2002.
44. A. K. Sikder, A. Acar, H. Aksu, A. S. Uluagac, K. Akkaya, and M. Conti, "IoT-enabled smart lighting systems for smart cities," 2018 IEEE 8th Annu. Comput. Commun. Work. Conf. CCWC 2018, vol. 2018, pp. 639–645, 2018.
45. Y.-J. Lin, H. A. Latchman, M. Lee, and S. Katar, "A power line communication network infrastructure for the smart home," *IEEE Wireless Commun.*, vol. 9, no. 6, pp. 104–111, 2002.

8 Internet of Things Applications in Marketing

W. Pizło
Warsaw University of Life Sciences, Warsaw, Poland

A. Kałowski
University of Social Sciences, Warsaw, Poland

A. Zarzycka
Warsaw School of Economics, Warsaw, Poland

CONTENTS

8.1 INTRODUCTION

Social, economic, political, cultural and, above all, technological changes force many entities operating in the market to constantly redefine their strategies and adopted competitive advantages, which often leads to complete reorientation in terms of the goals of long-term operations. In recent decades, despite the growing local criticism manifested in the form of customs blockades, at the continental level, there has been a significant acceleration of globalization interactions supported by the development of technology and emerging new models of competition between international corporations [1]. The increasing globalization of the economy made some parts of the international community aware of the need to counteract negative consequences, such as environmental pollution and migration of dirty industries from countries recognized as wealthy to countries described as poor. At the same time, the rapid development of modern means of communication took place, which revolutionized the interaction between organizations and consumers as well as the flow of

DOI: 10.1201/9781003219620-8

data between individual economic entities. In the era of the so-called domination of digitization, the use of new, innovative information technologies contributes to significant changes in the world, which applies not only to broadly understood sectors operating on a global scale but also to functional areas within an open organization predisposed to operate in the 21st century. Consequently, also marketers face new challenges balanced with opportunities offered by digital innovations in this area [2]. Hence, one of the newest technological solutions, the Internet of Things (IoT), deserves special attention. Although its business potential has already been explored in detail in the literature, the research on the practical consequences of the IoT phenomenon is still at the pioneering stage [3], especially in the area of marketing.

For decades, the field of marketing, in the application sphere, has been a kind of "operating philosophy" applied by business entities to adapt to the realities of the market and aimed at satisfying the needs of customers. On the other hand, in the theoretical sphere, marketing is a set of accepted theories, laws and principles as well as rational methods of operation developed by experts in this field. Marketing actions undertaken by various types of organizations are modified under the pressure of increasingly intense competition with regard to the goals they want to achieve. The modern theory of marketing highlights the importance of the building long-term relationship between the business operator and the consumer, defined as "an effort to create, maintain and enhance strong relationships with target customers and stakeholders" [4]. Marketing practitioners point out that relationship marketing requires the cooperation of all parties of the economic entity and creating customer value must take place at all levels, i.e. economic, social, technical and legal. Relationship marketing, as opposed to the traditional approach, focuses on building a long-term relationship between an organization and a target consumer (e.g. a customer with an enterprise, a resident with a non-governmental organization, a voter with a political party or a fan with a celebrity). Acquiring knowledge about consumers is a key element in building and maintaining the relationships, as it enables the company to provide an offer tailored to customers' requirements and needs. Another critical factor in creating long-term customer relationships is trust [5], which derives from the conviction that each organization building lasting relationships with consumers will create new added value. Trust arises on the basis of responsibility for the actions taken throughout the whole product life cycle.

The remainder of the chapter is organized as follows. Section 8.2 presents research objectives and methodology. Section 8.3 describes the basic communication system used in the IoT. Section 8.4 presents the characteristics of the IoT and its importance in marketing activities. Section 8.5 describes the specifics of IoT marketing. Section 8.6 focuses on best practices of IoT activities in marketing. The theoretical foundations and examples of practical applications of the IoT have been used to develop a model for using the IoT in marketing, which is presented in Section 8.7. The whole project ends with a summary (Section 8.8).

8.2 RESEARCH OBJECTIVES AND METHODOLOGY

The purpose of this chapter is to identify the basic characteristics of IoT and review its current applications in the area of marketing. This research covered selected main tools of machine-to-machine (M2M) communication and, in particular, individual marking of objects (including Radio Frequency Identification [RFID] technology). The research involved the literature review and an original attempt to develop a theoretical model. The knowledge about IoT comes from the analysis of the content of scientific publications from the following databases: Elsevier, SpringerLink and Scopus, which have been critically reviewed in order to develop an optimal model of operations in this area. Due to the dynamic changes in the area of technology and marketing, the literature review covered the years 2020–2021. The research is qualitative and refers to the methodology of grounded theory as well as the recommendations of Eisenhard and Goya. Therefore, not only preliminary research questions were formulated to build the foundations of the inductive regularity of IoT operations in marketing, but also a creative approach was presented. The formulation of the further research problem and the determination of its essential features took place on the basis of the intellectual achievements contained in the widely available literature [6]. In this study, the method of critical thinking was used, consisting in formulating questions within the research area and developing concepts. The following techniques of conducting literature review were used: selection of keywords, selection of databases (Springer, Elsevier and Scopus) and definition of concepts [7], although the latter is not always possible in inductive research.

8.3 COMMUNICATION TECHNOLOGY SYSTEM

RFID is a system that uses radio waves to transmit data. This system enables the identification of objects, people and animals. It is a universal system, widely used in business, including marketing activities. The Radio Frequency Identification system consists of a computer system containing a database, transponders – transmitting and receiving devices (also called tags) – and a reader with an antenna and software. With this system, an identifier with encoded information must be within the reader's range in order to be identified. Readers can identify multiple objects at the same time. The transponders can take the form of a card, key ring, wristband or they can be in the form of a microchip implanted under the skin of a human or animal (Figure 8.1).

Identification of People, Animals and Objects via RFID Technology Applied in Marketing

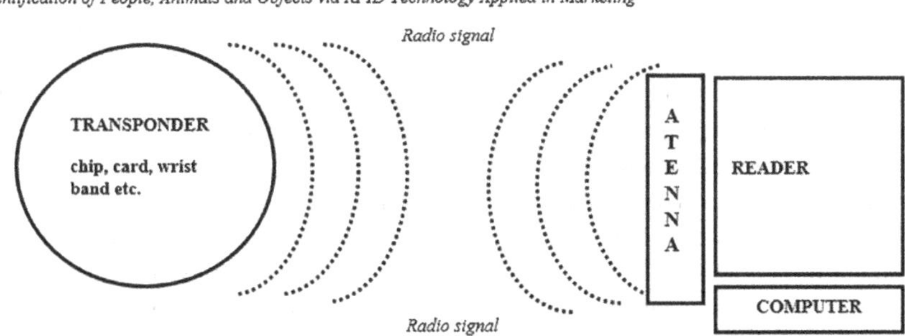

Note. [1]

FIGURE 8.1 Identification of people, animals and objects via RFID technology applied in marketing.

In the case of RFID technology, the transponder is not visible, and when it is within the range of an active reader, the data is transferred – the door is opened, access to the account granted or promotional offer activated.

This technology can be further enhanced by the use of artificial intelligence (AI), which is defined as the ability of machines and devices to perform cognitive functions, such as perception, reasoning, learning as well as independent problem-solving [8]. The greatest advances in AI have been made by applying machine learning to very large data sets. Machine learning algorithms, without prior software instructions, are able to detect patterns and learn how to create forecasts and directions for action by processing data and gaining new experiences. Algorithms also adapt to new data and situations and improve their effectiveness. Such development can lead to the creation of a computer with a fully independent programming ability. On the one hand, the human-machine relationship generates completely new problems, which will require a change in the doctrine and the operational philosophy. On the other hand, this situation creates great opportunities for optimization of operations, including those in marketing.

An important element of an RFID system is the software, which is a set of functions that underlies the interaction between the tag and the reader. The essence of the communication channel is usually the ability to process the radio signal. Hardware and software are required for proper marketing management of the data transmitted from the tag to the reader and back. Important features of the IT infrastructure are, apart from the hard infrastructure, also the RFID software system, the implementation of which requires technically adequate tags and readers. The use of this technology is applicable in various areas of marketing activities, i.e. it may refer to promotional activities and also the operation of supply and distribution chains. It is worth emphasizing that contemporary marketing reaches not only to the issue of sales channels (channels providing buyers of goods and services) but also to supply channels as an important element of not only marketing narrative, social marketing but also trade ethics. Supply chains are a distributed system where different counterparties may have unique RFID infrastructures with unique system needs. It is therefore important to create a uniform tool for formatting, processing and sharing data to facilitate the interaction of heterogeneous, unique systems. An Electronic Product Code (EPC) is a standardized numbering system that allows a unique identifier to be assigned to any physical object. The EPC can be considered the next generation of Universal Product Codes (UPC). The EPC is a way to assign a unique identifier to any product and make it easily traceable. Easy traceability thus becomes a tool that can be used for both marketing, social control, and the dangerous control of the totalitarian state. The information contained in the device becomes sensitive information, which must be the subject to strict protection. The EPC data format of Type One consists of a title (indicates the EPC version number), an owner number (indicates the entity that uses the EPC number), an object class (indicates the class or category of product, similar to a Stock Keeping Unit (SKU)), and a serial number (contains the unique code of the item being labelled). The 96-bit EPC1 code specification allows for the unique recognition of 268 million companies, each of which can have 16 million object classes and 68 billion serial numbers. EPCglobal is addressing this need by developing standards of the same name. To develop a solution that harnesses the power of RFID technology, it is necessary to understand both the internal organization of RFID components and how they interact. The distance from which a chip will be read will depend on the design, frequencies used, transmission power and environmental impact. In practice, all currently used RFID technologies can be divided into the following groups:

- Low frequency includes technologies and devices operating at frequencies from 125 to 134 KHz. The use of these frequencies allows the penetration of electromagnetic waves to a depth of 8 m in water, living tissues up to 2 m and metals within 2 mm. This frequency allows, for example, the tagging of livestock and pets and the use of technology in IoT communications, although data transfer at this frequency is too slow).
- High frequency includes technologies and devices designed to operate at 13.56 MHz. The ability to limit the reading area by a few centimetres allows them to be used in access control and payment systems – passes, cards, intercom keys, mobile phones etc.
- Ultra-high frequency includes technologies and devices designed to operate at 433 MHz, 860–960 MHz, 2.4–2.45 GHz and 5.2–5.8 GHz. Referred to as rain – Radio Frequency Identification with a frequency band around 866 MHz (ETSI – European Telecommunications Standards Institute) or 910 MHz (FCC – Federal Communications Commission) – which is designed to support several sensor platforms.

On the one hand, the high wavelength (high number of oscillations per second) means that more information is transmitted per second. On the other hand, the wavelength (distance from one "peak" to another) at high frequencies is noticeably shorter – such waves "break up" into much smaller obstacles and much more material is impermeable to them.

8.4 IoT IN MARKETING

Research and study in the field of IoT is in nascent stage as there is not even a single appropriate definition for IoT available yet. Three perspectives can broadly define IoT: Internet-oriented, semantics-oriented, and things-oriented. These views can be used jointly to achieve the goals of IoT [9]. The term IoT was first coined by Kevin Ashton presenting the possible use of RFID technology in the supply chain of the global company Procter & Gamble. According to the McKinsey Institute [10], IoT is defined as sensors and actuators connected by networks to devices that can monitor and manage the activities of connected devices. In turn, Porter and Heppelmann [11] identified the IoT as an information transmission mechanism in which things (objects and devices) have the ability to operate intelligently. Thus, the IoT means the digitization of physical objects by supplementing them with smart digital functionalities thanks to the use of new technologies at various stages of creating customer value, which allows for better satisfaction of customer's needs. Hence, the key feature of IoT is the integration of the physical and digital spheres [12, 13] (Figure 8.2).

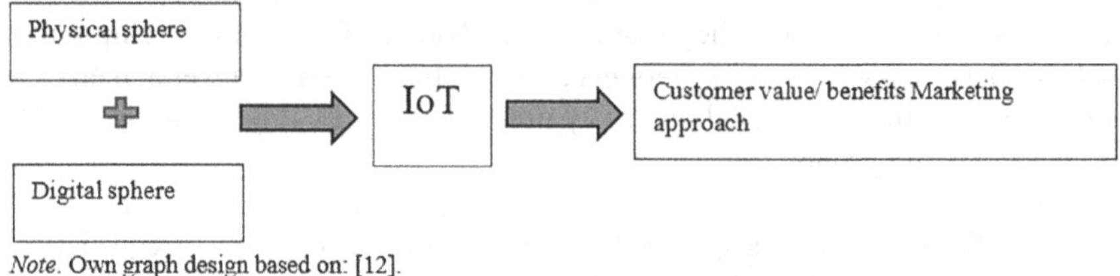

FIGURE 8.2 IoT spheres according to Fleisch-Weinberg-Wortmann – marketing approach.

To extend and modernize the identification of IoT by management classics [11], it should be stated that IoT consists of:

- the functionality characteristic of physical objects (mechanical and electrical parts of the product);
- "intelligence" enhancing the functionality of physical objects, through the use of, e.g., light, motion, sound or temperature sensors, as well as methods of collecting and sharing data and having a friendly interface equipped with a virtual agent (software that delivers results of observations and research of AI in a human-friendly manner);
- connectivity allowing, within the defined world of things and human needs, for the exchange of information and building added value over the physical existence of things.

The development of IoT in marketing is greatly facilitated by blockchain technology, defined as a chain of cryptographically linked timestamped blocks that operates as a distributed ledger, the data of which is shared among its peers [14]. The technology is based on a peer-to-peer network without central computers, management systems and transaction verification. Any computer connected to the network can participate in the transmission and authentication of transactions. Opportunities for its development have been identified in areas such as transport and decentralized markets (using the principles of the sharing economy), energy distribution and production, gaming systems and event prediction and identification and tracking of luxury goods [15].

IoT technology from the consumer's point of view is also reflected in the idea of the so-called smart city, as it contributes to improving the standard of living of residents and increases their participation in making important decisions. Smart city refers to the concept of applying all the available resources and technologies in a coordinated manner aiming to develop integrated, habitable and sustainable urban centres. Some well-known applications of smart city in modern societies include smart energy for optimizing consumption, smart building (home), smart mobility, and smart security [16]. A smart home is based on sensors, IoT devices, GPS, alarm systems, dedicated network connections and so on [17]. The decisive factor in the implementation of the IoT system in marketing is the trust of customers, because it determines the market success of digital products and services. Trust in this respect is important because it is often strictly related to the willingness (consent) to share information, which is one of the pillars of the contemporary business model [18]. The technological revolution has also brought about changes in the behaviour of consumers. Nowadays, consumers increasingly often make decisions taking into account organizations' social responsibility and ethics, including respect for animal rights and care for the natural environment. IoT concerns marketing management, because consumers are the "addressees" of what is being offered. The literature on this subject points to the fact that, for many years, consumption has exceeded the biological capacity of nature. The IoT offers an opportunity to balance many aspects of consumption, as it provides technology capable of limiting the waste of goods. Digitization in the use of new technologies, including IoT, requires change in consumer behaviour [19].

8.5 WHAT IS CONTEMPORARY DIGITAL MARKETING?

"Digital marketing" differs in its essence from "general marketing". It is presented as a modern form of communication between businesses and their stakeholders. Digital marketing means "a much wider choice of products, services and prices from different suppliers and a more convenient way to select and purchase items" [20]. It involves Ref. [21] identifying the needs of potential customers, who leave their footprint in the virtual space, with the use of, e.g., Big Data, Internet advertising, content marketing, influencers' activity and the virtual community. A characteristic feature of digital marketing is the focus on building customer loyalty and retaining the consumer as long as possible (preferably forever). Contemporary Internet marketing puts emphasis on developing strong customer loyalty by creating desired customer attitudes, e.g. through social media and influencers so that the consumer remains emotionally attached to the organization for as long as possible. When using digital marketing, organizations put the accent on creating customer value and customer benefits and not on the product itself [8]. Another characteristic feature of digital marketing is that companies' operations are aimed at increasingly narrow target groups of identical consumers, often dispersed across national borders.

The directions of development in digital marketing can be outlined as the following: digital marketing channels can be treated as an entirety, penetrated with algorithms, from which the best access channel to the customer will be selected; media planning and buying will be smart and automatic; marketing activities will be permanently updated with the use of Big Data; customer behaviour, including emotional behaviour in social media, will be automatically analysed based on monitored conversations, posted photos and messages; customer profiling based on the analysis of their activity in social media will enable targeting them with precisely selected media campaigns and promotional activities. Numerous studies have been devoted to applications of modern technologies, including their implementation in marketing activities [22–24]. The issue of managing the knowledge and customer needs is of interest to many researchers [25, 26]. IoT plays an important role in the area of managing consumer needs and creating recognizable customer value. The IoT is a functional integration between devices, a network to network (as a virtual exchange platform), and a driver of value creation through product and process innovation [27]. Therefore, looking from the customer's perspective in the aspect of broadly understood marketing instruments, the use of IoT is felt, (although probably in many cases unconsciously), mainly when they use or buy a given product (e.g. they can track the delivery of ordered goods, often thanks to RFID technology). As regards the product itself, the application of IoT leads to "smart products". Their "intelligence" consists of autonomy, adaptability, reactivity, multi-functionality, and the ability to cooperate and interact with humans, which translates into functional, emotional, social and epistemic benefits for the consumer [28]. Another important issue related to the customer and IoT is the so-called reverse distribution, which originates in the concept of circular economy (CE) [29], i.e. one in which the value of products, materials and resources is maintained in the economy as long as possible and waste production is minimized [30]. The IoT is considered part of it and its tools are designed to deal with its challenges.

8.6 SELECTED PRACTICES OF APPLYING IoT IN MARKETING

In order to demonstrate the significant benefits of using IoT activities in marketing, some practical cases should be presented. The selected cases originate in various industries, so as to demonstrate the universality of this solution and the application possibilities in the vast majority of business entities. In addition, the possibilities of using this technology in some specific operations of an organization, such as returns in retail, were also indicated, along with the demonstration of the benefits for the consumer, and ultimately for the organization, in the context of the correct application of this technology. Selected practices of IoT applications in marketing constitute a limited illustration of the implementation of this solution, which will be used in the construction of the theoretical model of IoT application in marketing presented in the next section of this chapter.

8.6.1 Tracking Customer Behaviour in the Fashion Industry using RFID

The RFID technology was applied in one of the luxury clothing stores in Oslo [31], owned by a Norwegian company that has been operating through both online and traditional outlets for 15 years. The three fitting rooms in this store were equipped with separately addressable RFID antennas, one per room, connected to a common reader (Figure 8.3). The RFID reader was used to continuously track consumers. A computer connected to the store network recorded the IDs detected by the three antennas and saved them in an online database. The data was collected for 46 days during the opening hours of the store. A total of 8014 situations of trying on clothes by customers, which lasted from 1 to 30 min, were recorded and analysed. Four patterns of behaviour in the fitting room were identified that required different treatment from a retail business perspective (be it upselling, visual merchandizing or customer experience customization). Based on the data set collected using RFID technology, it was possible to identify new customer typology/segmentation and work out new solutions [32].

Shop layout

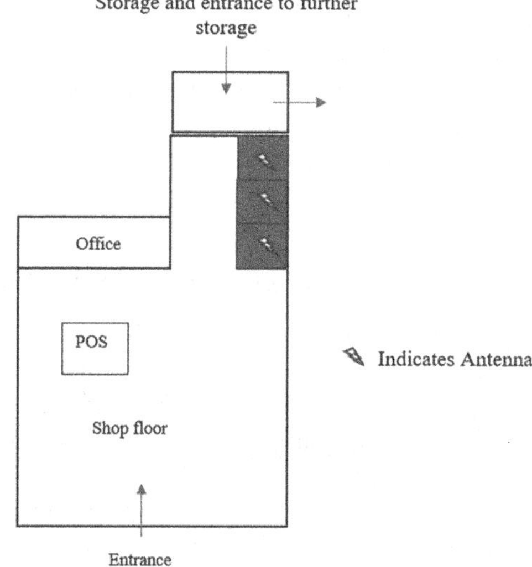

Note. [31].

FIGURE 8.3 Shop layout.

8.6.2 LG Smart Products

LG Corporation is a South Korean company operating in the international market, producing, inter alia, electronics and telecommunications products which are technologically advanced devices aimed at adding value to the consumer everyday life experience. Thanks to the use of AI and the LG ThinQ application – a smart home assistant – the technology offers such benefits as [33]:

- easy control of home devices via the LG ThinQ application and voice commands,
- personalized performance (LG ThinQ can learn the consumer's preferences to better meet their needs),
- proactive care (analysing habits and giving advice to help prevent problems before they occur),
- energy efficiency (optimization of product performance to minimize energy consumption and save money).

8.6.3 Smart Direct Trade Returns – 12Return

12Return platform deals with the management of product returns in a customer-centric and CE. It enables retailers and logistics service providers to deal with returns from anyone, anywhere. These returns have two dimensions: customer experience with returns and returns management operations. Customer expects the possibility to return a product at different stages of using it, seamless return with easy shipping, fast refund and accurate status [33]. The platform offers consumers returns across every distribution channel. This is important from a seller's perspective, as 97% of consumers say they will buy again if they have a positive experience with returns. The automation of processes and the use of smart solutions have allowed the platform to handle returns for customers from 149 countries, using global shipping and collection services from UPS, DHL and local carriers in all major European countries. In addition, the portal supports communication with the customer, who can create, send and track returns via a communication interface integrated into the shopping cart system.

8.7 MODEL OF IoT APPLICATION IN MARKETING

The logic behind the construction of the IoT model in marketing is based on the main areas of its application and the basic attributes of this technology (Figure 8.4). Identification of the characteristics of IoT technology (intelligence, including AI, increasing the functionality of physical things; integration of the physical and digital sphere; prioritizing consumer satisfaction) was used to identify the elements which, according to the authors of this chapter, are important in the area of marketing activities. This research shows that IoT technology has implications mainly in three scopes:

- The consumer – researching their behaviour and meeting their needs,
- The product – designing products meeting the consumer's requirements, leading to the creation of the so-called smart products, which can also constitute a competitive advantage,
- The distribution – creating cost-effective routes to reach the consumer.

In order to deepen the analysis of the model presented in Figure 8.4, it can be observed that the role of IoT is not only to increase the consumer portfolio but also

Model of IoT Applications in Marketing

Characteristics

Intelligence enhancig the functionality of physical objects

Integration of physical and digital spheres

Customer satisfaction

Internet of Things in marketing

Elements

Customer (Information about their needs)

Product and knowledge about it (directions of development)

Distribution (selection of optimal paths)

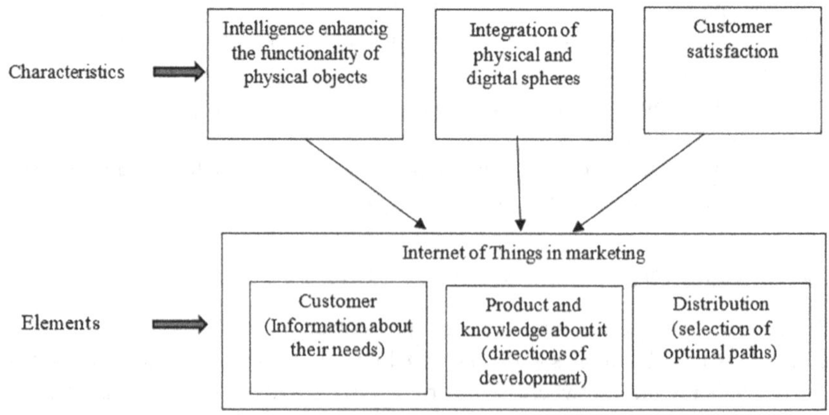

Note. Own graph design based on: [34] [18] [19] [12].

FIGURE 8.4 Model of IoT applications in marketing.

to build resources of knowledge about the consumer. Knowledge about the consumer can be used to create new products that will enhance consumer satisfaction to an even greater extent. In turn, consumer satisfaction will make it possible to build a long-term relationship between the organization and the consumer, which nowadays is the ultimate marketing goal.

8.8 CONCLUSIONS

The advancing digitization process is modifying the approach to traditional marketing principles and enables applications of more advanced technological solutions in marketing activities [35]. Moreover, the development of new communications technologies has direct impact on the needs and requirements of the consumers themselves. Having access to more information, consumers have become increasingly demanding and are now looking for unique products, tailored to their individual needs. In the economy where information becomes, or already is, dominant, the use of new technologies revolutionizes the functioning of many sectors of the economy. The IoT is a relatively new tool that enhances the added value of many so far "ordinary" household items. The RFID technology enables the transmission of data about users, their lifestyle, the way they use devices (sending information between devices without the users' awareness) and optimizing the consumption of, e.g., electricity. In addition, this system enables the identification of people and animals, it is a universal system, used in business, but its capabilities are greater than the current implementation. The IoT is defined as a mechanism of information transfer in which "things" (objects and devices, people and animals [equipped with appropriate devices]) gain the possibility to act and interact smartly. The exchange of information enables its further processing and ultimately creates an added value, which can be shared between the consumer and the "owners" of information. However, the necessary condition is the trust of customers, because it determines the possibility of undertaking marketing activities based on IoT. The intention of the authors of this chapter is to develop the theoretical model of IoT application in the field of marketing and identify the benefits that organizations can generate through the use of this technology. Based on the literature review and selected IoT applications in marketing, it was possible to develop a model of IoT implementation in marketing, consisting of both the elements derived from the key areas of its application in marketing and its generic features. In addition, the development of the theoretical model has showed the benefits of using this technology in marketing, which should encourage marketing practitioners to further popularize this solution.

M2M technology can be improved through the use of AI. The 21st-century economy is constantly changing and redefining the goals of international communities governments, and businesses. In the future, we can foresee innovative use of IoT and AI models in traditional industries such as agriculture, for example, and are also related to medicine and biotechnology. As machine learning algorithms are able to detect patterns, learn and make predictions by processing data, they will become the essence of a new business model. In the case of marketing and ecology, the relationship between the IoT and the consumer is particularly important, which can contribute to a significant reduction in pollution and consequently to the saving of raw materials, which is derived from the concept of CE.

REFERENCES

1. Pizło, W. (2019) *Marketing międzynarodowy [International marketing], wyd.* SGGW, Warszawa.
2. Urbach, N. (2020) Einleitung: Digitalisierung verändert die Geschäftswelt. In *Marketing im Zeitalter der Digitalisierung. essentials.* [Introduction: Digitization is changing the business world. In: Marketing in the Age of Digitization. Essentials] Springer Gabler, Wiesbaden.
3. Sestino, A., Prete, M. I., Piper, L., & Guido, G. (2020), Internet of Things and Big Data as enablers for business digitalization strategies, Technovation 98, 102173.
4. Kotler, Ph., Armstrong, G., Saunders, J., & Wong, V. (2002) Marketing. Podręcznik europejski [Marketing. European textbook], wyd. PWE, Warszawa.
5. Mazurkiewicz-Pizło, A., & Pizło, W. (2017). Marketing. In *Wiedza ekonomiczna i aktywność na rynku [Marketing. Economic knowledge and market activity], wyd.* PWN, Warszawa.
6. Eisenhardt, K. M. (1989). Building theories from case study research. Academy of Management Review, 14:4, 537.
7. Creswell, J. W. (2009). *Projektowanie badańnaukowych. Metody jakościowe, ilościowe i mieszane, Wyd.* Uniwersytetu Jagiellońskiego, Kraków. [Scientific Research Design. Qualitative, quantitative and mixed methods].
8. Pizło, W. & Filipowicz, A. (2019). Konsekwencje rozwoju sztucznej inteligencji [Consequences of the development of artificial intelligence], [w:] Problems of Economics and Law, 3:12, 1–14.
9. Bhushan, B., Sahoo, Ch., Sinha, P., & Khamparia, A. (2021). Unification of Blockchain and Internet of Things (BIoT): requirements, working model, challenges and future directions. Wireless Networks, 27:55–90.
10. McKinsey (2018) The Internet of Things: How to capture the value of IoT. https://www.mckinsey.com.
11. Porter, M. E. & Heppelmann, J. E. (2014) How smart, connected products are transforming competition. Harvard Business Review, 92, 64–88.
12. Fleisch, E., Weinberger, M. & Wortmann, F. (2015) Geschäftsmodelle im Internet der Dinge [Business models in the Internet of Things]. Schmalenbachs Z betriebswirtsch Forsch, 67, 444–465. https://doi.org/10.1007/BF03373027.
13. Schumacher C. (2017) Marken im Internet der Dinge [Brands on the Internet of Things]. In: Theobald E. (ed) *Brand Evolution*. Springer Gabler, Wiesbaden.
14. Saxena, S., Bhushan, B., & Ahad, M. A. (2021), Blockchain based solutions to secure IoT: Background, integration trends and a way forward. Journal of Network and Computer Applications 181.
15. Deptuła, E., Co to jest blockchain i jakie może mieć znaczenie z punktu widzenia ekonomii? [What is blockchain and what can it mean in terms of economics?] https://www.lazarski.pl Aceessed Julay 2021.
16. Bhushan, B., Khamparia, A., Sagayam, K. M., Sharma, S. K., Ahad, M. A., and Debnath, N. C. (2020), Blockchain for smart cities: a review of architectures, integration trends and future research directions, Sustainable Cities and Society, 61.
17. Haque, A. K. M. B., Bhushan, B., & Dhiman, G. (2021). Conceptualizing smart city applications: Requirements, architecture, security issues, and emerging trends. Expert Systems. 10.1111/exsy.12753.

18. Müller-ter Jung, M. (2021). Datenschutz im Internet der Dinge [Data protection in the Internet of Things]. Datenschutz Datensich, 45, 114–119.
19. Jacob, M. (2020). Nachhaltige Digitalisierung in Unternehmen [Sustainable digitalisation in companies]. Wirtschaftsinformatik & Management 12, 224–229.
20. Chaffey, D. & Ellis-Chadwick, F. (2019). Digital marketing. In *Strategy, Implementation and Practice*, England New York Pearson, Harlow.
21. Świerczyńska-Kaczor, U. (2012), *e-Marketing. Przedsiębiorstwa w społeczności wirtualnej, wyd.* Difin, Warszawa, pp. 13–43 [e-Marketing. Enterprises in the virtual community].
22. Daoud, M. K. & Trigui, I. T. (2019), Smart packaging: consumer's perception and diagnostic of traceability information. In *Digital Economy. Emerging Technologies and Business Innovation*, Springer.
23. Lydekaityte, J. & Tambo, T. (2020) Smart packaging: definitions, models and packaging as an intermediator between digital and physical product management, The International Review of Retail, Distribution and Consumer Research, 30:4, 377–410.
24. Smith, K. T. (2020) Marketing via smart speakers: what should Alexa say?, Journal of Strategic Marketing, 28:4, 350–365.
25. Hibbard, J. D., Kacker, M., & Sadeh, F. (2019). Performance impact of distribution expansion: a review and research agenda. In Handbook of Research on Distribution Channels. Edward Elgar Publishing, Cheltenham.
26. Guitart, I. A., Hervet, G., & Hildebrand, D. (2019). Using eye-tracking to understand the impact of multitasking on memory for banner ads: the role of attention to the ad, International Journal of Advertising, 38:1, 154–170.
27. Visconti, M. R. (2020) Internet of Things valuation. In *The Valuation of Digital Intangibles*. Palgrave Macmillan, Cham.
28. Kaldewei, M. & Stummer, C. (2018). Der Einfluss der Produkt intelligenz auf den Konsumentennutzen und die Produktnutzung [The influence of product intelligence on consumer benefit and product use]. Schmalenbachs Z betriebswirtsch Forsch 70, 315–349.
29. Kristoffersen, E., Blomsma, F., Mikalef, P., & Li, J. (2020). The smart circular economy: a digital-enabled circular strategies framework for manufacturing companies, Journal of Business Research 120, 241–261.
30. Barreiro-Gen, M. & Lozano, R. (2020). How circular is the circular economy? Analysing the implementation of circular economy in organisations. Business Strategy and the Environment, 29:8, 3484–3494.
31. Landmark, A. D. & Sjøbakk, B. (2017), Tracking customer behaviour in fashion retail using RFID, International Journal of Retail & Distribution Management, 45:7/8, 844–858.
32. Jones, R. P. & Runyan, C. R. (2016), Conceptualizing a path-to-purchase framework and exploring its role in shopper segmentation, International Journal of Retail & Distribution Management, 44:8, 776–798. https://www.lg.com/pl/o-lg.
33. 12Return | Returns Management Platform, www.12return.com, Accessed 21.05.21.
34. Kubach, M. & Mihale-Wilson, C. (2021). Kundenorientierte strategische Auswahl von Geschäftsmodellen für IoT-Ökosysteme [Customer-oriented strategic selection of business models for IoT ecosystems]. Wirtsch Inform Manag.
35. Crick, J. M. (2021) Qualitative research in marketing: what can academics do better? Journal of Strategic Marketing, 29:5, 390–429.

9 Internet of Things (IoT) for Sustainable Smart Cities

E. Fantin Irudaya Raj
Dr. Sivanthi Aditanar College of Engineering,
Tamil Nadu, India

M. Appadurai
Dr. Sivanthi Aditanar College of Engineering,
Tamil Nadu, India

S. Darwin
Dr. Sivanthi Aditanar College of Engineering,
Tamil Nadu, India

E. Francy Irudaya Rani
Francis Xavier Engineering College,
Tamil Nadu, India

CONTENTS

DOI: 10.1201/9781003219620-9

9.1 INTRODUCTION

The world's population is rapidly increasing. Only 16% of the world's population lived in cities at the starting of the 20th century. In 1950, it was increased to 31% and 45% in 2009. This portion is further increased to 55% in the year 2020. By 2050, it is expected to reach 70%. With the ongoing economic downturn, this unprecedented rate of urbanization has created a significant problem for politics and city governance. Human consumption has accelerated environmental and natural resource depletion on the planet. The towns and cities struggle to provide essential urban services. Governments and city planners face major challenges in providing more reliable, decent facilities to sustain peak demand, provide clean and affordable energy, and meet CO_2 emissions reductions.

Additionally, governments, businesses, and city planners will face challenges such as integrating public transit, providing more safe and good-quality water, and optimizing the use of existing assets. Following the introduction of modern digital technology and computers, new opportunities for managing urban transitions and fostering the era of sustainability have emerged. Along with environmental changes, cities' health is impacted by fragmented, deteriorating, insufficient, and aging infrastructure. City infrastructure and services operate inefficiently due to a lack of communication between their system components in the network, restricting the proliferation of new value-added services, increasing transportation costs, and weakening existing economic models and logistics chains. Infrastructure problems have frequently aggravated the numerous challenges related to urban life regarding housing, healthcare, transportation, the rule of law, utility services, education, waste management, and delivery of essential public amenities. Policymakers must balance urban development, economic growth, industrialization, environmental regulations, and geographic sprawl to create sustainable smart cities. Cities are particularly interested in intelligent technologies that will improve the quality of urban services while lowering operating expenses.

Through smart city design, planning, and construction, digital technologies present a new array of prospects to alleviate a few of these effects and strike a balance of social, environmental, and economic benefits. The Internet of Things (IoT) is a new archetype that brings together wireless sensor networks, sensor technology, communication technologies, Internet communication protocols, ubiquitous computing, and embedded systems. The IoT is used to regulate, offer additional knowledge and insight, interconnect, and interact with the various warehouses of splintered systems found throughout city areas. The vast network of connected devices, combined with the vast volume of data collected by them, opens up emerging technological options for resolving urban problems. These components are designed and integrated with the city systems to create a dynamic coexistence of the physical and digital worlds. This ubiquitous and intelligent environment serves as the foundation for connected, sustainable smart cities. Figure 9.1 shows the IoT-enabled smart environment, which is the combination of smart industry, smart transport, smart health, smart city, smart home, and so on.

The current work discusses IoT-based smart home automation, which aids in the remote monitoring of various equipment in residences from faraway places. It also discusses the importance of air quality monitoring and management system. The system

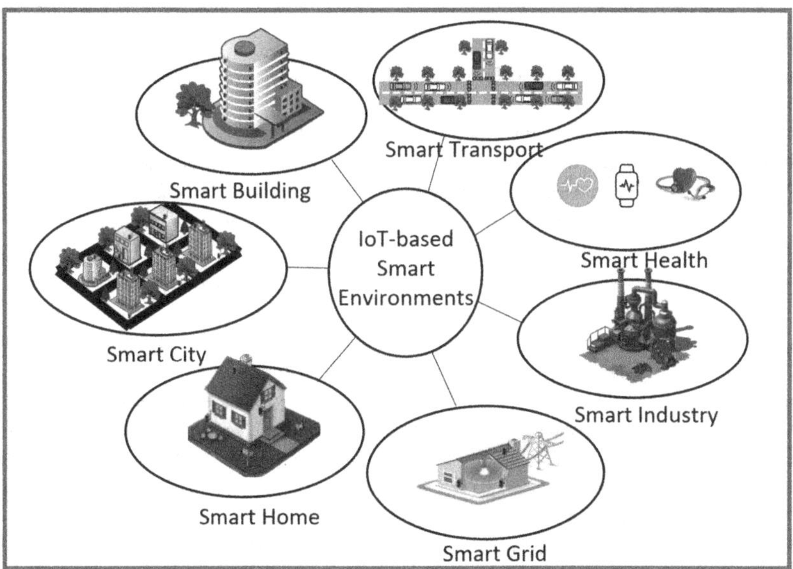

FIGURE 9.1 IoT-enabled smart environment.

will alert the government officials and residents to take precautionary steps to overcome the situation. Healthcare is another important factor to be considered in smart cities. The present work provides the classification of a smart health-care system. In addition, detailed architecture and hardware prototypes of the IoT-based health monitoring system are provided. IoT-based traffic management and vehicle parking systems are also explained.

Furthermore, Section 9.2 provides a detailed review of the literature. An IoT-based smart home environment that provides safety and security to the inhabitants of smart homes is discussed in Section 9.3. Section 9.4 discusses IoT-based air quality management. IoT-based modern health-care system and its importance and effectiveness are discussed in detail in Section 9.5. Sections 9.6 and 9.7 discuss the IoT-based smart traffic management system and vehicle parking system, respectively. Finally, Section 9.8 covers the IoT-based smart waste management system in smart cities.

9.2 RECENT WORKS

In recent years, smart city initiatives have become increasingly widespread around the world. Because of the population of cities and the complexities of town planning, local governmental bodies were turning to technology to improve the efficiency of metropolitan areas and the delivery of government services. The allure of smart cities is in their ability to connect advanced technology, a sustainable environment, and citizens' comfort regardless of their culture or geographic location. Many works related to smart cities and their implementation are presented in the literature in recent times. Some of the few works are listed in Table 9.1.

TABLE 9.1
Recent and Important Notable Works in the Literature

Author and Year	The Focus of the Study	Findings
Bakıcı et al. [1], 2013	Utilizing Information and Communication Technologies (ICT) to transform the cities, urban growth, and policymaking of smart cities	A unique model of the smart city is proposed. The benefits and challenges in adopting ICT are also explained.
Venolo [2], 2014	Discussed the smart city from a critical perspective and focused more on knowledge and power implications	A new way to organize, manage, and imagine the smart city and its flow is proposed. Identified technical parameters to know the city is good or bad.
Neirotti et al. [3], and Angelidou [4], 2014	Discussed the importance of the policymakers' agendas and provided valuable guidelines for city managers to describe and proceed with their smart city strategy.	Uncovers the spread of smart initiatives through an empirical study. The ratio of domains covered by a city's best strategies to the total number of potential domains for the smart initiative is determined.
Tai-Hoon Kim et al. [5], 2017	Focuses on the Internet of Things (IoT) and its implantation in smart city creation and management.	IoT's new protocols, services, and architectures for developing scalable, reliable, and efficient smart cities are proposed.
Gaur et al. [6], and Krylovskiy et al. [7], 2015	Focus on the adoption of IoT in various domains of smart cities.	A new architecture of the smart city is developed and designed a smart city IoT platform using the microservice architectural style.
Chakrabarty et al. [8], 2016	The focus of this study was secure IoT architecture for smart cities.	A set of IoT architectural blocks was proposed to safeguard smart city data. The new architecture's efficiency is meticulously detailed.
Basford et al. [9], 2020	LoRaWAN is a low-power wide area network architecture developed and optimized for IoT deployment in smart cities like urban environments.	The network was set up to facilitate the installation of air quality monitors and test the LoRaWAN capabilities. The results obtained are satisfactory.
Suri et al. [10], 2018	The authors examined the possibility of utilizing smart city IoT capabilities to assist with disaster recovery operational activities.	Resources, such as personnel and supplies, can be prioritized to have the most impact and help those who are in need. It can be greatly aided by data gathered from IoT devices, particularly in a smart city setting.
Peneti et al. [11], 2021	The importance of IoT and 6G technologies and their adoption in smart cities like intelligent environments are discussed.	The system's efficiency is then assessed using simulation results, showing that the system provides low latency and great security compared to a multilayer perceptron network.
Al Kindhi et al. [12], 2021	Discussed in detail IoT and fuzzy-logic-based lighting systems suitable for smart cities.	The remote monitoring of lights and their state from the control center is proposed. Moreover, it is more economical compared with any other conventional methodologies.

(Continued)

TABLE 9.1 *(Continued)*
Recent and Important Notable Works in the Literature

Author and Year	The Focus of the Study	Findings
Saxena et al. [13], 2021	Introduced a new data fusion system in which data is obtained from various smart city sub-applications and sensors.	IoT-based new framework is deployed, and the data is gathered through various sensors and networks. A unique data fusion technique is applied and used to manage the information effectively.
Napolitano et al. [14], 2021	Discussed the necessity of collecting vast data in real time in a smart city like the urban environment.	Insisted on using IoT devices and wireless networks that can sense, react, and process the data acquired from the environment without any centralized sources.
Mukhopadhyay et al. [15], 2021	Artificial intelligence (AI)-based modern sensors for future generation IoT applications in smart cities are discussed.	AI skills that can detect, identify, and avert performance degradation and uncover new patterns have recently evolved. Thus, it enables the IoT to collect data and make smart decisions.
Shafali Jain et al. [16], 2020, and Raj [17], 2016	Improved IoT-based control system combined with an advanced control management server-based system for smart city environment is explained.	Recent advancements in modern control systems and IoT sensors are identified. As a result, the proposed methodology shows good performance compared with the conventional one.
Goyal et al. [18], 2021	The focus is on IoT-enabled technologies in smart healthcare and their applications, difficulties, and future directions.	IoT networks for healthcare are depicted and the various characteristics of IoT confidentiality and safety, including security requirements.
Khamparia et al. [19], 2020	The use of transfer learning to detect and classify skin cancer is discussed. The authors were utilizing an Internet of Health Things-driven deep learning framework.	The proposed framework outscored existing conventional architectures to classify and detect skin cancer using skin lesion images in terms of accuracy, recall, and precision.
Sharma et al. [20], 2020	WSN and biometric models' applicability in the field of healthcare to be examined.	A comparison table is created that lists all of the benefits and drawbacks of various biometric-based models utilized in healthcare.

Many similar works in the literature also explore smart cities with IoT technology in depth. In the present work, we will discuss the IoT-based smart home environment in smart cities, air quality management system, modern health-care system, traffic management system in smart cities, vehicle parking system, and waste management system.

9.3 IoT-BASED SMART HOME ENVIRONMENT IN SMART CITIES

In today's world, electronic devices and machines take over tasks previously performed by humans. Because this enables us to free up time for other activities, we reorganize our lives. Our schedules are reset. Now that we have more time to create, we can do so. However, relying on an excessive number of devices eventually necessitates controlling them without exerting additional effort or wasting time. As a result, we've begun discussing novel concepts such as the smart home. Having a smart home allows us to access all of our devices remotely via a network system that allows us to assign tasks to be completed later.

When considering the smart home system, we can see how each step forward in living with machines has helped us get to this stage. Human beings settled down and have been upgrading their way of life ever since. They've been looking for a simpler way to accomplish tasks. Indeed, laziness took the lead and continues to do so. As a result, it was unavoidable to develop an invention for each. And this is how we became acquainted with household appliances. The first one was introduced during the Industrial Revolution. Following that, the evolution of appliances commenced.

Before creating Echo IV, around the 1960s, the need for home automation technology was recognized. However, it was not popular until 1998. Humans took a much longer duration to complete a task until the advent of modern appliances. Cleaning the kitchen, doing the dishes, laundry, mixing and cooking food ingredients, and so on took up even more time in a day than they should have. Given that so many of these activities are considered routine housework, it's safe to assume that housekeeping is a huge burden for families. We needed a smart home system capable of controlling the devices we own to avoid this and free up time for more worthwhile endeavors. Figure 9.2 shows the smart homes with IoT-based home automation.

Smart homes encompass all aspects of domestic life, including lighting, air conditioners, and dishwashers. Smart homes operate on the Internet, which has access to the home's content and commands and its applications, such as the music system. For instance, consider the smart television, which operates via visual, voice, and gesture consolidation. Phillips recently launched a customizable intelligent lighting system that detects whether or not the owner is at home and switch on and off the light based on the situation. In addition to the sensors, data can also be manipulated using smart devices like smartphones, personal computers, and tablets. The addition of

FIGURE 9.2 Smart home with IoT-based automation.

thermo-controlled devices, such as nest labs that can be governed through a sophisticated Wi-Fi connection, allows home temperatures to be scheduled and monitored in the housing area. It not only gives the user warmth, but it can also shift a person's attitude toward energy conservation with a single button press.

IoT systems have now transformed the field of pet care and automatic sprinklers that water plants on a routine in residents' gardens. In short, from smart refrigerators to coffee makers, our homes are becoming self-sufficient and controlled by the owners' commands. Everything is becoming possible thanks to IoT devices: an automated washing machine that takes care of our laundry, a cup of coffee made by an automated coffee maker, and autonomous systems capable of generating our grocery list. As a result, we may conclude that the IoT is changing the way our homes look and changing several business modules.

A smart, automated device operates based on the home's infrastructure. They are built up during the construction process. While older homes can be converted to smart homes, the process is more expensive and complex; on the other hand, prefitted smart homes are easy to install and include significantly more features. The majority of smart homes are still powered by Bluetooth-enabled devices and X10 or Insteon protocols. This one has increased in popularity due to its affordability. In 2018, Zigbee had been the most widely used protocol for home automation communication systems. Compared to a mesh network, this one is more limited in range, has a lower frequency, and has complications with radio signals. The range may be between three and ten meters.

A smart home is fundamentally based on the coordination of inter- and intra-device systems. It works in conjunction with hundreds of other small or large devices to establish a remotely controlled network system that operates according to predefined preferences. Typically, the master home automation controller, also referred to as the smart home hub, controls the devices in the smart home. The hardware device communicates with the central point system and can wirelessly sense, process, and act on the data. It is made abundantly clear by the recent introductions of a wink, Insteon Pro, Amazon Echo, Amazon Dot, and Google Homes.

While some smart homes can be easily constructed using Raspberry Pi or prototyping, which is significantly less expensive and simpler, others are more complicated and need bulk assembly, such as the smart home kit. They are primarily time-consuming and contain small data packets. The module's operation is often based on time management, such as triggering an action within a specified time frame. For example, if your home is automated with lights, it should be able to turn on the lights when the owner's smart device approaches the appropriate time. Thus, judgment must be maintained by making one aspect of time incompetent, such as turning on the lights at 6 p.m. and whether or not the smart device somehow doesn't approach the house.

Not only does the IoT make it simple to modulate and control your own home from the comfort of your chosen device, but it can also assist in preserving as much energy as possible through all these technological efforts. According to a recent survey, the United States of America could preserve energy and cut consumer spending by 30% simply by implementing smart homes and sensors. These sensors enable automatic pressure and temperature adjustment, ensuring that your air conditioner does not overcool and thus consumes less energy. In your office, on the other hand, your automated system can detect your absence and turn off the lights, conserving energy until someone manually switches them off. Furthermore, with the introduction of smart homes, maintenance has become quite simple, as generators can be switched on instantly when the power goes out.

Mohamed Asadul Hoque et al. [21] proposed Arduino compatible Elegoo Mega 2560 microcontroller board-based architecture and Raspberry Pi 2 board for communicating with a web server for IoT-enabled smart-city-based smart home environment. Yonghee Kim et al. [22] examined smart home devices in smart cities like urban environments. At the same time, the author described the new architecture and explained various new protocols for safe communication between the devices. It will enable an intelligent and secure home environment in urban conditions. Zaidan et al. [23] discussed various features of the IoT-based evolving approach in their review article. In this, they made precious recommendations for the approval and utilization of the intelligent process for IoT-based smart home applications.

Bhushan et al. [24] discussed blockchain for smart cities. They also reviewed new architecture and future trends of adopting blockchain in smart cities. The study provides basic information while examining the utility of blockchain in various smart communities, including data center networks, financial systems, supply chain management, smart grid, transportation, and healthcare. Haque et al. [25] conceptualized the applications of a smart city in the context of emerging trends, security issues, architecture, and requirements. Sustainability, device-level vulnerability, illegal access, authentication, and data security have all been highlighted due to smart cities' growing popularity. Various smart city applications, as well as prominent implementations, are also shown in order to comprehend the quality of living standards. Bhushan et al. [26] explained the concept of combining blockchain and the IoT (BIoT) is discussed. Future directions, obstacles, the working model, and requirements are also explored. Future directions, barriers, the working methodology, and requirements are all investigated. The proposed system's decentralization, security, and auditable are all described in depth. Its architecture and security features are also highlighted.

9.4 IoT-BASED AIR QUALITY MANAGEMENT SYSTEM

According to a global study, one of the most concerning issues in modern cities is air quality, with air pollution causing 120 deaths per 1 million people per year. The World Health Organization (WHO) indicated that 97% of cities in low- and middle-income countries with more than a million population do not meet WHO air quality standards. Due to poor air quality, potential health risks such as asthma, lung cancer, heart disease, and stroke will increase. As a result, cities must install an air quality monitoring system to ensure that the air is not contaminated. It can be accomplished by deploying sensors to monitor sulfur dioxide, carbon monoxide, nitrogen dioxide, carbon dioxide levels, and dust particles. The insights are received from the sensors sharing with the public via smartphones, where the smartphone app allows users to monitor the current air quality level in the area in real time. As a result of these implementations, a higher standard of living can be achieved.

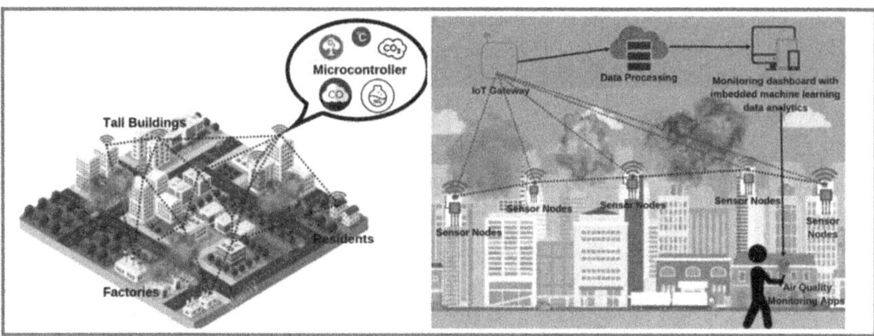

FIGURE 9.3 Deployment of air quality sensors and IoT framework to measure air quality index.

The main objectives of the present systems are (1) to develop an innovative air quality monitoring system capable of covering an urban area of one square kilometer, (2) supervise the air quality level via smartwatch and smartphone applications using sensors installed on buildings ranging in height from three to six meters, and (3) recommend safety procedures via smartwatch and smartphone applications when air quality levels exceed predefined thresholds.

Air quality sensors are placed on traffic areas, industrial areas, top of buildings, and residential areas in the targeted area, as shown in Figure 9.3. These detectors are interfaced with the microcontroller, which is used to manage the network of sensors. The microcontroller sends the information it collects to the cloud-based database. Finally, the analyzed data is made available to the public via a smartphone application.

Sulfur dioxide, dust particles, nitrogen dioxide, carbon monoxide, carbon dioxide, and dust particles are all detected in the air by wireless sensors mounted in strategic locations. This data is transmitted via Bluetooth or Wi-Fi communication to a gateway, which then forwards to a cloud database. The data is analyzed in the cloud to provide information on air quality. A mobile app is used to exchange information about air quality. It encourages both the relevant authorities and communities to take corrective action and preventative measures. It allows citizens to monitor real-time air quality in their area quickly. The workflow of the process is shown in Figure 9.4.

It makes use of devices, including a gas sensor, nitrogen dioxide sensor, carbon dioxide gas sensor, carbon monoxide gas sensor, and dust sensor, which are all inexpensive and readily available. Microcontrollers monitor these sensors, and they often serve as transmitters, sending data to a cloud database. In addition, air quality information can be obtained in real time through a smartphone app.

Many cities around the world are experiencing poor air quality. Every year, people died from contaminated air, and their health deteriorates due to their exposure to poor air quality. The government will take precautionary measures if they are aware of the polluted air on time. It will also enable the appropriate authority to take corrective action.

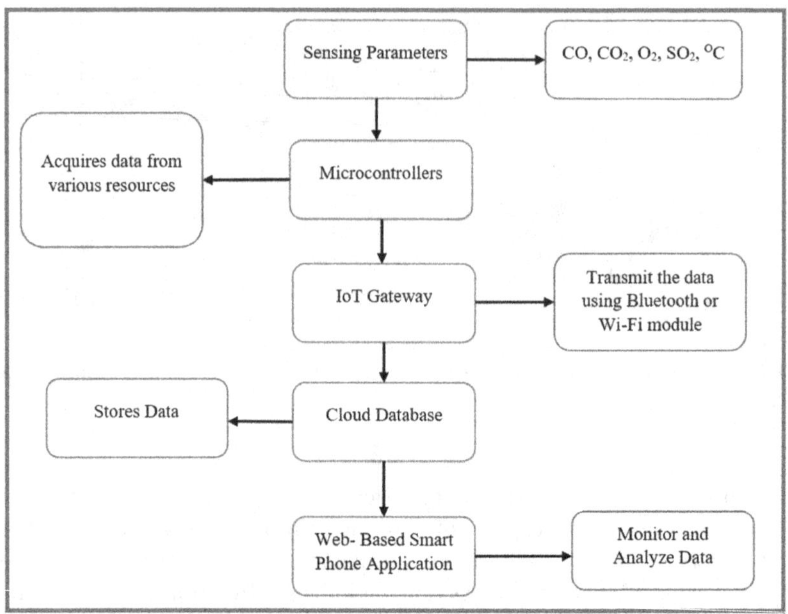

FIGURE 9.4 Workflow of the IoT-based to measure air quality index.

Nasution et al. [27] discussed the ThingSpeak cloud-service-based IoT-assisted air quality monitoring system. If compared with the conventional system, it is more economical and affordable. Xiaojun et al. [28] explained the various forecasting techniques in pollution monitoring using machine learning and IoT-based systems. The authors use a radial basis functional neural network to predict the nearby future values more precisely. In Refs. [29, 30], the researchers detailed indoor air quality monitoring using IoT devices. A novel indoor air quality detector is proposed, and information retrieved from the detector is passed to the microcontroller to take necessary action through a short-range communication network. The main advantage of these proposed works is that the devices used in the process consume only a small amount of energy. So, in terms of energy consumption, the proposed work is superior. Sai et al. [31] explained the low-cost Arduino-based IoT-dependent air quality monitoring system for a smart-city-like urban environment.

9.5 IoT-BASED MODERN HEALTH-CARE SYSTEM FOR SMART CITIES

The IoT, which has a lot of potential for smart medical treatment applications, will help hospitals achieve smart medical care and intelligent medical material management. It also allows for the transfer and exchange of internal medical data, staff data, digital data collection, retrieval, recording, management data, medication data, and equipment data. It may also meet intelligent supervision and management demands in health-care data, hospital instruments, and pharmaceutical products.

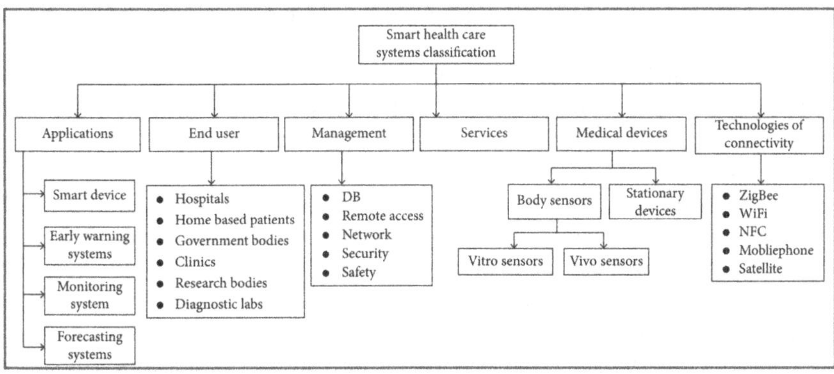

FIGURE 9.5 Classification of smart health-care system.

The smart health-care system is classified as (1) application, (2) end user, (3) management, (4) services, (5) medical devices, and (6) technologies of connectivity. It is also further classified into various subcategories. Figure 9.5 shows the detailed classification of smart health-care systems.

The smart health-care system architecture is shown in Figure 9.6. In this, we have three different layers, (1) sensor, (2) network access, and (3) services access layers.

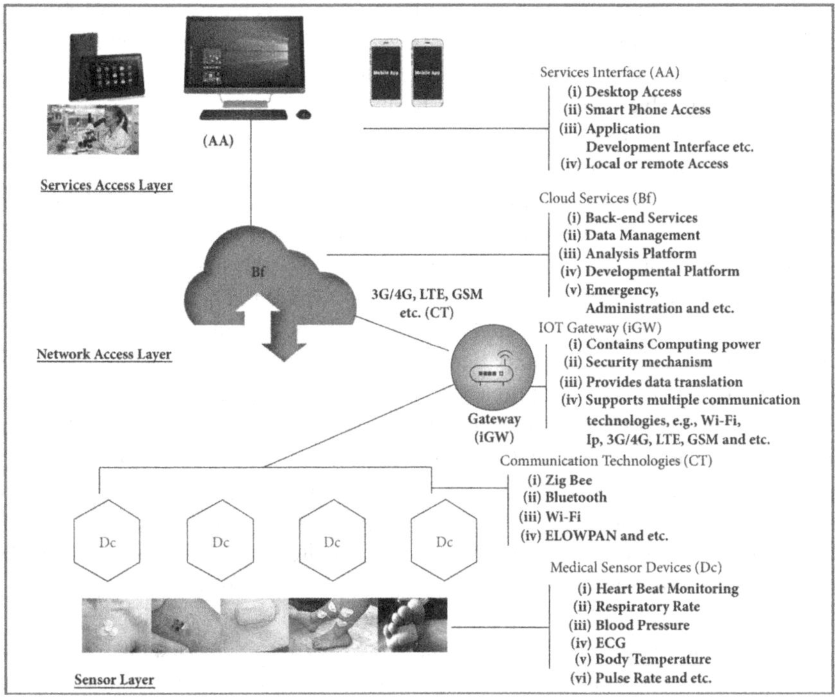

FIGURE 9.6 Architecture of smart health-care system.

Each layer of the architecture is equipped with various technologies that enable the doctors to monitor the patient's condition even in remote locations regularly. It is fundamental behind all types of e-health methodologies nowadays proposed [32, 33]. In the present pandemic time, even this type of system will help older adults and children get medical advice from the doctors in the home itself.

The sensor layer is directly connected with the patients or the person who needs to monitor. There are various types of sensors and devices placed in this layer. The important sensors and devices are heartbeat monitoring, respiratory rate monitoring, oxygen content monitoring, blood pressure monitoring, echocardiogram, body temperature, pulse rate devices, etc. These medical sensor devices are placed over the patients who need to monitor continuously. It will sense the patient's actual physical data, convert it to digital data, and pass it to the next layer of the network access layer.

In the network access layer, there are different types of communication technologies adopted. Zigbee, Wi-Fi, and Bluetooth are the few crucial technologies used in modern health-care system data transfer. The information received from the sensors is transmitted via various IoT gateways that contain security mechanisms, provide data translation, contain computing power, and support multiple communications. Cloud services are also adapted to store an enormous amount of data storage. By storing all such data in the cloud, we can retrieve it at any time. It holds the information about the patient's present and past conditions. Based upon the information, the doctor can analyze the patient's response to the treatment and take suitable corrective actions. This storage also acts as a permanent record for reference. It is beneficial in solving some complicated cases. The hospital management data, existing medical equipment available, health-care workers' details, and doctors' details are also stored for permanent reference in the cloud storage. This data is precious in crisis time to make suitable actions and better hospital management.

The next layer of this architecture is the service interface layer. It is one of the vital layers. Through this layer, only the physician or the patient can able to monitor his status. In this, mostly a computer with web application written using any software tool is used for viewing the details. Nowadays, the mobile phones have also become smart. Using smartphones also we can monitor the status of the person who is monitored using a smartphone-based web application. The details can be used for local access or remote access. The doctors or patients inside the hospital or in a local area can only sense the patient's actual data and condition in local access. According to that, the doctors prescribe medicine suggesting suitable remedies. In remote access, the information is shared via a wireless network. In this, either doctor or patient can be anywhere. Continuous monitoring is possible in this case. The doctors can suggest proper medication from remote places even. In this pandemic situation, these facilities are helpful for many peoples [34–36].

A simple prototype of the smart health-care monitoring system is developed. The entire system architecture of the prototype of a health-care monitoring system is depicted in Figure 9.7. The hardware prototype created is shown in Figure 9.8.

In this, Arduino-based microcontroller is used. It offers a less expensive and more economical IoT-based environment. Zigbee model is used for communication purposes. The plethysmography theory is used to develop the heartbeat sensor. It determines how much blood volume has changed. The timing of the pulses is a more vital

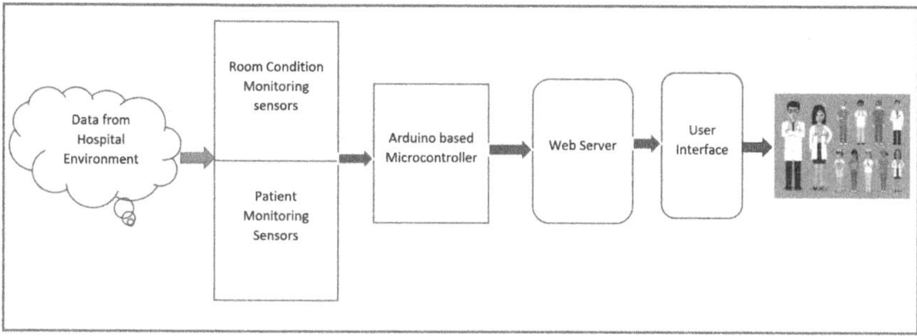

FIGURE 9.7 Prototype system architecture of the entire health-care monitoring system.

parameter to monitor because it allows the heart pulse rate to be accurately tracked. LM35 series body temperature sensor is used. It provides accurate, optimized temperature output voltage in the circuit. DHT11 sensor is also used to measure the room temperature. This sensor comes with a dedicated temperature measurement kit and an 8-bit microcontroller to process humidity and temperature values in series. MQ-9 sensor is normally used to monitor the CH4, CO, and LPG. Here, it is used for measuring CO. It has high sensitivity in nature and quick response time. MQ-135 sensor is used to measure CO_2. This sensor can work without the use of a microprocessor.

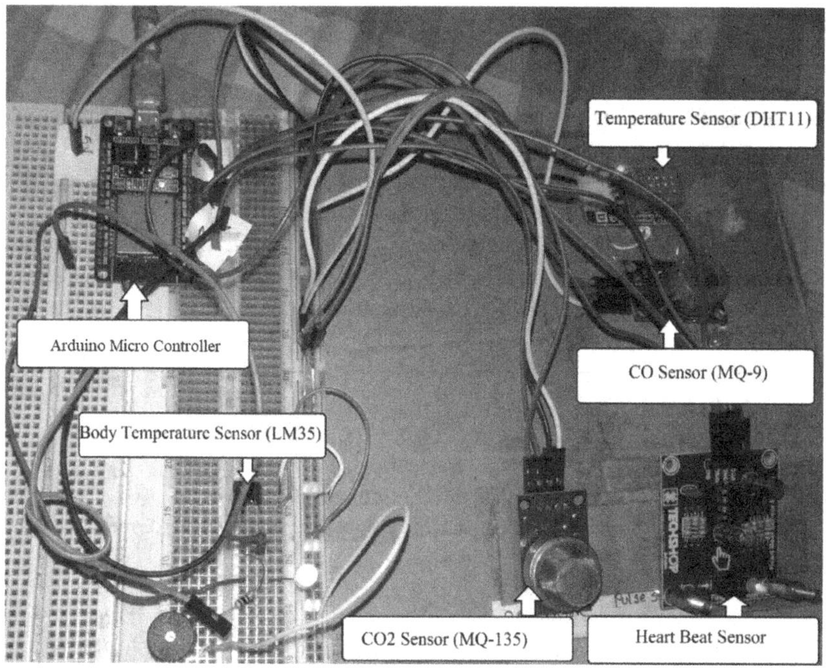

FIGURE 9.8 Hardware prototype of the IoT-based health-care monitoring system.

TABLE 9.2
Room Humidity Data

No of Experiment	Actual Data by Analog Machine (%)	Observed Data by Proposed System (%)	Error (%)
1	62	64	3.22
2	61	62	1.63
3	70	72	2.85
4	66	63	4.54
5	65	64	1.53

TABLE 9.3
Heartbeat Rate Data

No of Samples	Actual Data by Analog Machine (bpm)	Observed Data by the Proposed System (bpm)	Error (%)
1	68	69	1.47
2	70	72	2.85
3	74	77	1.29
4	73	72	1.47
5	78	76	2.56

It is more useful for specific gas detection. The humidity data collected by a conventional and the proposed system is shown in Table 9.2. Similarly, Tables 9.3 and 9.4 show the information of heartbeat rate and body temperature data collected by the conventional analog system and the proposed system.

The values observed using analog machine and observed data by the proposed system are shown graphically in Figure 9.9 (room humidity data), Figure 9.10 (heartbeat rate data), and Figure 9.11 (body temperature data).

TABLE 9.4
Body Temperature Data

No of Samples	Actual Data by Analog Machine (°C)	Observed Data by Proposed System (°C)	Error (%)
1	36.27	36.55	0.77
2	36.88	36.5	1.03
3	36.72	37	0.76
4	36.05	36.38	0.91
5	36.77	36.11	1.79

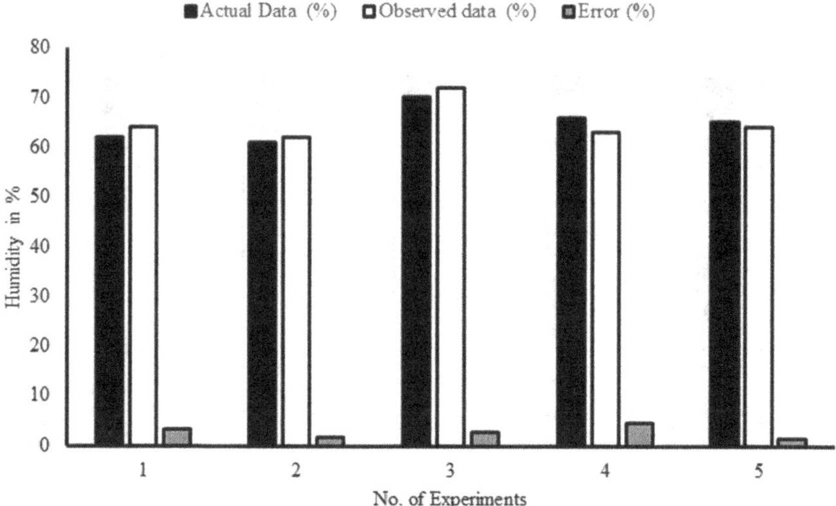

FIGURE 9.9 Room humidity data – actual, observed, and error in percentage.

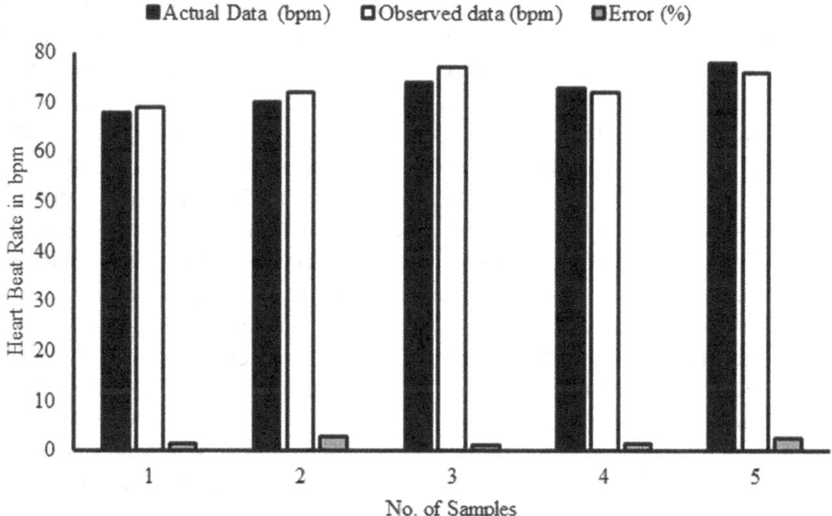

FIGURE 9.10 Heartbeat rate data – actual, observed, and error in percentage.

9.6 IoT-BASED TRAFFIC MANAGEMENT SYSTEM IN SMART CITIES

In large cities, the population of the people and automotive day by day gradually increases. There is a need for proper urban management systems to handle densely populated cities. Real-time handling of traffic is one of the tedious tasks. Intelligent

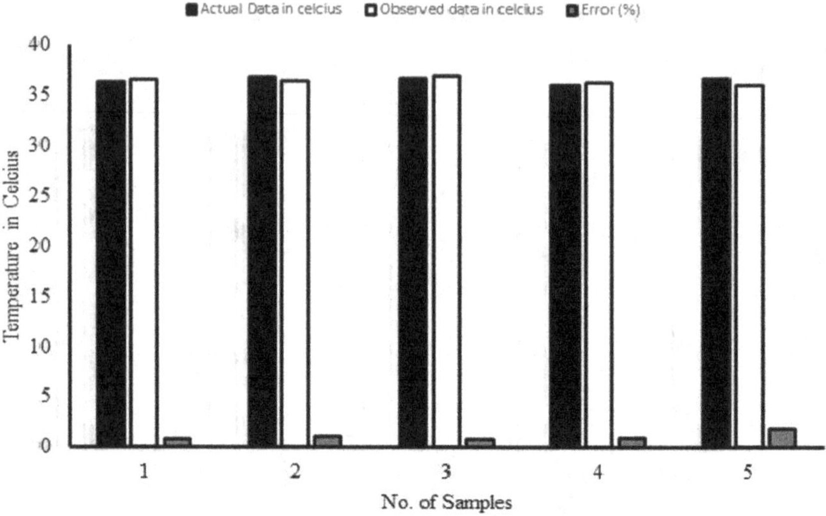

FIGURE 9.11 Body temperature data – actual, observed, and error in percentage.

smart techniques should be implemented to handle heavy traffic. The IoT device and other sensors are installed throughout the cities to connect traffic statistics with the central traffic management structure. The data from the IoT and sensors is handled automatically to adjust the traffic lights within a fraction of seconds. The past time data is utilized to guess the traffic in advance without human involvement [37, 38].

The traffic flow is measured by the sensors available in all prime areas of the cities. Modern wireless sensors and automated sensing and handling systems are the critical technology used to monitor, predict, and control all heavy traffic routes. Heavy traffic is first identified to handle traffic congestion. Then, real-time video imaging is analyzed to provide priority to the right road lanes. The traffic signals are altered and updated in real time. Based on the data, the dynamic handling of the infrastructure reduced the traffic density. Doing the work empowers more efficient usage of installed IoT devices, sensors, and modern technologies effectively. Figure 9.12 shows the display of real-time traffic for drivers.

Connected vehicles are another trustworthy technology in real-time traffic handling systems. The individual automotive is monitored continuously for efficient management of traffic flows. The fire engines, police vehicles, and ambulance handling are challenging tasks in the urban area dense traffics. The fraction of seconds' delay for every emergency vehicle produced higher negative impacts. The actual real-time schedule is made for immediate response for every event in the vehicular network [39–41]. This type of technique is primarily used for handling highway traffics. Figure 9.13 represents an optimization of traffic using IoT-based smart traffic control.

The real-time and upcoming traffic intensity is displayed graphically for the public on the roadside using signs, colors, or phone messages. The graphical display is mounted on tunnels, bridges, or toll gates. The pictorial data is also uploaded on the

FIGURE 9.12 Display of real-time traffic for drivers.

Internet continuously. The dynamic communication is immediately responded to by the drivers based on the situations.

The key advantages of IoT-based traffic management system are as follows:

- Efficient control of city lighting systems by automated sensors.
- It enhanced the urban area's traffic management quality.
- Provide public transport facilities to the people at their convenient time.

FIGURE 9.13 Optimization of traffic using IoT-based smart traffic control.

- Used to observe the entire city in the centralized form.
- Accidents and traffics are reduced by timely decisions based on centralized data.
- Higher road safety.
- City infrastructure is utilized efficiently.
- People safety and life standard are improved.

9.7 IoT-BASED VEHICLE PARKING SYSTEM IN SMART CITIES

The smart city concept is discussed worldwide in recent days. The innovation in IoT devices makes the concept of forming a smart city into a real one. The urban infrastructure is modernized with IoT equipment to enhance the yield and reliability of the invested infrastructure. The smart parking concept contains IoT devices that gather the data from every parking area and signalize the free space available in them [42–44]. The free parking slot is dynamically updated in the mobile app or Internet resources. Figure 9.14 shows smart traffic management using IoT.

In city life, the traffic density and car parking space are the main problems. The private car user's count is dramatically increased. Searching for a free slot in the parking area is very tedious and harder for every driver in high-density traffic. The implementation of IoT modulus in the cities gives solutions to these difficulties. The IoT devices gave a way to increase the parking area utilization. The traffic flow and parking area demand are continuously monitored. The alert is given to the public in advance about the destination parking space availability.

Smart infrastructure in parking regions provides the details of empty parking spaces for automotive at various public locations. The densely populated areas required the advanced parking system to study the free space available in the parking space of the required destination. The electronic modules are laid on each commercial buildings and parking area to gather and report their state. Through Internet resources and mobile apps, the driver is guided with real-time data for their requirement. The accurate location of the parking space and count of the automotive in that region, coming up cars down to that path, is updated for the users to reduce the

FIGURE 9.14 Smart traffic management using IoT.

resistance of the traffic flow near the parking region. Thus, the location and time are precisely updated to the drivers. Also, the smart parking system minimized vehicle driver stress, vehicle emission, and fuel cost. The pollutants emission is reduced at the idling condition of the vehicle during the parking area searching in the city-centered malls or offices. The smart parking concept does not need huge capital cost investment. The technology is implemented with a moderate-level capital investment with huge benefits as returns.

9.8 IoT-BASED SMART WASTE MANAGEMENT SYSTEM FOR SMART CITIES

The rapid increase of population produces a massive amount of garbage waste, which pollutes the surroundings quickly. The garbage spillout in the open places causes serious health issues to the living beings. So, sanitation of surroundings is essential for human health. The waste collection bins are appropriately supervised, and waste is gathered effectively for maintaining hygienic conditions in cities [45]. The conventional method of garbage waste collection and disposal is complicated. Trash bins are collected at a certain interval of time. Some waste collection bins are quickly attained full level. The delay in removing bins caused severe environmental problems due to the spilling of garbage in the surroundings. Festivals and national holidays are the special reasons for the rapid collection of waste within a shorter time. But the bin is dispersed in the scheduled time in traditional techniques of waste management. In the computer era, using a computer and modern IoT modules for day-to-day applications gives automated data gathering, analyzed statistical reports, and accurate decision-making results [46]. Figure 9.15 denotes the smart waste handling using IoT.

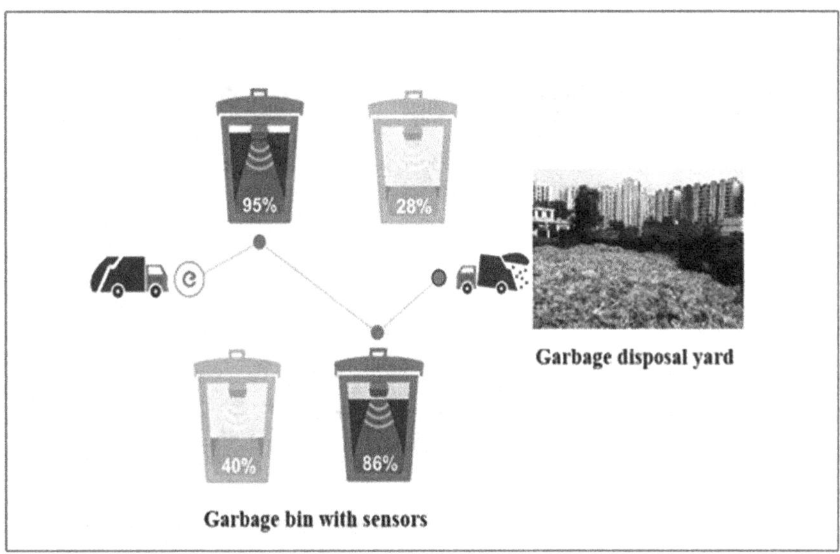

FIGURE 9.15 Smart waste handling using IoT.

The different types of waste handling techniques are used in smart cities. They are, namely, battery-based, RFID-based, and IoT-based waste collection systems. In IoT-based waste collection system, user-friendly service is the ultimate goal. This type of waste collection solution provided an efficient gathering system at minimal operational cost. While the waste bin got a signal from the level sensor, the truck driver is notified with the help of a smartphone to handle that particular bin by the management platform. The employment of IoT devices for real-time garbage collection signalized the management workers about the complete details of all bins in the city. Supervising the level of garbage in a bin is very easy. The IoT devices efficiently updated the information of trash bin through the Internet. The stack holders got alerts through a mobile app or graphical monitoring devices. The garbage is segregated according to physical nature. The real-time monitoring of waste without man workforce involvement provides the chance to react to any deviations. The data is sent to the central server computing stations for the decision-making process. The sanitary department continuously monitors the location, level of garbage in the bin, and count of the waste bin. The data is visualized to the management workers in graphical form. Thus, smart waste management techniques with the application of IoT devices gave a route plan for proper garbage collections and eliminated the possibility of a filled bin in the city over a week. Figure 9.16 displays the segregation of waste in smart waste management.

The sensors are instilled on every garbage bin to monitor the level of a trash bin and pinpoint the location of the bin in the city. The information collected from all the trash bins is stored in the central server. The collected data is analyzed, and reports are sent to the substations. The IoT connects the waste collection bin, vehicle, and computer sever of the sanitary department. The details from IoT devices are properly

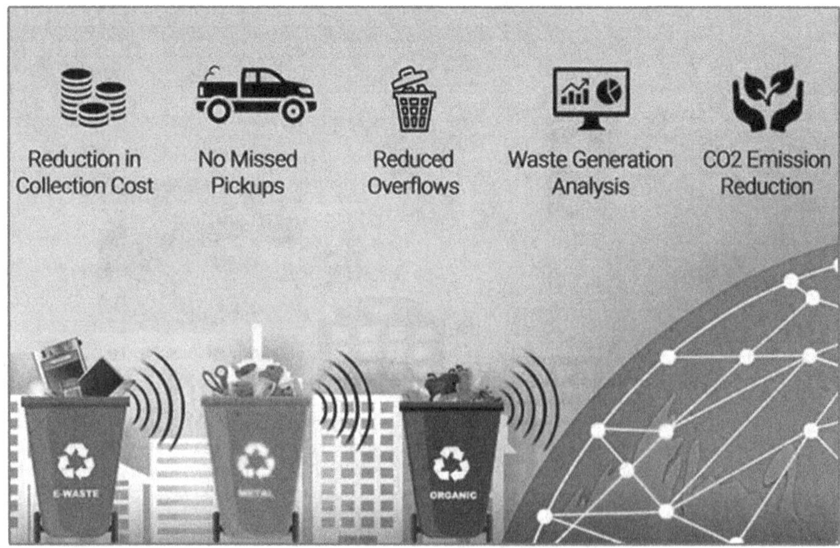

FIGURE 9.16 Segregation of waste in smart waste management.

analyzed to find the optimal route to handle all the trash bins with minimal fuel cost. Thus, IoT-based smart waste management system is very popular nowadays.

9.9 KEY BENEFITS OF IoT-BASED APPLICATIONS IN SMART CITIES

The advantages of IoT-based smart cities applications are discussed below:

- The service providers dynamically analyze the demand and shortage of gas, water, and electricity.
- An automated timer gives control for switching on or off the streetlight. The smart grid modules adjust the luminous power based on environmental conditions.
- Driverless automotive is implemented in smart cities to reduce accidents by human mistakes.
- Smart parking reduces the searching time and energy for parking slots in denser populated cities.
- The emergency needs are predicted rapidly by the advanced sensors integrated with the computer modules.
- The drivers are alerted about the city traffics by visual graphics on the roadsides.
- Effective immediate removal of filled trash bins leads to maintaining the environment healthy.
- Smart garbage disposal provides optimal routes for waste handler vehicles, which provides less fuel usage of trucks.
- Traffic lights are controlled for effective managing the traffic ways.
- Tenants have informed the states of the common government services electronically.

9.10 CONCLUSION

IoT applications will serve the same purpose as other distributed networks in the city. For example, both IoT and conventional distributed network systems can control energy systems and traffic lights remotely. When considering what IoT can do that no other system can, the true potential of the IoT in smart cities and its economic benefits become apparent. For example, the fragmented systems that operate and manage cities' physical infrastructure are frequently prohibitively expensive to implement and maintain. Those conventional systems are frequently too complex to operate by a nonexpert, possibly requiring engineers and thus incurring engineering costs. Additionally, these systems require human intervention and frequently depend on manual processing data, incorporation, and parameterization, limiting buildings' ability to operate at the desired condition. The IoT addresses these issues by going beyond simple preprogrammed settings and enabling the development of intelligent and context-aware applications that can adapt to changing conditions and ensure that city services operate at their optimum point.

In the present work, we discussed the importance of IoT-based smart homes in a smart city environment. The importance and implementation of the air quality monitoring system are discussed. In addition to that, the implementation of a sensor-based air quality monitoring system is also discussed. The IoT-based health-care monitoring system for the smart city environment is also discussed. In this, typical architecture for the IoT-based monitoring of the health-care system is prescribed, its prototype is created, and the results are also discussed. From the results, we can conclude that the error percentage in the prediction is significantly less. If it can be implemented in real time, it brings good efficiency to the monitoring system. Traffic management and vehicle parking are also the primary concern in smart cities like the urban environments. It is also explained in detail. In addition to that, IoT-based waste management system is also discussed. From this, we can conclude that the deployment of IoT-based devices in our smart-city-like environment brings a more convenient, more comfortable, and traffic-free environment to the citizens. It also provides a chance to improvise the business model and reduce the environmental impact by providing a pollution-free, nature-assisted lifestyle.

REFERENCES

1. Bakıcı, Tuba, Esteve Almirall, and Jonathan Wareham. "A Smart City Initiative: The Case of Barcelona." Journal of the Knowledge Economy 4, no. 2 (January 28, 2012): 135–148. doi:10.1007/s13132-012-0084-9.
2. Vanolo, Alberto. "Smartmentality: The Smart City as Disciplinary Strategy." Urban Studies 51, no. 5 (July 11, 2013): 883–898. doi:10.1177/0042098013494427.
3. Neirotti, Paolo, Alberto De Marco, Anna Corinna Cagliano, Giulio Mangano, and Francesco Scorrano. "Current Trends in Smart City Initiatives: Some Stylised Facts." Cities 38 (June 2014): 25–36. doi:10.1016/j.cities.2013.12.010.
4. Angelidou, Margarita. "Smart City Policies: A Spatial Approach." Cities 41 (July 2014): S3–S11. doi:10.1016/j.cities.2014.06.007.
5. Kim, Tai-Hoon, Carlos Ramos, and Sabah Mohammed. "Smart City and IoT." Future Generation Computer Systems 76 (November 2017): 159–162. doi:10.1016/j.future.2017.03.034.
6. Gaur, Aditya, Bryan Scotney, Gerard Parr, and Sally McClean. "Smart City Architecture and Its Applications Based on IoT." Procedia Computer Science 52 (2015): 1089–1094. doi:10.1016/j.procs.2015.05.122.
7. Krylovskiy, Alexandr, Marco Jahn, and Edoardo Patti. "Designing a Smart City Internet of Things Platform with Microservice Architecture." 2015 3rd International Conference on Future Internet of Things and Cloud (August 2015). doi:10.1109/ficloud.2015.55.
8. Chakrabarty, Shaibal, and Daniel W. Engels. "A Secure IoT Architecture for Smart Cities." 2016 13th IEEE Annual Consumer Communications & Networking Conference (CCNC) (January 2016). doi:10.1109/ccnc.2016.7444889.
9. Basford, Philip J., Florentin M. J. Bulot, Mihaela Apetroaie-Cristea, Simon J. Cox, and Steven J. Ossont. "LoRaWAN for Smart City IoT Deployments: A Long Term Evaluation." Sensors 20, no. 3 (January 23, 2020): 648. doi:10.3390/s20030648.
10. Suri, Niranjan, Zbigniew Zielinski, Mauro Tortonesi, Christoph Fuchs, Manas Pradhan, Konrad Wrona, Janusz Furtak, et al. "Exploiting Smart City IoT for Disaster Recovery Operations." 2018 IEEE 4th World Forum on Internet of Things (WF-IoT) (February 2018). doi:10.1109/wf-iot.2018.8355117.
11. Peneti, Subhashini, M. Sunil Kumar, Suresh Kallam, Rizwan Patan, Vidhyacharan Bhaskar, and Manikandan Ramachandran. "BDN-GWMNN: Internet of Things (IoT) Enabled Secure Smart City Applications." Wireless Personal Communications 119, no. 3 (March 12, 2021): 2469–2485. doi:10.1007/s11277-021-08339-w.

12. Kindhi, Berlian Al, and Ilham Surya Pratama. "Fuzzy Logic and IoT for Smart City Lighting Maintenance Management." 2021 3rd East Indonesia Conference on Computer and Information Technology (EIConCIT) (April 9, 2021). doi:10.1109/eiconcit50028.2021.9431917.
13. Saxena, Monika, Vaibhav Vyas, and C. K. Jha. "A Framework for Multi-Sensor Data Fusion in the Context of IoT Smart City Parking Data." IOP Conference Series: Materials Science and Engineering, vol. 1099, no. 1, p. 012011. IOP Publishing, 2021.
14. Napolitano, Rebecca, Wesley Reinhart, and Juan Pablo Gevaudan. "Smart Cities Built with Smart Materials." Science 371, no. 6535 (March 18, 2021): 1200–1201. doi:10.1126/science.abg4254.
15. Mukhopadhyay, Subhas Chandra, Sumarga Kumar Sah Tyagi, Nagender Kumar Suryadevara, Vincenzo Piuri, Fabio Scotti, and Sherali Zeadally. "Artificial Intelligence-Based Sensors for Next Generation IoT Applications: A Review." IEEE Sensors Journal (2021): 1. doi:10.1109/jsen.2021.3055618.
16. Shafali Jain, Balarengadurai Chinnaiah, M P Rajakumar, E.Fantin Irudaya Raj, Gujar, Anantkumar Jotiram. "Improved IoT-Based Control System Combined with an Advanced Control Management Server-Based System." Journal of Green Engineering, 10, no. 10 (2020):8488–8496.
17. Raj, E. Fantin Irudaya. "Available Transfer Capability (ATC) under Deregulated Environment." Journal of Power Electronics & Power Systems 6, no. 2 (2016): 85–88.
18. Goyal, Sukriti, Nikhil Sharma, Bharat Bhushan, Achyut Shankar, and Martin Sagayam. "IoT Enabled Technology in Secured Healthcare: Applications, Challenges and Future Directions." Studies in Systems, Decision and Control (October 20, 2020): 25–48. doi:10.1007/978-3-030-55833-8_2.
19. Khamparia, Aditya, Prakash Kumar Singh, Poonam Rani, Debabrata Samanta, Ashish Khanna, and Bharat Bhushan. "An Internet of Health Things-driven Deep Learning Framework for Detection and Classification of Skin Cancer Using Transfer Learning." Transactions on Emerging Telecommunications Technologies 32, no. 7 (May 4, 2020). doi:10.1002/ett.3963.
20. Sharma, Nikhil, Ila Kaushik, Bharat Bhushan, Siddharth Gautam, and Aditya Khamparia. "Applicability of WSN and Biometric Models in the Field of Healthcare." Advances in Information Security, Privacy, and Ethics (2020): 304–329. doi:10.4018/978-1-7998-5068-7.ch016.
21. Hoque, Mohammad Asadul, and Chad Davidson. "Design and Implementation of an IoT-Based Smart Home Security System." International Journal of Networked and Distributed Computing 7, no. 2 (2019): 85. doi:10.2991/ijndc.k.190326.004.
22. Kim, Yonghee, Youngju Park, and Jeongil Choi. "A Study on the Adoption of IoT Smart Home Service: Using Value-Based Adoption Model." Total Quality Management & Business Excellence 28, no. 9–10 (April 12, 2017): 1149–1165. doi:10.1080/14783363.2017.1310708.
23. Zaidan, A. A., and B. B. Zaidan. "A Review on Intelligent Process for Smart Home Applications Based on IoT: Coherent Taxonomy, Motivation, Open Challenges, and Recommendations." Artificial Intelligence Review 53, no. 1 (July 23, 2018): 141–165. doi:10.1007/s10462-018-9648-9.
24. Bhushan, Bharat, Aditya Khamparia, K. Martin Sagayam, Sudhir Kumar Sharma, Mohd Abdul Ahad, and Narayan C. Debnath. "Blockchain for Smart Cities: A Review of Architectures, Integration Trends and Future Research Directions." Sustainable Cities and Society 61 (October 2020): 102360. doi:10.1016/j.scs.2020.102360.
25. Haque, A. K. M. Bahalul, Bharat Bhushan, and Gaurav Dhiman. "Conceptualizing Smart City Applications: Requirements, Architecture, Security Issues, and Emerging Trends." Expert Systems (June 11, 2021). doi:10.1111/exsy.12753.

26. Bhushan, Bharat, Chinmayee Sahoo, Preeti Sinha, and Aditya Khamparia. "Unification of Blockchain and Internet of Things (BIoT): Requirements, Working Model, Challenges and Future Directions." Wireless Networks 27, no. 1 (August 6, 2020): 55–90. doi:10.1007/s11276-020-02445-6.
27. Nasution, T H, M A Muchtar, and A Simon. "Designing an IoT-Based Air Quality Monitoring System." IOP Conference Series: Materials Science and Engineering 648 (October 18, 2019): 012037. doi:10.1088/1757-899x/648/1/012037.
28. Xiaojun, Chen, Liu Xianpeng, and Xu Peng. "IOT-Based Air Pollution Monitoring and Forecasting System." 2015 International Conference on Computer and Computational Sciences (ICCCS) (January 2015). doi:10.1109/iccacs.2015.7361361.
29. Zhao, Liang, Wenyan Wu, and Shengming Li. "Design and Implementation of an IoT-Based Indoor Air Quality Detector With Multiple Communication Interfaces." *IEEE Internet of Things Journal* 6, no. 6 (December 2019): 9621–9632. doi:10.1109/jiot.2019.2930191.
30. Barot, Virendra, Viral Kapadia, and Sharnil Pandya. "QoS Enabled IoT Based Low Cost Air Quality Monitoring System with Power Consumption Optimization." Cybernetics and Information Technologies 20, no. 2 (June 1, 2020): 122–140. doi:10.2478/cait-2020-0021.
31. Sai, Kumar, Kinnera Bharath, Subhaditya Mukherjee, and H Parveen Sultana. "Low Cost IoT Based Air Quality Monitoring Setup Using Arduino and MQ Series Sensors With Dataset Analysis." *Procedia Computer Science* 165 (2019): 322–327. doi:10.1016/j.procs.2020.01.043.
32. Yasnoff, William A. "Health Information Infrastructure." Biomedical Informatics (2021): 511–541. doi:10.1007/978-3-030-58721-5_15.
33. Gampala, Veerraju, M. Sunil Kumar, C. Sushama, and E. Fantin Irudaya Raj. "Deep Learning Based Image Processing Approaches for Image Deblurring." Materials Today: Proceedings (December 2020). doi:10.1016/j.matpr.2020.11.076.
34. Ch, Gangadhar, S. Jana, Sankararao Majji, Prathyusha Kuncha, Fantin Irudaya Raj E., and Arun Tigadi. "Diagnosis of COVID-19 Using 3D CT Scans and Vaccination for COVID-19." World Journal of Engineering ahead-of-print, no. ahead-of-print (June 30, 2021). doi:10.1108/wje-03-2021-0161.
35. Krishna, BrahmadesamViswanathan, G. Amuthavalli, D. StalinDavid, E. FantinIrudaya Raj, and D. Saravanan. "Certain Investigation of SARS-COVID-2-Induced Kawasaki-Like Disease in Indian Youngsters." *Annals of the Romanian Society for Cell Biology* (2021): 1167-1182.
36. Agarwal, Parul, Mohammad Akram Ch, Dinesh Sheshrao Kharate, E. Fantin Irudaya Raj, and S. Balamuralitharan. "Parameter Estimation of COVID-19 Second Wave BHRP Transmission Model by Using Principle Component Analysis." *Annals of the Romanian Society for Cell Biology* (2021): 446–457.
37. Guillen-Perez, Antonio, and Maria-Dolores Cano. "Intelligent IoT Systems for Traffic Management: A Practical Application." IET Intelligent Transport Systems 15, no. 2 (January 5, 2021): 273–285. doi:10.1049/itr2.12021.
38. Priyadarsini, K., E. Fantin Irudaya Raj, A. Yasmine Begum, and V. Shanmugasundaram. "Comparing DevOps Procedures from the Context of a Systems Engineer." Materials Today: Proceedings (October 2020). doi:10.1016/j.matpr.2020.09.624.
39. Shinde, Sunita Sunil, Ravi M. Yadahalli, and Ramesh Shabadkar. "Cloud and IoT-Based Vehicular Ad Hoc Networks (VANET)." Cloud and IoT-Based Vehicular Ad Hoc Networks (April 22, 2021): 67–82. doi:10.1002/9781119761846.ch4.
40. Chouhan, Arun Singh, Nitin Purohit, H. Annaiah, D. Saravanan, E. Fantin Irudaya Raj, and D. Stalin David. "A Real-Time Gesture Based Image Classification System with FPGAand Convolutional Neural Network." International Journal of Modern Agriculture 10, no. 2 (2021): 2565–2576.

41. Ye, Zhoujing, Guannan Yan, Ya Wei, Bin Zhou, Ning Li, Shihui Shen, and Linbing Wang. "Real-Time and Efficient Traffic Information Acquisition via Pavement Vibration IoT Monitoring System." Sensors 21, no. 8 (April 10, 2021): 2679. doi:10.3390/s21082679.
42. Sadhukhan, Pampa. "An IoT-Based E-Parking System for Smart Cities." 2017 International Conference on Advances in Computing, Communications and Informatics (ICACCI) (September 2017). doi:10.1109/icacci.2017.8125982.
43. Rane, Sagar, Aman Dubey, and Tejisman Parida. "Design of IoT Based Intelligent Parking System Using Image Processing Algorithms." 2017 International Conference on Computing Methodologies and Communication (ICCMC) (July 2017). doi:10.1109/iccmc.2017.8282631.
44. Thorat, S. S., M. Ashwini, Akanksha Kelshikar, Sneha Londhe, and Mamta Choudhary. "IoT Based Smart Parking System Using RFID." International Journal of Computer Engineering In Research Trends 4, no. 1 (2017): 9-12.
45. Medvedev, Alexey, Petr Fedchenkov, Arkady Zaslavsky, Theodoros Anagnostopoulos, and Sergey Khoruzhnikov. "Waste Management as an IoT-Enabled Service in Smart Cities." Internet of Things, Smart Spaces, and Next Generation Networks and Systems (2015): 104–115. doi:10.1007/978-3-319-23126-6_10.
46. Roshan, Rakesh, and Om Prakash Rishi. "Effective and Efficient Smart Waste Management System for the Smart Cities Using Internet of Things (IoT): An Indian Perspective." Advances in Intelligent Systems and Computing (October 2, 2020): 473–479. doi:10.1007/978-981-15-6014-9_54.

10 An Integration of IoT and Machine Learning in Smart City Planning

Ritu Khandelwal, Hemlata Goyal, and Rajveer Singh Shekhawat
Manipal University Jaipur, Jaipur, India

CONTENTS

DOI: 10.1201/9781003219620-10

10.1 INTRODUCTION

An integration of Internet of Things (IoT) with machine learning (ML) in smart city planning aims to design environmentally friendly applications with optimal use of space and resource along with the promise to give the optimal distribution of benefits in an effective manner. The smart city idea can be defined as a sustainable framework, which contains information, data and communication technologies to build and manage practice to support urbanization smartly. Smart technologies like IoT integrated with ML have the potential to tame the stresses of urbanization to implement new and smart experiences to make day-to-day life more comfortable. These technologies are capable of absorbing the various demands of urbanization through the delivery of innovative and smarter options aimed toward smart city.

10.1.1 Paper Organization

The organization of the chapter is followed by five sections in which Section 10.1 covers the introduction of the integration of IoT and ML in smart city planning. A detailed background knowledge of IoT and ML applications, motivation and major contribution is given in Sections 10.1.2, 10.1.3 and 10.1.4, respectively.

Technical aspects of IoT and ML in the context of smart city segment, ML algorithms, techniques, use cases, benefits and combination of ML algorithms with IoT in particular smart city application have been carried out in Section 10.2.

Section 10.3 covers the applications of IoT and ML in various dimensions of smart city planning with suitable matching of ML algorithm and IoT devices carried out references.

Section 10.4 gives the insights of current challenges of IoT infrastructure and devices with ML algorithms. Finally, conclusion with future research is pursued by readers in Section 10.5 followed with references in the last section.

10.1.2 Background

Smart city requires the active deployment of IoT and ML technologies to work in an effective manner. IoT with ML combines as smart technology that plays an important role in intelligent parking system, public safety, smart traffic administration, water management plan, smart street lamps, track of air purity level, smart planning of schools, hospitals, roads etc. to examine, manage and control the resources of the smart city. A brief knowledge about IoT and ML is given in subsequent sections.

10.1.2.1 Internet of Things (IoT)

The **IoT** defines the networking of physical objects, embedded with sensors, software and combining technologies to serve the purpose of data collection, pre-processing of the data, machine-to-machine connection and exchange the data with other systems and devices over the Internet [1]. It is ripened with a set of commonly used technologies and deployment approaches. Things are needed to communicate over the network with each other, initiating the requirement for communication in between machine to machine (M2M). This communication may be of short range with the help of wireless technologies, i.e. Wi-Fi, Bluetooth and ZigBee, or wide range with the help of mobile network like WiMAX, LoRa, Sigfox, CAT M1, NB-IoT, GSM, GPRS, 3G, 4G, LTE and 5G.

An IoT solution comprises sensors, actuators, communication channel, data collection on the cloud and applications having its own challenges of scale, diversity, connectivity, privacy, security and regulatory compliance. Sensors can include basic temperature, pressure, flow, voltage, vibration, frequency, optical, navigational, inertial or many more [2].

There is a lot to be done for standardization in terms of IoT infrastructure and technology. As discussed so far, IoT devices turn out to be smart enough with the amalgam of embedded processing, remote communication with the smart device for easy life.

10.1.2.2 IoT in Smart City

IoT is coupled to many aspects of day-to-day life application for everyone, such as smart farming, smart medical health-care assistance, smart transportation, environmental conditions monitoring, logistics and supply chain management and smart city planning. Smart city has advanced sensors that manage city assets and plays an important role in the grouping of significant information that can be later used in several applications. Sensors range from electronic sensors such as parking and speedometer sensors; chemical sensors that include carbon dioxide sensor or oxygen sensor and catalytic bead sensor; biosensors, to detect the components of biomedicine; and smart grid sensors for efficient generation, transmission and distribution of power from the point it is generated to its users [3].

Following are the details of the most necessary applications that occur in the planning of smart city.

10.1.2.2.1 Smart Homes

It comprises day-to-day important household devices, i.e. refrigerators, laundry machines, fans or tube lights, which are able to communicate and enable monitoring, energy consumption optimization, smart doors locks, smart home assistants etc. with authorized users using Internet.

10.1.2.2.2 Health-care Assistance

In order to assist patient, health-care assistance wireless sensors monitor the state of wound and report the observations to the doctors without physical presence. Apart from this, wearable device can trace and report a huge variety of measurement like heart rate, blood oxygen level, blood sugar level or temperature.

10.1.2.2.3 Smart Transportations

In order to achieve smart transportation, vehicles are embedded with sensors, or mobile devices that are used to provide smart and optimized route decision, reservations in the parking area, automatic street lighting with temperature sensor for economy and autonomous driving and prevention of accident.

10.1.2.2.4 Environmental State Monitoring

Environmental condition monitoring assesses the wide variety of the perfect infrastructure for a smart city development. With the advancement of Geographical Information System (GIS), and wide variety of wireless sensor like barometer, humidity sensors, or ultrasonic wind sensors, temperature sensors are able to help and generate advanced weather stations. Besides these, smart sensor is used for monitoring the air quality and water pollution levels across the city.

10.1.2.2.5 Logistics and Supply Chain Management

In order to use the smart RFID tags, a logistic can be easily traced from the production store within timeframe. Apart from this, smart packaging offers feature like brand protection, quality assurance and client personalization.

10.1.2.2.6 Security and Surveillance Systems

Smart cameras can capture pictures/videos through the streets network. In addition to real time, smart security systems can recognize the visual objects and determine suspects or avoid dangerous situation.

The Internet Version Protocol (IPv6), dropping in sensors price, edge computing, improvements in network and GIS are the major factors in contribution to the growth of IoT [4].

10.1.2.3 Machine Learning

The availability of intelligent machines through ML has been crucial in escalating the concept of smart cities.

If started with a canonical definition according to Tom Mitchell in 1997, “Machine learning is a leaning from experience with respect to some class of tasks, and a performance measure P, if [the learner’s] performance at tasks in the class, as measured by P, improves with experience.” To define learning with respect to a specific class of tasks, it could be answering exams in a particular subject or it could be diagnosing patients of a specific illness. We should be very careful while defining the set of tasks with appropriate performance criteria P to measure the learning in right direction

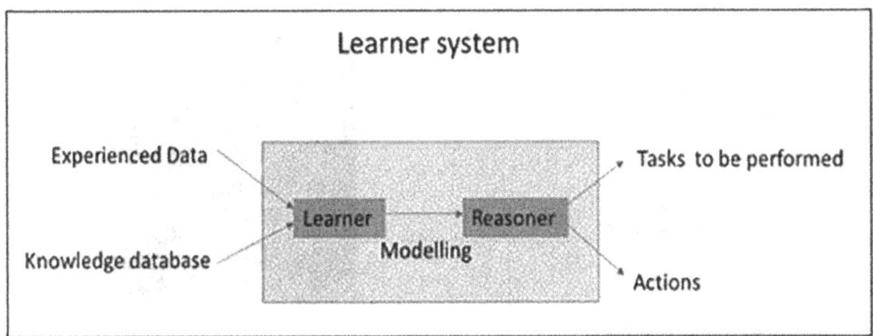

FIGURE 10.1 Block diagram of learning model.

or vague learning. For example, if we study the illness diagnosis of the patient, then performance measure P should be the number of patients or we can compute the number of patients who do not have opposing reactions to the medicine that is prescribed to him or it can be a variety of ways to define the performance measure P depending on what criteria we are looking for and the next most important component is experience. The following block diagram depicts the ML system, as shown in Figure 10.1, and how a basic learning model can be developed is discussed in the following steps.

- Select training dataset/training experience.
- Select the target function to represent the model for learning, which wants to learn. For example, if we are writing an ML system for playing the tic-tac-toe game, then the target function can be defined by giving a board position, the number of moves and what move we want to take. After this, we have to define the class of function and the task, move on the board position as an input function.
- Select the learning algorithm to infer the target function, which is based on training experience as the best function.

Training experience can be defined in terms of features of the domain to decide the target function for choosing the appropriate class of function to infer decision-making. Now we are able to define more precisely that ML is the field of computer science and a domain of Artificial Intelligence (AI), which provides intelligent system having the ability of learning without explicit programming. As discussed above, learning algorithm is used as an input function for training dataset. Learning algorithm is divided into three main categories according to their characteristic nature:

- **Supervised learning:** the training dataset comprises sampling of known input/source features together, which relate to fitted target feature, also called Label.
- **Unsupervised learning:** the training dataset comprises sampling of unknown input/source features together, which relate to fitted target feature, without knowing the Label. Here, no label is required for the training dataset.
- **Reinforcement learning:** a kind of maximized learning payoff.

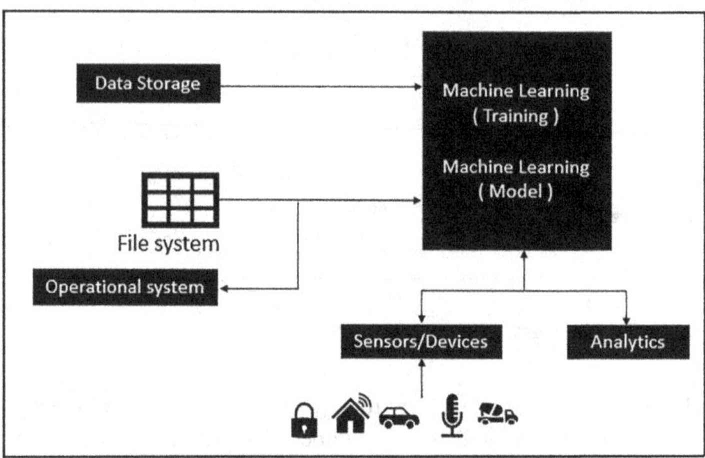

FIGURE 10.2 IoT in machine learning architectural view.

Supervised/unsupervised learning is widely used and compared to reinforcement learning in IoT smart data analysis on the basis of computationally fast and intense algorithm.

10.1.2.4 Machine Learning in IoT

The IoT produces a huge volume of dataset, which is gathered from millions of devices. ML is empowered with hidden data and produce insights from it. ML practices past behavior to recognize patterns and construct model, which is helpful in the prediction of future behavior and event state. With the integration of ML and IoT resulting in the projection of future trends, anomalies detection, and augmented intelligence by different types of datasets nature. Figure 10.2 discusses the architecture of IoT with ML.

With the integration of ML and IoT, we could be able to:

- Receive and transformation of dataset into a consistent format
- Building an ML model with learning algorithms
- Deployment of this ML model on cloud, edge and device

With the advancement of techniques in machine intelligence, now massive volume of IoT sensor's dataset, mining is possible and gives insight for the real-world problems. Hence, IoT and ML must complement each other in real-world problem solving with optimized computing and communicating manner. Nowadays, IoT with ML has added significant importance and can be understood, as shown in Figure 10.3.

Smart data analysis is necessary due to the following reasons:

- Distributed IoT devices produced a large volume of dataset.
- Heterogeneous data points generated a huge variety of data formats.
- IoT data streams are uncertain.
- IoT with ML enables balanced and efficient scalability.

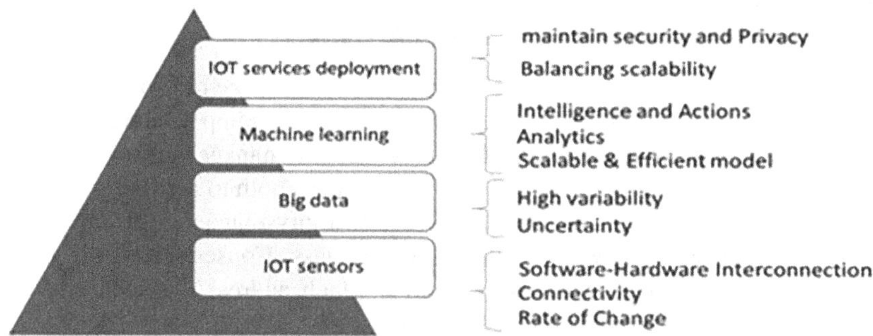

FIGURE 10.3 Challenges in IoT application from analytics perspective.

10.1.3 Motivation

Transforming any city into a "smart city" requires the active deployment of IoT technologies. Several practical examples can pick up across different industries like manufacturing, healthcare, home, environment, and transportation. The extensiveness of the Internet across global regions is the core of IoT applications. Apart from the reduced cost of connection, the introduction of better gadgets with sophisticated sensors and Wi-Fi connectivity contributes to smarter involvement of the IoT.

According to the research, it is estimated that, by year 2025, 75.44 billion devices would be linked over the Internet throughout the world and approximately 30% of data creation, capturing or replicating in real-time scenario. IoT is tightly coupled with Big Data, Cloud Storage, Embedded System and ML applications. In addition, IoT devices are immensely used and integrated into various technologies and day-to-day applications. So, it is required to keep a balance among cost, processing power and energy consumption in terms of design or selection of an IoT device. Due to the lack of established standards and mechanisms, most of the data coming from the sensors is wasted without the extraction of potentially useful information. ML can be utilized in the best possible way to find optimal decision.

10.1.4 Major Contribution

The objective of this chapter is to answer the following:

- To better understand ML algorithms in the field of IoT-based applications of smart city planning.
- With the taxonomy of ML algorithms with IoT, a depth literature review has been studied, revealing that eight ML algorithms are applied to IoT data along the way to deal generated data from IoT sensors.
- To solve the technical aspects of real-life application in smart city planning.
- Challenges that have been facing out with IoT implementation in the real world.
- In order to derive the smart city planning, this chapter is able to direct the best machine algorithm to handle smart data characteristics.

10.2 TECHNICAL ASPECTS OF IoT AND MACHINE LEARNING IN SMART CITY PLANNING

Smart city planning plays an exceptionally well-known space of exploration, since it experiences numerous regular issues, with a tremendous impression in a cutting-edge shrewd city. Furthermore, the idea of the issues it manages favors the utilization of both IoT and ML advances. This survey means both to recognize the latest thing in the utilization of ML and IoT in shrewd city management and inspect exploration inclusion in every single keen city class. Consequently, the review centers around the latest examination works, which address the keen city classifications (Modular Infrastructure, smart home, smart farming, smart health, smart environment and smart transportation), utilize IoT and additionally ML methods. Section 10.2 is focused on ML algorithms that are applied to smart city planning discussed in subsections.

10.2.1 Machine Learning Algorithms

In order to plan the smart city with appropriate integration of IoT and ML algorithm, hereby the main objective is to focus on efficient eight ML techniques of Bayesian-Naïve Bayes, Markov Model, decision tree, Ensemble methods, Regression Analysis, Clustering technique, Artificial Neural Network (ANN) and deep learning. The detailed knowledge of the abovesaid ML techniques is given in the following subsections.

10.2.1.1 Ensemble Techniques

Ensemble learning generated much accurate solutions than an individual model. The frameworks built by ensemble technique can utilize the frail classifiers that are effortlessly built and furthermore acquire the nature of a robust classifier. The researchers expect road mishaps by making a structure to recognize the consciousness of the driver; this information is received by a camera. At that point, these features are chosen from an extricated fundamental picture provided by the AdaBoost ensemble technique.

Random Forest (RF) is utilized for performing classification, collection of multiple decision trees. Four models are created to conjecture present moment and long-term traffic stream in work zones. The consequences of this exploration showed that RF beats the carried out benchmark indicator. To recognize street surface irregularities, the considered boundaries are the number of repetition, features, regression and Feed Forward Neural Network, given all estimations for long-short learning. The outcomes show that each classifier performed enough with more than 99% precision. RF examined with Support Vector Machine (SVM) and ANN to distinguish street mishaps. SVM uses the hyperplane for making predictions, which is considered by hypothesis function as defined in Equation (10.1).

$$h(x_i) = \begin{cases} +1, \omega \cdot x + b \geq 0 \\ -1, \omega \cdot x + b < 0 \end{cases} \tag{10.1}$$

The point above or on the hyperplane will be classified as class +1, and the point below the hyperplane will be classified as class –1.

At last, five ML techniques are prepared utilizing a named set of information for vehicle-to-vehicle correspondence. Among these, the logistic Regression marginally outflanked different strategies. The ML techniques referenced over that don't have a place with the ensemble classification and will be audited independently in the accompanying areas.

10.2.1.2 Bayesian-Naïve Bayes

It is a probabilistic classifier, which works by assuming no pair of features are dependent, which executes a likelihood appropriation for a given set of irregular factors. Using Bayesian classifier every one of the factors and their relationship can be depicted. In this way, the perception of certain factors contributes the information of other arrangements of factors [5]. Bayes relates the conditional and marginal probabilities of stochastic events C and A as shown in Equation (10.2).

$$P(C/A) = \frac{P(A/C)P(C)}{P(A)} \propto L(A/C)P(C) \tag{10.2}$$

where P is the probability of the variable; $L(A|C)$ is the likelihood of A, given fixed C; $P(C)$ is the prior probability or marginal probability of C; $P(C|A)$ is the conditional probability of C, given A and also called posterior probability because of already incorporated outcome of event A; $P(A|C)$ is the conditional probability of A, given C; $P(A)$ is the prior or marginal probability of A, normally the evidence; ratio $P(A|C)/P(A)$ is the standardized likelihood; posterior = likelihood × prior/normalizing constant.

In Ref. [6], the author utilizes a Bayesian Network technique to accomplish transient traffic anticipating the coupling of time series of ARIMA model and Bayesian network. The strategies were thought about in contrast to the MAE, RMSE and MAPE measurements. The combined approach of deep learning and the Bayesian technique come out with comparative output in momentary rush hour gridlock anticipating [7].

10.2.1.3 Markov Models

It is dependent on likelihood circulation. In a Markov chain, a variable changes its worth arbitrarily over the long run that relies just upon the dissemination of the past state. The Hidden Markov Model is prepared by utilizing speed information of driving style of a vehicle [8]. Mathematical expression has been derived in Equation (10.3).

$$P(Y) = \sum_{x}^{n} P(Y,X) = \sum_{x}^{n} P(Y \mid X)P(X) \tag{10.3}$$

where P is the probability of the variable, Y is the observed events, $P(C)$ is the prior probability or marginal probability of C, $P(Y|X)$ is the emission probability of Y where X is given, $P(X)$ is the transition probability, which is already calculated, n is the total no. of states.

To execute a parking spot identification framework, the authors take the advantage of an SVM and a Markov model. The input information has been gathered from cameras fixed at parking areas. After capturing pictures, SVM is utilized to group parking spaces to free or involved. To distinguish street surface irregularities, SVM has been utilized in numerous studies [3].

In the examination over, the MDP utilizes a bunch of activities to settle on a few choices, the activities choice approach is performed utilizing a support learning method, named *Q*-learning. The algorithm learns by executing every one of the potential activities over and over and assessing the subsequent state. Eventually, the algorithm has taken in the ideal arrangement of activities [9].

10.2.1.4 Decision Tree

Decision tree is a classification technique, which addresses relationship of dependent variable and a set of independent variable. It comprises nodes to represent a class leading a test set to a target label in top-down manner.

Decision tree is defined as "a database $D = \{t_1,\ldots, t_n\}$ where $t_i = \{t_{i1},\ldots, t_{ih}\}$ and the database schema contains the attributes $\{A_1, A_2,\ldots, A_h\}$, set of classes $C = \{C_1,\ldots, C_m\}$." It divides the entire search space into rectangular regions and the classification of a tuple is done on the basis of the region in which it is falling in the following two steps.

- Decision tree induction: decision tree construction is done using training of the dataset.
- For each $t_i \in D$, apply the decision tree to assign its class.

Statistical measure can be defined in terms of entropy, where P_i is the frequentist probability of an element of class i of state s and n is total class as shown in Equation (10.4).

$$Entropy(s) = -\sum_{i=1}^{n} P_i * \log(P_i) \tag{10.4}$$

A comparison is done of decision tree with other ML techniques to anticipate traffic crowding with consideration of precision, recall and accuracy.

The authors tried a Regression model of decision tree to predict short long-term traffic stream. The decision tree was less powerful in the traffic forecasts contrasted with the Random Forest or FF-NN. Regression and RF trees are the ML techniques, which play the role of getting value of fresh water level [10].

10.2.1.5 Clustering Technique

Clustering is a task to arrange components into different clusters based on their similarity or dissimilarity [11]. A Fuzzy *C*-Means strategy is utilized to make short-term traffic prediction [12]. While in a standard clustering technique, a record needs to have in one analyzed class where each occurrence has a place with an offered class in a limited way. The author experimented for utilizing a traffic system with genuine traffic information via street network of Japan. They analyze consequences of the Fuzzy *C*-Means strategy to the *K*-Nearest Neighbor (*K*NN) technique. The proposed technique shows about 26% more modest mistake rate contrasted with the *K*NN. In clustering, proximity measure could be defined in terms of Euclidean distance as given in Equation (10.5).

$$J = \sum_{j=1}^{k} \sum_{i=1}^{n} \left\| x_i^{(j)} - c_j \right\|^2 \tag{10.5}$$

where J is the objective function, k is the total clusters, n is the number of cases, $x_i \times c_j$ is the distance function, c_j is the centroid for cluster j.

The authors utilize *K*-Means to streamline the traffic network, to limit the cost [12]. Preceding the *K*-Means clustering, the ongoing information get from GIS gadgets are pre-handled with a Deep Belief Network (DBN) algorithm.

10.2.1.6 Artificial Neural Networks (ANN)

It is the arrangement and connections of the neurons with weight. It is having three basic layers – input, hidden and output layers. Input variable is sent to the input layer where each neuron represents a variable and passing to hidden and output layers where neuron is assigned a label, shown in Figure 10.4.

The activation function used in it can be expressed as shown in Equation (10.6).

$$\sigma(x) = 1 + e - x1 \tag{10.6}$$

where $\sigma(x)$ is sigmoid function for input x.

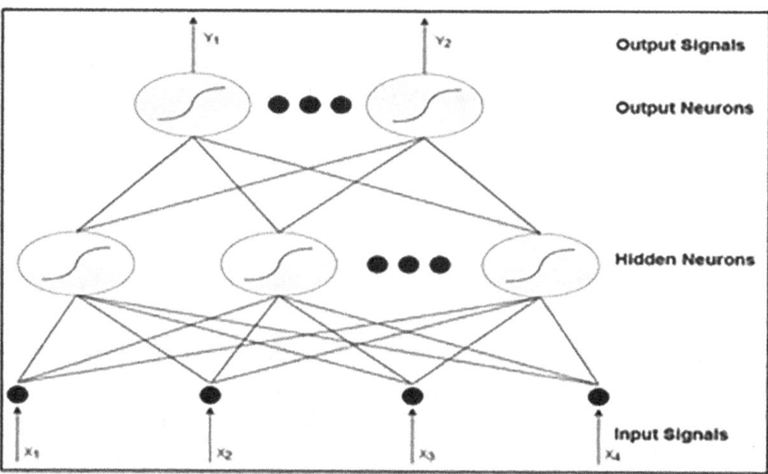

FIGURE 10.4 Interconnected layered architecture of neural network. (Adapted from Jaihar and John, 2020.)

The FFNN uses the straight blend of the information factors to extract information, which is the combination to forecast the traffic flow to utilize a function. The FFNN strategy generated the second smallest error metrics after the RF technique. In Ref. [13], FFNN is used in the anticipation of the varieties in timing movements. Street sensors generate the information related to speed meter, travel timing. The ANN is trained with the back-propagation strategy to anticipate street mishaps progressively so that error can be minimized [14, 15]. The model accepts input information from recorded information gathered from street sensors, with data like the quantity of vehicles, speed, path inhabitance proportion and so forth. The outcomes show that every one of the models can recognize mishaps with incredible precision. Nonetheless, there are countless false-positive estimations. ANNs, clustering and SVMs are effective ML techniques to perform the different types of analytics [16].

10.2.1.7 Deep Learning

Deep learning strategies are appropriate for large information dealing with and computationally exceptional cycles like image pattern recognition, speech recognition and amalgamation, and so on. "Deep" word means the number of hidden layers that have been increased in neural network, which required computation power increments. It can be done on unstructured data. The number of layers is directly proportional to the number of features, and extraction is not required [17].

Convolutional Neural Network (CNN) is most popular in deep learning based on neural network layered architecture. Architecture of hidden layers in a CNN is categorization into pooling, convolutional, subsampling, or Fully Connected (FC) layers. The objective of CNN technique used here is to depend solely on smart cameras for it working [18]. Also, an all-encompassing audit is dependent on CNNs and Deeper CNNs (DCNNs) [19]. The CNN and DCNN are widely applying Google maps, cell phone pictures, and 3D pictures built with specific equipment [20] and performance is very well in asphalt picture classification. Deep Neural Network (DNN), CNN, Recurrent Neural Network (RNN), Long Short-Term Memory (LSTM) and DBNs are being tested for different smart city applications that have been discussed in the next section [16]. LSTM uses sigmoid function to generate the output using gates. Gate equations are defined in Equations (10.7)–(10.9).

$$i_t = \sigma\left(\omega_i\left[h_{t-1}, x_f\right] + b_i\right) \tag{10.7}$$

$$f_t = \sigma\left(\omega_F\left[h_{t-1}, x_t\right] + b_F\right) \tag{10.8}$$

$$O_t = \sigma\left(\omega_0\left[h_{t-1}, x_t\right] + b_0\right) \tag{10.9}$$

where i_t is the input gate, f_t is the forget gate, o_t is the output gate, σ is the sigmoid function, w_x is the weight for the respective gate(x) neurons, h_{t-1} is the output of the previous LSTM block (timestamp of $t-1$), x_t is the input for current timestamp, b_x is the biases for the respective gates.

As an advancement of a vulnerable side identification framework for smart transportation, it utilizes contribution from cameras settled on a vehicle [21]. As referenced before, the motivation is to perform ongoing traffic network tasks [12].

This examination utilizes a DBN to measuring input information. DBNs are likewise utilized in Ref. [22], to help advanced guide creation via programmed road components recognition, for example, traffic signals, traffic circles, and so forth. The input information for the framework are gotten exclusively by users GPS information. The initial phase in their strategy is the execution of an exception identification method to make assessments for the road components position on time. Deep learning techniques are used to classify different object by taking input in image form [8].

The Stacked Auto Encoder (SAE) is introduced in Ref. [23] to empower traffic stream predictions dependent on enormous information using logistic regression algorithm.

10.2.1.8 Regression Analysis

Regression technique attempts to develop numerical models and is ready to portray, or distinguish, the connection between at least two factors and is efficient in forecasting. Logistic regression is appropriate where the depending variable has binary qualities. A linear relationship exists between the independent and the response variables as shown in Equation (10.10) where Y is the dependent variable for the X independent variable, W_1 is the slope and W_0 is the intercept.

$$Y = W_0 + W_1X \tag{10.10}$$

Semi-supervised ML strategies address the shortage of explained information in large information streams. Reinforcement learning methods produce the most promising outputs of the framework when feedback is taken from environment, used to increase the overall output rather than a class label in account of other supervised learning strategies. These strategies address IoT-related issues and convey total versatile solutions [24, 25].

10.3 APPLICATIONS OF IoT AND ML IN SMART CITIES

Smart city requires the active deployment of IoT with ML technologies. IoT with ML combines as smart technology that plays a significant role in Modular Infrastructure, home automation, smart farming, smart environment, social health, crime monitoring, smart transportation, intelligent parking system, public safety, smart traffic administration, water management plan, smart street lamps, track of air purity level, smart planning of schools, hospitals and roads etc. to examine, manage and control the resources of the smart city [26].

ML is applied to automated systems to achieve less manual operations. Table 10.1 explores the different dimension of emerging ML techniques with IoT in respective smart city application.

10.3.1 Modular Infrastructure

Roads, underpasses, air route, water route, railway network, communication power supply, controlling and maintaining activities are helping in optimization of the resources with security [27].

TABLE 10.1
Smart City Applications with IoT and Machine Learning Methods

S. No	Smart City Applications	Machine Learning Methods	References	State of IoT
1	Modular infrastructure	Deep learning	[27]	Wireless sensor network Adhoc network
2	Smart home	Reinforcement learning	[4,28–32]	Location sensing Environment sensing Secure communication
3	Smart farming	Naive Bayes, Random Forest, and decision trees, SVM	[32]	Environment sensing Remote sensing Secure network Wireless network
4	Smart weather monitoring and prediction	Hidden Markov Model	[4,33]	Location sensing Environment sensing Secure communication
5	Smart health	Deep learning	[34,35]	Location sensing Environment sensing Adhoc network Secure communication
6	Smart environment	ANN, SVM, Random Forest, CNN, naïve Bayes algorithm	[36,37]	Location sensing Environment sensing Secure communication
7	Crime monitoring and public safety	Clustering, Random Forest	[38]	Location sensing Environment sensing Secure communication
8	Smart water and waste management plan	ANN, SVM	[3,39–41]	Remote sensing Adhoc network
9	Smart energy management	Clustering	[16]	Remote sensing
10	Smart transportation	SVM, *K*NN, Markov random field, regression tree, clustering, DNN, DBN	[4,5,38,42]	Location sensing Secure communication

10.3.2 Smart Home

Smart home is the centralized automated controlling system, including electrical appliances like smart TV, refrigerator, air conditioner, fan, light system, door locks, window blinds accessing control, fire detection, temperature monitoring security and interrelated unit, to make the home as smart behavior.

The sensors of the appliance monitor the condition and send surveillance dataset to the centralized controlling at home, which enables monitoring and remote controlling the home [4].

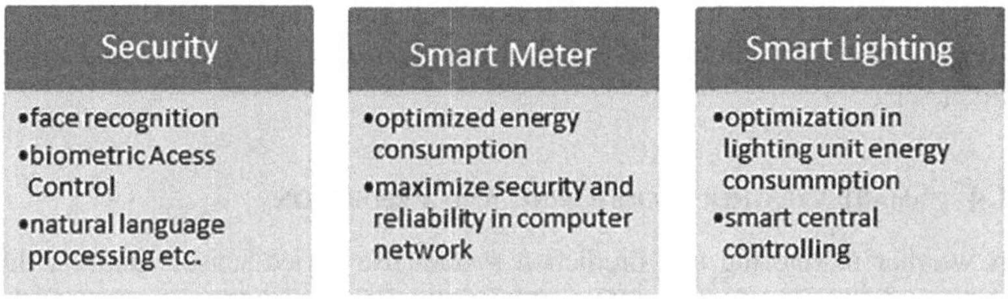

Security
•face recognition
•biometric Acess Control
•natural language processing etc.

Smart Meter
•optimized energy consumption
•maximize security and reliability in computer network

Smart Lighting
•optimization in lighting unit energy consummption
•smart central controlling

FIGURE 10.5 Smart home automation.

ML module identifies the patterns from the training dataset of raw sensor readings on the basis of training-testing-validation [28]. It can be developed in two ways of data-driven and emotion-recognition approaches. For object-based face detection, SVM and neural network are used [29].

As shown in Figure 10.5, smart home automation comprises security, smart meter, smart lighting etc. Mostly security systems integrate the face-recognition techniques into linked visual cameras for smart home automation. Recognition of the face unit like eye, cheekbones, nose, chin and other instances of the persons' picture, matching actual data with image generated using visual cameras, can be done. ML model of a neural network recognizes points on a face, for instances – eye, cheekbone, nose and chin in a persons' photograph – and matches the data with the image generated by visual camera. With the help of AI, security component is able to recognize the face of a smart home person, give notification for the doubtful movement.

Biometric door locks like "August, Kwikset (News – Alert) Kevo and Samsung" are coupled with "Amazon, Google or Samsung" to maintain the ecosystem of smart home. They regularly capture imagery of a user's fingertip using optical scanner and store for similar by computer vision or two-step authentication fingerprint and password.

Smart home automation is usually based on voice recognition and natural language processing (NLP) tools like "Amazon Transcribe and Azure Custom Speech Service or intelligent personal assistants like Siri, Alexa or Google (News – Alert) Home." These systems isolate a user's voice from background noise, transform the analog voice into digital mode for NLP analysis.

IoT device collects the electricity requirement of different appliances configured in smart home and conveys to smart meter. The centralized control unit in smart grids schedule energy consumption of appliances to the user's preference in a plan to minimize electricity bill and maximize the security and reliability of the network [30].

Lighting control popularly explored nowadays. The aim of the smart house is to decrease the electricity consumption, whereas heat, ventilation and air conditioning increase the consumption [30, 31].

10.3.3 Smart Farming

Smart farming system enables to monitor the crop field using sensors related to irrigation like light, humidity, temperature, soil moisture, etc. It uses robot, drone, remote sensor and imagery etc. with ML and analysis tools with ML and analytical tools. It helps to monitor crop, survey and map field and send analyzed results to farmers for managing the farming plan in terms of time and monetary benefits [32].

10.3.4 Smart Weather Monitoring and Prediction

Smart weather monitoring and prediction system use varied sensors and satellites to obtain the temperature, water runoff, vegetation, precipitation and wind speed data [4, 33].

10.3.5 Smart Health

Smart health applications are able to track patients' status, staff people, ambulances, automatic data gathering, sensing and hospital monitoring. The position of the objects, blood products and organs' transplantation can be observed and the results can be obtained online [34]. Patient-based deep brain close loop simulator can be created using ML to simulate patient behavior and mental health according to the use of computers, social media and smartphones. ML can identify simply the status of the mental health [10]. Bio-signal monitoring is able to investigate the patients' condition or obtain the patients' data from anywhere via diverse wireless accessing methods [35].

10.3.6 Smart Environment

Smart environment parameters can be considered like temperature, humidity, moisture, ambient CO_2 level, harmful gases, noise and pollution. An IoT is able to monitor the noise in place like rivers and parks and can reduce the noise at city centers at specific hours [36, 37].

10.3.7 Crime Monitoring and Public Safety

Smart crime monitoring is able to track people at night place, with the help of infrared camera based on temperature mechanism and video-sequence framework is applied to motion works. In order to provide the structure for smart surveillance and public safety, CCTV cameras have been used. This CCTV camera is linked to the video recorder, not having the power of intelligent processing [38]. It is possible to monitor the activity and can alarm any violence with the smart surveillance system.

10.3.8 Smart Water and Waste Management Plan

Water and waste management planning is an important activity of the smart city. In order to deploy sensors at suitable point location, it behaves as an intelligent one to detect faults, leakage in the pipeline etc. The leakage in the pipeline can be detected with three types of sensors: "vibration (using dual-axis accelerometers), pressure (piezo-resistive sensor) and sound (ultrasonic sensor) monitoring." Smart water distribution comprises water source (lake/river), storage service (reservoir), and distribution network (under or aboveground pipelines) [39].

In order to improve water conservation and allocation of the scarce resource in a more effective manner, address flooding and drought should be prevented and wastewater management should be done. With the coupling of IoT and ML, local authorities are able to plan guidelines regarding water, for example, to control a large volume of raw sewage and efficient management for storms [4, 40]. This system is dealing in the prediction for the demand for water and is helpful in future planning requirement for authorities [41].

10.3.9 Smart Energy Management

Smart energy management is capable of sensing and network for increasing the likelihood of optimum progress of energy providers in emergency situations like fault location, isolating and service restoration. It is also helpful to meter readers to send the bill summary directly to the bill server [16].

10.3.10 Smart Transportation

Smart transportation comprises major subsystem as the following (Figure 10.6).

10.3.10.1 Intelligent Parking Systems

Intelligent parking system enables arrival and departure time of various vehicle trace within city. Road sensors and intelligent displays lead drivers in the selection of the best route of parking.

10.3.10.2 Smart Traffic Administration

ML algorithms – logistic regression, decision tree, RF, SVM and MLP – enable to predict accurate and early prediction of the traffic-congestion for a given static road network [5]. Logistic Regression is most accurate and can use hybrid algorithms to get more accurate results. In order to establish sensor utilities and GPS devices, traffic-monitoring can be possible for authority and locality to manage the discipline.

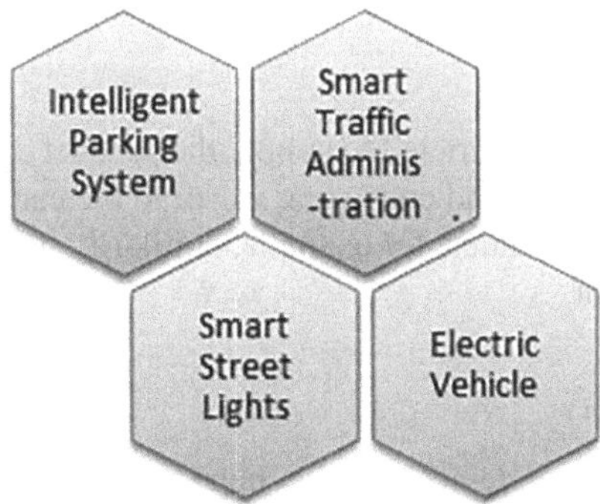

FIGURE 10.6 Smart transportation.

10.3.10.3 Smart Street Lights

Smart street lights play an important role in the smart transportation in terms of reduction in energy consumption, dynamic operations and management. GPS is used in monitoring the lamps their location and status. The light sensors are used to trigger the lamp with on/off signals according to sunrise/sunset. Apart from this, these lamps are capable of detecting passing vehicles, pedestrians and immediately switching the lamp and reduction in energy consumption in dynamic manner. The smart street lights enable cost reduction significantly by switching on/off according to the nearby atmosphere with the help of Wi-Fi hotspot centralized server. Lamps are integrated with camera and environment sensor for the safety measure of the public [38].

10.3.10.4 Electric Vehicle

In order to save energy, pollution-free transportation efficient energy storage devices also known as Electric Vehicles (EVs) can be used. Efficient scheduling algorithms and ML have the potential to charge/discharge, shave peak-load, reducing emission, increasing the use of renewable energy sources [42] to make a green smart city with focus on smart grids and EVs [4].

10.4 CHALLENGES OF IoT

IoT architecture involves with wide-range of interconnected gadgets attached with servers on cloud. Server is based on various interconnected administrations and applications. Building a major, complex arrangement always been a challenging task. By understanding the architecture of an IoT framework, exceptional challenges of an IoT framework can be examined.

Sensing technologies impact the quality of the data fed into the IoT system. Some of the important factors to be considered in the selection of sensors are accuracy, reliability, range, resolution and ability to deal with noise and interference. Some of the other operational challenges include power consumption, security and interoperability [43]. In a particularly different field, there are numerous product, firmware, protocols and equipment stage variations. With these dissimilar frameworks on different scale and variety, and inferring significant results is a huge task. Following are the details about to give the major challenge that has to face with IoT devices.

10.4.1 Compatibility

An IoT arrangement is not restricted to simply the product application or the gadget. It is about both cooperating and conveying worth to the end-client. Software installation presents novel difficulties, for example, similarity issues, device accessibility and model development.

10.4.2 Connectivity

Thousands of associated devices put a huge burden on the overall network. Difficulties from temperamental network equipment and Internet associations could affect gadget execution and at last the IoT architecture.

10.4.3 Storage

The device state changes quickly. For example, the wearable gadget, the state could change rapidly – detached, halting, resting, awakening, due to area, time and speed setting. Such changes need to be made at the server and create a lot of data.

10.4.4 Analytics

The genuine worth of an IoT comes from analyzing the gathered IoT information and determining significant experiences. In any case, its huge size, exactness, information, and exceptions give a challenging issue. An erroneous information sent and recorded on the cloud could prompt mistaken investigation, which may affect dynamic.

10.4.5 Intelligence and Actions

There are challenges in automating actions. It is not uncommon for machines to be in an unpredictable situation, which is not learnt by the model. The ability to programmatically interact (take action) with legacy machines, which are mostly hardwired, could also pose a challenge. Unless, algorithms that take action are well designed, and too many or too few actions could also diminish the value of intelligent analysis and automated recovery action.

10.4.6 Maintain Security and Privacy

Networked gadgets and applications opened for public are susceptible being hacked. With IoT development, programmers are continually attempting to discover framework shortcomings. Security and testing is an unquestionable requirement in any IoT deployment. Security improvement has been achieved by integrating blockchain technology with IoT [44, 45].

10.5 CONCLUSIONS

A study on ML with IoT procedures investigated for keen city applications has been introduced. This study focused on the assortment of ML and IoT strategies that have been assessed for smart city applications. Then again, given the current applications and framework in regard to IoT and ML, a relatively more modest ML inclusion for the shrewd city arranging applications is identified. This chapter investigates the capability of ML to use big in aspect of customized benefits in smart city planning. The capacity to comprehend residents' needs to oblige customized needs as indicated by the conditions is the objective of a smart city planning. Concerning IoT approaches for the smart city planning application categories, smart health, device security, smart farming and energy management, smart transportation, smart home, and smart environment have proven to be the most popular among them. Accordingly, there is a distinct requirement for advantageous inclusion in those spaces, from the ML viewpoint.

With assistance of IoT and ML, smart city planning has been accomplished with incredible advancement. In order to raise the extent of IoT devices, the variety, volume etc. will also increase; hence, ML could be able to think of numerous significant applications in forthcoming years.

ABBREVIATIONS

The following abbreviations are used in this manuscript:

ML	Machine Learning
SC	Smart Cities
AI	Artificial Intelligence
SVM	Support Vector Machine
SGs	Smart Grids
ANN	Artificial Neural Network
SCP	Smart City Planning
CNN	Convolutional Neural Network
HMM	Hidden Markov Model
SAE	Stacked Auto Encoder
RL	Reinforcement Learning
DNN	Deep Neural Networks
RNN	Recurrent Neural Network
LSTM	Long Short Term Memory
DBN	Deep Belief Network
DCNN	Deeper CNNs
MLP	MulitLayer Perceptron
ANN	Artificial Neural Networks
IP	Internet Protocol
RF	Random Forest
NLP	Natural Language Processing
EVs	Electric Vehicles

REFERENCES

1. R. Sethi, B. Bhushan, N. Sharma, R. Kumar, and I. Kaushik, "Applicability of Industrial IoT in Diversified Sectors: Evolution, Applications and Challenges." In Multimedia Technologies in the Internet of Things Environment (pp. 45–67). Springer, Singapore, 2021.
2. S. Chen, H. Xu, D. Liu, B. Hu, and H. Wang, "A Vision of IoT : Applications, Challenges, and Opportunities with China Perspective." IEEE Internet of Things Journal, vol. 1, no. 4, pp. 349–359, 2014.
3. M. Badar, "Machine Learning Approaches in Smart Cities Motivation," 2020.
4. "Big Data Analytics for Electric Vehicle Integration in Green Smart Cities," *IEEE J. Mag.* | IEEE Xplore https://ieeexplore.ieee.org/document/8114543 (accessed May 28, 2021).
5. S. Devi and T. Neetha, "Machine Learning based Traffic Congestion Prediction in a IoT based Smart City," Int. Res. J. Eng. Technol., pp. 3442–3445, 2017 [Online]. Available: www.irjet.net.

6. "Short-Term Traffic Predictions on Large Urban Traffic Networks: Applications of Network-Based Machine Learning Models and Dynamic Traffic Assignment Models." IEEE Conference Publication | IEEE Xplore. https://ieeexplore.ieee.org/abstract/document/7223242 (accessed May 31, 2021).
7. S. Iyengar, S. Lee, D. Irwin, P. Shenoy, and B. Weil, "WaHome: A Data-Driven Approach for Energy Efficiency Analytics at City-Scale." Proceedings of the ACM SIGKDD International Conference on Knowledge Discovery and Data Mining, vol. 18, pp. 396–405, July 2018, doi: 10.1145/3219819.3219825.
8. J. Ba, V. Mnih, and K. Kavukcuoglu, "Multiple Object Recognition with Visual Attention," 3rd Int. Conf. Learn. Represent. ICLR 2015 – Conf. Track Proc., December 2014 [Online]. http://arxiv.org/abs/1412.7755 (accessed June 12, 2021).
9. C. J. C. H. Watkins and P. Dayan, "Q-Learning," *Mach. Learn.*, vol. 8, no. 3–4, pp. 279–292, 1992, doi: 10.1007/bf00992698.
10. B. N. Mohapatra and P. P. Panda, "Machine Learning Applications to Smart City," *Accents Trans. Image Process. Comput. Vis.*, vol. 5, no. 14, pp. 1–6, 2019, doi: 10.19101/tipcv.2018.412004.
11. A. Ghosh, T. Chatterjee, S. Samanta, J. Aich, and S. Roy, "Distracted Driving: A Novel Approach towards Accident Prevention," *Adv. Comput. Sci. Technol.*, vol. 10, no. 8, pp. 2693–2705, 2017.
12. J. Yang, Y. Han, Y. Wang, B. Jiang, Z. Lv, and H. Song, "Optimization of Real-Time Traffic Network Assignment based on IoT Data Using DBN and Clustering Model in Smart City," *Future Gener. Comput. Syst.*, vol. 108, pp. 976–986, 2020, doi: 10.1016/j.future.2017.12.012.
13. J. Yu, G. L. Chang, H. W. Ho, and Y. Liu, "Variation Based Online Travel Time Prediction Using Clustered Neural Networks." In 2008 11th International IEEE Conference on Intelligent Transportation Systems (pp. 85–90). IEEE, October 2008. https://ieeexplore.ieee.org/abstract/document/4732594 (accessed June 12, 2021).
14. P. J. Werbos, "Backpropagation Through Time: What It Does and How To Do It." Proceedings of the IEEE, vol. 78, no. 10, pp. 1550–1560, 1990. https://ieeexplore.ieee.org/abstract/document/58337 (accessed June 12, 2021).
15. M. Ozbayoglu, G. Kucukayan, and E. Dogdu, "A Real-Time Autonomous Highway Accident Detection Model Based on Big Data Processing and Computational Intelligence." In 2016 IEEE International Conference on Big Data (Big Data) (pp. 1807–1813). IEEE, December 2016. https://ieeexplore.ieee.org/abstract/document/7840798 (accessed June 12, 2021).
16. K. Soomro, M. N. M. Bhutta, Z. Khan, and M. A. Tahir, "Smart City Big Data Analytics: An Advanced Review," *Wiley Interdiscip. Rev. Data Min. Knowl. Discovert*, vol. 9, no. 5, pp. 1–25, 2019, doi: 10.1002/widm.1319.
17. G. Hinton, "Deep Learning."
18. A. Krizhevsky, I. Sutskever, and G. E. Hinton, "ImageNet Classification with Deep Convolutional Neural Networks." [Online]. http://code.google.com/p/cuda-convnet/ (accessed June 12, 2021).
19. K. Gopalakrishnan, "Deep Learning in Data-Driven Pavement Image Analysis and Automated Distress Detection: A Review," *Data*, vol. 3, no. 3. MDPI AG, p. 28, 2018, doi: 10.3390/data3030028.
20. D. J. Daniels, "Ground Penetrating Radar." In *Encyclopedia of RF and Microwave Engineering*. John Wiley & Sons, Inc., Hoboken, NJ, 2005.
21. D. Kwon, S. Park, S. Baek, R. K. Malaiya, G. Yoon, and J. T. Ryu, "A Study on Development of the Blind Spot Detection System for the IoT-Based Smart Connected Car." In 2018 IEEE International Conference on Consumer Electronics (ICCE) (pp. 1–4). IEEE, January 2018. https://ieeexplore.ieee.org/abstract/document/8326077 (accessed June 12, 2021).

22. M. Munoz-Organero, R. Ruiz-Blaquez, and L. Sánchez-Fernández, "Automatic Detection of Traffic Lights, Street Crossings and Urban Roundabouts Combining Outlier Detection and Deep Learning Classification Techniques based on GPS Traces While Driving," *Comput. Environ. Urban Syst.*, vol. 68, pp. 1–8, 2018, doi: 10.1016/j.compenvurbsys.2017.09.005.
23. Y. Lv, Y. Duan, W. Kang, Z. Li, and F. Y. Wang, "Traffic Flow Prediction with Big Data: A Deep Learning Approach." IEEE Transactions on Intelligent Transportation Systems, vol. 16, no. 2, pp. 865–873, 2014. https://ieeexplore.ieee.org/abstract/document/6894591 (accessed June 12, 2021).
24. G. Fusco, C. Colombaroni, L. Comelli, and N. Isaenko, "Short-Term Traffic Predictions on Large Urban Traffic Networks: Applications of Network-Based Machine Learning Models and Dynamic Traffic Assignment Models." In 2015 International Conference on Models and Technologies for Intelligent Transportation Systems (MT-ITS) (pp. 93–101). IEEE, June 2015. https://ieeexplore.ieee.org/abstract/document/7223242 (accessed June 11, 2021).
25. M. Mohammadi and A. Al-Fuqaha, "Enabling Cognitive Smart Cities Using Big Data and Machine Learning: Approaches and Challenges," *IEEE Commun. Mag.*, vol. 56, no. 2, pp. 94–101, 2018, doi: 10.1109/MCOM.2018.1700298.
26. B. Bhushan, A. Khamparia, K. M. Sagayam, S. K. Sharma, M. A. Ahad, and N. C. Debnath, "Blockchain for Smart Cities: A Review of Architectures, Integration Trends and Future Research Directions," *Sustain. Cities Soc.*, vol. 61, p. 102360, 2020, doi: 10.1016/j.scs.2020.102360.
27. S. Joshi, S. Saxena, T. Godbole, and Shreya, "Developing Smart Cities: An Integrated Framework," *Procedia Comput. Sci.*, vol. 93, no. September, pp. 902–909, 2016, doi: 10.1016/j.procs.2016.07.258.
28. S. Talari, M. Shafie-Khah, P. Siano, V. Loia, A. Tommasetti, and J. P. S. Catalão, "A Review of Smart Cities based on the Internet of Things Concept," *Energies*, vol. 10, no. 4, pp. 1–23, 2017, doi: 10.3390/en10040421.
29. J. Jaihar, N. Lingayat, P. S. Vijaybhai, G. Venkatesh, and K. P. Upla, "Smart Home Automation Using Machine Learning Algorithms," *2020 Int. Conf. Emerg. Technol. INCET 2020*, pp. 16–19, 2020, doi: 10.1109/INCET49848.2020.9154007.
30. T. F. Abiodun, F. Temitope, A. Abiodun, O. Jamaldeen, and A. F. Adebola, "Environmental Problems, Insecurity in the Sahel Region and Implications for Global Security," 2020 [Online]. https://www.researchgate.net/publication/349553954 (accessed May 25, 2021).
31. S. Talari, M. Shafie-Khah, P. Siano, V. Loia, A. Tommasetti, and J. P. S. Catalão, "A Review of Smart Cities based on the Internet of Things Concept," *Energies*, vol. 10, no. 4. MDPI AG, p. 421, 2017, doi: 10.3390/en10040421.
32. E. Adi, A. Anwar, Z. Baig, and S. Zeadally, "Machine Learning and Data Analytics for the IoT," *Neural Comput. Appl.*, vol. 32, no. 20, pp. 16205–16233, 2020, doi: 10.1007/s00521-020-04874-y.
33. M. F. McCabe et al., "Hydrological Consistency Using Multi-Sensor Remote Sensing Data for Water and Energy Cycle Studies," *Remote Sens. Environ.*, vol. 112, no. 2, pp. 430–444, 2008, doi: 10.1016/j.rse.2007.03.027.
34. L. Atzori, A. Iera, and G. Morabito, "The Internet of Things: A Survey," *Computer Networks*, vol. 54, no. 15, pp. 2787–2805, 2010, doi. 10.1016/j.comnet.2010.05.010.
35. D. Niyato, E. Hossain, and S. Camorlinga, "Remote Patient Monitoring Service Using Heterogeneous Wireless Access Networks: Architecture and Optimization," *IEEE Journal on Selected Areas in Communications*, vol. 27, no. 4, pp. 412–423, May 2009, doi: 10.1109/JSAC.2009.090506.
36. E. A. S. Ahmed and M. E. Yousef, "Internet of Things in Smart Environment: Concept, Applications, Challenges, and Future Directions," *World Sci. News*, vol. 134, no. 1, pp. 1–51, 2019, [Online]. Available: www.worldscientificnews.com.

37. A. K. M. B. Haque, B. Bhushan, and G. Dhiman, "Conceptualizing Smart City Applications: Requirements, Architecture, Security Issues, and Emerging Trends," *Expert Syst.*, 2021, doi: 10.1111/exsy.12753.
38. F. Zantalis, G. Koulouras, S. Karabetsos, and D. Kandris, "A Review of Machine Learning and IoT in Smart Transportation," *Future Internet*, vol. 11, no. 4, pp. 1–23, 2019, doi: 10.3390/FI11040094.
39. G. P. Hancke and B. de Carvalho e Silva, "The Role of Advanced Sensing in Smart Cities," *Sensors (Switzerland)*, vol. 13, no. 1. MDPI AG, pp. 393–425, 2013, doi: 10.3390/s130100393.
40. Y. Nam, S. Rho, and J. H. Park, "Intelligent Video Surveillance System: 3-Tier Context-Aware Surveillance System with Metadata," *Multimed. Tools Appl.*, vol. 57, no. 2, pp. 315–334, 2012, doi: 10.1007/s11042-010-0677-x.
41. "Identification of Road Surface Conditions Using IoT Sensors and Machine Learning | springerprofessional.de." https://www.springerprofessional.de/en/identification-of-road-surface-conditions-using-iot-sensors-and-/16885118 (accessed May 29, 2021).
42. B. Zhou et al., "Smart Home Energy Management Systems: Concept, Configurations, and Scheduling Strategies," *Renew. Sustain. Energy Rev.*, vol. 61, pp. 30–40, 2016, doi: 10.1016/j.rser.2016.03.047.
43. B. Russell and D. Van Duren, Practical Internet of Things Security, 2016.
44. B. Bhushan, C. Sahoo, P. Sinha, and A. Khamparia, "Unification of Blockchain and Internet of Things (BIoT): Requirements, Working Model, Challenges and Future Directions," *Wireless Network*, vol. 27, no. 1, 2021. Springer US.
45. S. Saxena, B. Bhushan, and M. A. Ahad, "Blockchain based Solutions to Secure IoT: Background, Integration Trends and a Way Forward," *J. Network Comput. Appl.*, vol. 181, p. 103050, 2021, doi: 10.1016/j.jnca.2021.103050.

11 The Internet of Medical Things for Monitoring Health

Rehab A. Rayan
High Institute of Public Health, Alexandria University, Alexandria, Egypt

Christos Tsagkaris
Novel Global Community Educational Foundation, Hebersham, NSW, Australia

Andreas S. Papazoglou
Athens Naval Hospital, Athens, Thessaloniki, Greece

Dimitrios V. Moysidis
Hippokration University Hospital, Aristotle University of Thessaloniki, Thessaloniki, Greece

CONTENTS

DOI: 10.1201/9781003219620-11

11.1 INTRODUCTION

Conventional techniques of security cannot be readily applied to the Internet of Things (IoT) for its special needs. Implementing information and communication technologies (ICTs) in health records should guarantee the various major protection measures along with credibility, privacy, accessibility, approval, and verification tasks to facilitate healthcare delivery without affecting the streamlining of services and confidentiality of patients' records. The main issue here is that patients primarily live in remote areas where there is a shortage of physicians and medications in critical conditions. Today, adopting novel technologies such as IoT for healthcare monitoring have addressed these issues. IoT may not completely ensure patients' health and safety; however, it could improve the delivery of healthcare. Medical IoT could engage the patient by enabling them to have extra time with their physicians. The IoT is a broad context in medical care that could be called the Internet of Medical Things (IoMT) in a broader set of digital health and clinical care (Kaur, Atif, and Chauhan 2020).

The Radio Frequency Identification (RFID) organization defines the IoT as the global connectivity of networked devices. It comprises traditional themes like embedded systems, digital and control systems, and networks of wireless sensors to enable interconnected communication among devices (D2D). Designing mini devices was derived via developing and integrating micro-electromechanical technologies, non-wired connections, and appliances that could sense, compute, and communicate remotely. These mini devices are known as wireless sensor networks (WSNs) nodes (Bagci et al. 2016). IoT facilitates gathering, listing, and exploring the new flow of information quickly and accurately through appliances that rapidly gather and exchange statistics with the cloud. IoT has many areas of application that could be classified primarily according to accessibility, security, scope, variety, authenticity, usability, and significance (Bandyopadhyay and Sen 2011).

Overall, the aim of this chapter is to provide a critical overview, elaborating on the concept and status of IoMT research, applications, and limitations of adoption. It seeks to present the emerging IoMT evidence in the field along with real-world perspectives from the industry, the market, and the policymaking sector.

This chapter starts by describing the concept of IoMT, introducing the methodology of selecting included research items, then presents the landscape of IoMT research via providing a catalog of selected studies that summarize the IoMT research. Next, it discusses applying IoMT on body, at home, in the community, and in clinics. Furthermore, it reports a case study highlighting the application of the IoMT. Ultimately, it concludes by outlining the opportunities and limitations of adoption.

11.2 METHODOLOGY

The authors conducted a literature search in PubMed/Medline and Cochrane with keywords (IoMT, healthcare, applications, research, digital health, wearable, devices, and software). Peer-reviewed original studies (preclinical animal research or clinical trials), reviews, and meta-analyses published in English, French or German between

2000 and 2021 were included. Studies with a significant conflict of interest were excluded. The authors identified the included studies through title and abstract screening and assessed the full text of the studies that fulfilled the inclusion criteria. In total, fifteen studies were reported based on these criteria.

Additionally, product licensing reports, policy briefs, and case studies were included in the discussion subsections. Their selection was based on relevance and (recently) time of publication. The authors declare no financial or other ties guiding to the inclusion of reports and case studies. The study technique applied is descriptive, specifically content analysis, which is meant to comprehensively discuss a message. The structure of this study is not meant to test a specific hypothesis or evaluate the correlation among elements.

11.3 THE INTERNET OF MEDICAL THINGS

The IoMT has been described as an amalgamation of clinical gadgets and usages compatible with healthcare informatics systems via networking modalities. Although the IoMT is not as popular as the IoT in terms of research and entrepreneurial activity, it has grown significantly during the last few years. In 2016, the IoMT market accounted for 22.5 billion dollars. With a growth compound rate of more than 20%, the value of the IoMT market is expected to surpass 70 billion dollars in 2021 (Joyia et al. 2017).

The IoMT has the potential of transforming healthcare, enough to manage patients with chronic conditions timely and effectively from distance. This feature has become highly relevant, and it may be the starting point for the IoMT to unravel its potential. One of the most popular and typical examples is wearables with embedded analytics and real-time communication systems. Such gadgets bring the concept of continuous monitoring to almost any disease and contribute to tertiary prevention through the early detection of relapses or deterioration in individuals with known medical history (Goyal et al. 2021). Collected data can be incorporated into the machine and deep learning frameworks, with the potential for automated diagnosis. Although automated diagnosis cannot substitute appropriate assessment and diagnosis from a physician, it may alert healthcare providers to investigate earlier.

Khamparia et al. (2020) have recently proposed a framework of IoMT- enabled deep learning system for the detection and classification of skin cancer (Khamparia et al. n.d.). Their approach is transferable to other communicable and noncommunicable diseases, ranging from herpes zoster infection to cutaneous lupus erythematous. Nonetheless, such a volume of biometric data requires several safety valves against cyberattacks and discriminatory behavior of insurance companies and employers. Sharma et al. (2020) have presented a safety framework for the use of biometric data under a two-factor identification system (Sharma et al. 2020). Moreover, IoMT-driven systems can decrease unneeded clinical encounters and hospitalization length. The load on health sectors and the healthcare workforce can be alleviated as well by linking patients to healthcare providers and facilitating the exchange of confidential information through a guarded connection (Joyia et al. 2017; Vishnu, Jino Ramson, and Jegan 2020).

11.3.1 The Research Landscape of IoMT

Several review studies are currently available providing a comprehensive overview of IoMT, its applications and perspectives. Sadoughi et al. (2020) have provided a detailed mapping of IoMT research and applications. Their findings suggest that IoMT applications are mostly used in neurology, cardiology, and psychiatry/psychology, either at home or in healthcare facilities. These applications appear to have mostly been studied in the United States, China, and India, which collectively account for 40% of the available evidence in the field (Sadoughi, Behmanesh, and Sayfouri 2020). Table 11.1 presents some studies summarizing the IoMT research between 2015 toward 2021.

Between 2015 and 2021, the majority of IoMT reviews attempted to capture and assess a wide range of relevant applications, mechanistic frameworks, concerns, recommendations for improvement, and future trends. Data protection was a common ground of concern for almost all the studies. A limited number of studies focused on mechanistic frameworks such as fog computing or health issues relevant to the COVID-19 pandemic. It would be pivotal for future research in the field to dig deeper into IoMT solutions for health issues (infectious outbreaks, access to preventive medicine services, diabetes, rhythm control, rehabilitation, etc.). Scoping reviews could help physicians and informatics specialists to set clear goals for relevant original studies in the future. Many of the above-listed studies have also emphasized the need for policymaking and effective and transparent decision-making not only at a regulatory level but also, within health bodies.

11.3.2 The IoMT Market and Capacity

To this date, the IoMT industry includes intelligent gadgets (wearables and clinical monitors) and supportive services focusing on data processing, real-time location detection, and telehealth. Such devices can be used on the body, in the house, in public, in clinics, and in hospitals. In most cases, they are strictly for medical use, but considerable effort is paid in making them part of lifestyle applications compatible with devices used in everyday life such as smartphones and tablets (Gomes et al. 2017).

11.4 APPLICATIONS OF IoMT

The applications of IoMT can be categorized in the following segments:

1. On-Body Segment
2. In-Home Segment
3. Community Segment
4. In-Clinic Segment

Figure 11.1 provides an overview of the evolving landscape of IoMT applications.

TABLE 11.1
A Catalog of Studies Summarizing the IoMT Research between 2015 and 2021

Topic	Type of the Study	Date	Key Content	Reference
The Internet of Things for Health Care: A Comprehensive Survey	Survey/Overview	2015	IoMT network architectures have a major potential in ambient intelligence and wearables. Health policy interventions need to focus on data protection in IoMT	Islam et al. (2015)
Internet of Medical Things and Big Data in Healthcare	Systematic mapping study	2016	An overview of devices, wearables, and mHealth applications with a reflection on contemporary challenges of IoMT (cost and implementation, safety concerns)	Dimitrov (2016)
Internet of Things for Smart Healthcare: Technologies, Challenges, and Opportunities	Survey paper on latest advances across the area	2017	Cloud computing and data storage is necessary to support IoMT-based wearable devices. This study suggests an IoMT operational paradigm	Baker, Xiang, and Atkinson (2017)
Advanced Internet of Things for Personalized Healthcare System: A Survey	Systematic literature review	2017	An overview of four-layer IoT architecture for personalized healthcare systems (PHS), classification of applications, and successful conclusion of case studies in IoMT	Qi et al. (2017)
Internet of Medical Things (IOMT): Applications, Benefits and Future Challenges in Healthcare Domain	Narrative literature review	2017	Emerging IoMT applications and future challenges with a focus on medical services delivery and spreading IoMT awareness among healthcare practitioners	Joyia et al. (2017)
Toward fog-driven IoT eHealth: Promises and Challenges of IoT in Medicine and Healthcare	Narrative review	2018	Overview of IoMT applications for patient-centered healthcare A suggestion of a three-pronged architecture for the IoMT eHealth ecosystem focusing on (i) devices, (iii) fog computing, (iii) cloud services	Farahani et al. (2018)
Internet of Medical Things: A Review of Recent Contributions Dealing with Cyber-Physical Systems in Medicine	Comprehensive literature review	2018	Efforts to improve the IoMT by means of formal methodologies developed by the cyber-physical systems community Democratization of IoMT-based medical devices for both patients and healthcare providers applications and implications	Gatouillat et al. (2018)

(Continued)

TABLE 11.1 *(Continued)*
A Catalog of Studies Summarizing the IoMT Research between 2015 and 2021

Topic	Type of the Study	Date	Key Content	Reference
The application of Internet of Things in healthcare: a systematic literature review and classification	Systematic review 2000–2016	2018	Contemporary state and future directions for the role of cloud-based IoMT architecture Data protection and interoperability implications and challenges of the IoMT	Ahmadi et al. (2019)
A Survey on Internet of Things and Cloud Computing for Healthcare	A survey focusing on research conducted between 2015 and 2018	2019	IoMT framework regarding topologies, platforms, and structures. Relevant data protection strategies. Challenges for the integration and implementation of IoMT and cloud computing in healthcare. Global policymaking efforts and recommendations oriented toward IoMT	Dang et al. (2019)
Internet of Things for Healthcare Using Effects of Mobile Computing: A Systematic Literature Review	Systematic review between 2011 and 2019	2019	The role of mobile computing in the IoMT infrastructure. Data protection and privacy related to the IoMT	Nazir et al. (2019)
Enabling technologies for fog computing in healthcare IoT systems	Systematic Review between 2007 and 2017	2019	Fog computing as an essential component of the IoMT. Evaluation of fog computing performance in healthcare services Three crucial factors for resource management regarding the IoMT: (i) computation offloading, (ii) load balancing, and (iii) interoperability	Mutlag et al. (2019)
IoT-Based Healthcare Applications: A Review	Systematic review between 2000 and 2017	2019	IoMT functional and nonfunctional requirements. Classification of IoMT protocols regarding communication (network protocols) and application (data transfer protocols)	de Morais Barroca Filho and Soares de Aquino Junior (2017)
Internet of Things (IoT), Applications and Challenges: A Comprehensive Review	Comprehensive literature review	2020	Evaluation of original studies about the application domain-specific parameters in IoMT Tools to assess the current state of the IoMT and future research directions	Khanna and Kaur (2020)

(Continued)

TABLE 11.1 *(Continued)*
A Catalog of Studies Summarizing the IoMT Research between 2015 and 2021

Topic	Type of the Study	Date	Key Content	Reference
The Internet of Things: Impact and Implications for Health Care Delivery	Narrative review and viewpoint	2020	IoMT enabled decision-making. IoMT to increase the accessibility of preventive healthcare services	Kelly et al. (2020)
Internet of Things for Current COVID-19 and Future Pandemics: an Exploratory Study	Exploratory review	2020	IoMT-based technologies in COVID-19 including infrastructures, platforms, applications, industrial IoMT products against COVID-19 Implementation of IoMT-based strategies in three steps, (i) early diagnosis, (ii) quarantine time, (iii) post recovery monitoring/long COVID-19	Nasajpour et al. (2020)
Internet of Things (IoT) enabled healthcare helps to take the challenges of COVID-19 Pandemic	Scoping review	2021	Achievements of the IoMT during the COVID-19 pandemic Seven major IoMT technologies with a major potential against the COVID-19 pandemic Sixteen IoMT applications with a relevant potential against the COVID-19 pandemic	Javaid and Khan (2021)

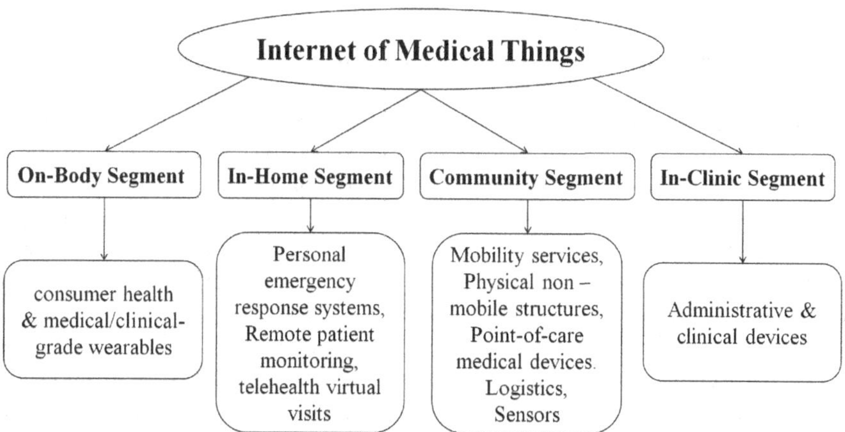

FIGURE 11.1 An overview of the evolving landscape of IoMT applications (Jain et al. 2021; S. Rubí and L. Gondim 2019; Siddiqui 2021).

11.4.1 On-Body IoMT Modalities

They can be sorted into client health wearables and clinical wearables. Consumer health wearables consist of gadgets for in-person exercise and wellbeing, like exercise monitors, bands, athletic watches, and intelligent clothes. The majority are non-licensed through a standard procedure by health authorities. In most cases, experts testify for such applications based on small-scale consumer studies. Companies such as Misfit (Fossil Group), Withings, Samsung Medical, and Fitbit have launched such products (Vishnu, Jino Ramson, and Jegan 2020).

Medical/clinical-grade wearables include regulated devices and support networks licensed by legal agencies, such as the US Food and Drug Administration or the European Medicines Agency. Most of these devices are prescribed by a physician and/or used in accordance with expert advice. Examples include but are not limited to (Madanian and Parry 2019):

- An intelligent bang from Active Protective detecting slips and deploying hip safeguarding for old people.
- The Halo Sport headset by Halo Neuroscience. If this helmet were worn while exercising and physical fitness, it could trigger cognitive sites controlling muscular retention, resilience, and durability.
- The Quell of Neuro-Metrix is a wearable neuro-modulation gadget. It is placed on sensory nerves, alleviating chronic pain.
- A wearable, wireless biosensor by Phillips. This device is positioned on the chest of COVID-19 in-patients, monitoring their heart and respiratory rate and detecting signs of early deterioration (Cuffari 2020).

11.4.2 The In-Home Devices

They consist of personal emergency response systems (PERS), remote patient monitoring (RPM), and telehealth virtual visits. PERSs rely on the interconnection of wearable devices/relay units and an on-call medical/support center to monitor home-bound or limited-mobility seniors and increase their self-reliance. The package provides users with a constantly available line to communicate and receive medical care in case of an emergency. RPM is a broader version of PERS including all home monitoring devices and sensors used in the context of chronic or relapsing conditions. RPM involves continuous monitoring of physiological parameters in the frame of lengthy care in the house as well as acute home monitoring. RPM objectives are to decrease the time of hospitalization, a burden for the patient, their carers, and the healthcare system, while securing the patient's proper monitoring at home (Guan et al. 2019).

Continuous observation of discharged patients in a safe and familiar environment can decrease convalescence duration and avoid further admissions. Medication control can contribute significantly to patients' compliance and boost clinical research as well. From providing patients with drug alerts and dosage data to rewarding compliance and results or notifying physicians in case of side effects, there is room for improvement and research. On top of these, telehealth virtual visits/consultations can enhance patients to control their health while obtaining medications or reassessing care. These are camera call-based appointments and assessment of conditions through video supervision and electronic examinations (Bielli et al. 2004).

Some examples include, but are not limited to:

- The Sensemedic® pill dispenser by Evalan. This device is connected to a real-time communication system and aims to improve adherence to medication (Evalan 2017).
- A smart ingestible pill intake sensor by Proteus Digital Health. The sensor's size is 1 mm^2. It is made with silicon and food ingredients, enabling it to be embedded in pills and swallowed. It is activated by gastric acids, notifying an app about the successful intake of the pill (Hutter 2012).
- An IoT-enabled smart yoga mat, helping monitor the patterns of physical exercise at home and providing feedback about its duration and intensity. This has been manufactured by SmartMat (SmartMat 2014).
- The TrackR Bravo® items locator. This is a coin-sized sensor/tracker manufactured by TrackR. It is essential for individuals with dementia, who might experience hardship or even risk in case of loss of essential items such as keys, glasses, and medical devices (TrackR 2016).

11.4.3 The Community-Oriented IoMT

There are five components of the community-oriented IoMT. First, motion services enable riders in cars to monitor status on transit. Such vehicles can be equipped with urgency-responding intelligence developed to notify and help rapid responders, paramedics, and healthcare providers. Such a system is under development by the Seattle-based Airbiquity. Although this is not a direct medical system, its contribution to decreasing the risk of car accidents can be an asset to community health (Airbiquity 2006). Second, physical nonmobile structures such as kiosks equipped with computer touchscreen displays can relate to care providers. Third, point-of-care medical devices can be placed in public spaces and buildings (Baker, Xiang, and Atkinson 2017). Defibrillators connected to a support line could be an example of community-oriented IoMT. Such a device was manufactured by Lantronix (Lantronix 2016).

Furthermore, community-oriented IoMT involves logistics in terms of transferring medical goods and services. Finally, yet importantly, such shipments can be accompanied by sensors monitoring adverse environmental conditions (temperature, humidity, and tilt). The latter could make end-to-end visibility for personalized medicine possible. This could be achieved using RFID barcodes and drones. Drones serve the needs of delivery and end-to-end visibility and have already been used in remote healthcare settings (Miorandi et al. 2012). An RFID tracker of food delivery has been launched by Dole Foods, and this technology can be extrapolated to medical equipment and medication among others (Maras 2015).

Finally, yet importantly, community IoMT can contribute to detecting air pollution and notifying residents and concerned authorities. Such solutions have become quite relevant in the context of smart cities, given the health-related effects of air pollutants in large urban centers around the world. Air Quality Internet of Things

(AQIoT) device platforms have been used in the United Kingdom and in Romania so far (Toma et al. 2019).

11.4.4 In-Clinic Implemented IoMT

It pertains to administrative or clinical devices. Such devices can be used either in physical healthcare structures or in telehealth modalities. They can also be used by emergency on-site responders. Some recent models involve Rijuven's Clinic in a Bag, which is an online testing framework for healthcare providers to evaluate patients timely; ThinkLabs' electronic stethoscope; and Tytocare's broad telemedicine patient testing tool for the heart, ears, lungs, skin, abdomen, and throat that could assess temperature as well (Rajasekaran et al. 2019).

In the context of a large-scale healthcare structure, such as a hospital, IoMT can be widely implemented. Examples include but are not limited to asset management monitors and tracking high-value equipment deployed throughout various departments, like wheelchairs and infusion pumps, throughout the facility. At the same time, logistics services can rely on personnel, and patients' management devices can be used to ameliorate patients' and procedures' flow. In the background, management streamlines assessing and ordering supplies, consumables, pharmaceuticals, and medical devices (Vishnu, Jino Ramson, and Jegan 2020). Telekom Healthcare solutions have provided a variety of programs encompassing aspects of the functionalities (Telekom Healthcare Solutions 2019). Finally, energy supply and environmental factors (e.g., electricity consumption, temperature, and humidity) can be addressed via IoMT (Michalakis and Caridakis 2017).

11.5 CASE STUDY

A research group from the Fujian University of Technology in China has recently published a study related to a sleep apnea IoMT device. The device consists of an SpO_2 sensor monitoring the cardiac rate and the gas blood concentration and a cloud-based support system. The device analyzes its input, correlating alterations of the cardiac rate diversity and oxygen saturation with apnea incidences. This analysis is facilitated through the cloud-based support center, which can diagnose and warn remote patients. Data and warnings become available to both the smartphone and computer of the patients (Abdel-Basset, Ding, and Abdel-Fatah 2020; Haoyu et al. 2019).

In recent years, it has been proved that appropriate home care has a significant contribution in preventing complicated sleep apnea, which includes hypertension and coronary heart disease. Home care devices can serve as a reliable monitoring modality, detecting exacerbations that require a reassessment of the therapeutic approach. The automatic portable sleep apnea detector of this study seems to be a noninvasive, low cost, and universally accessible solution. The algorithms of the device were assessed with the St. Vincent's University Hospital/University College Dublin sleep apnea dataset. Besides, the user data are collected from 10 apnea conditions. The function of the suggested algorithm reached a mean precision, specificity, and sensitivity of 98.54%, 98.95%, and 97.05% respectively (Haoyu et al. 2019).

11.6 DISCUSSIONS

IoMT has a major potential to support evidence-based decision-making and improve patients' quality of life. Interconnected health solutions and networked technologies could solidify and enhance therapeutic effectiveness, where various medical devices could alert patients to administer timely their prescribed medicines. Unique IoT techniques could strengthen managing emergencies. Many unique sensors and various equipment could facilitate physicians in examining the patient's health. IoT would enable healthcare providers to consult experts in urgent cases from all around the world.

However, keeping privacy and security of the daily produced big data is highly challenging. The deficiency of unique standards could lead to medical devices congesting connections for the overload and hence slow the speed of exchanging data. A questionable level of functionality would put incorporating IoMT in contemporary medicine on hold, given the increased cost of IoT infrastructure for both individuals and healthcare systems.

11.7 FUTURE INSIGHTS

IoMT technologies are gaining ground, with more than 70% of global healthcare organizations having endorsed them so far. Today, a change in basic assumptions to digital transformation is occurring in traditional healthcare. Although encouraging, this shift deepens existing inequalities between developed and developing countries (Gopal et al. 2019). The same gap applies between wealthy individuals and those in need. Careful observation of case studies from the IoMT sector may reveal crucial standpoints for further development in a tech-savvy and sustainable manner.

11.8 CONCLUSIONS

The IoMT can make significant contributions to healthcare. Including IoT-based devices and services in medical facilities and everyday patients' life can pave the way toward more effective, affordable, and patient-centered care. Such applications enable detailed patient documentation and remote real-time monitoring. Both are essential for preventive medicine and timely interventions. However, data protection is a major concern, given the detrimental impact of cybersecurity breaches on individuals, companies, and institutions. Integral improvements in the IoMT infrastructure, in combination with firm policymaking, transparency within and beyond healthcare, and health technology literacy among healthcare workers, patients, and carers, are pivotal. In the future, healthcare facilities would apply the IoMT techniques to build their medical capacities. In any case, rigorous assessment is necessary for the constant adaptation of IoMT to contemporary needs in either administrative or research context.

REFERENCES

Abdel-Basset, Mohamed, Weiping Ding, and Laila Abdel-Fatah. 2020. "The Fusion of Internet of Intelligent Things (IoIT) in Remote Diagnosis of Obstructive Sleep Apnea: A Survey and a New Model." *Information Fusion* 61 (September): 84–100. https://doi.org/10.1016/j.inffus.2020.03.010.

Ahmadi, Hossein, Goli Arji, Leila Shahmoradi, Reza Safdari, Mehrbakhsh Nilashi, and Mojtaba Alizadeh. 2019. "The Application of Internet of Things in Healthcare: A Systematic Literature Review and Classification." *Universal Access in the Information Society* 18(4): 837–69. https://doi.org/10.1007/s10209-018-0618-4.

Airbiquity. 2006. "Vehicle Telematics, Geofencing & Fleet Tracking Systems for the Connected Car." Text. Airbiquity. October 9, 2006. http://www.airbiquity.com/.

Bagci, Ibrahim Ethem, Shahid Raza, Utz Roedig, and Thiemo Voigt. 2016. "Fusion: Coalesced Confidential Storage and Communication Framework for the IoT." *Security and Communication Networks* 9(15): 2656–73. https://doi.org/10.1002/sec.1260.

Baker, Stephanie B., Wei Xiang, and Ian Atkinson. 2017. "Internet of Things for Smart Healthcare: Technologies, Challenges, and Opportunities." *IEEE Access* 5: 26521–44. https://doi.org/10.1109/ACCESS.2017.2775180.

Bandyopadhyay, Debasis, and Jaydip Sen. 2011. "Internet of Things: Applications and Challenges in Technology and Standardization." *Wireless Personal Communications* 58(1): 49–69. https://doi.org/10.1007/s11277-011-0288-5.

Bielli, Emilia, Fabio Carminati, Stella La Capra, Micaela Lina, Cinzia Brunelli, and Marcello Tamburini. 2004. "A Wireless Health Outcomes Monitoring System (WHOMS): Development and Field Testing with Cancer Patients Using Mobile Phones." *BMC Medical Informatics and Decision Making* 4(1): 7. https://doi.org/10.1186/1472-6947-4-7.

Cuffari, Benedette. 2020. "The Philips Wearable Biosensor for COVID-19 Patient Monitoring." AZoSensors.Com. November 10, 2020. https://www.azosensors.com/article.aspx?ArticleID=2087.

Dang, L. Minh, Md Jalil Piran, Dongil Han, Kyungbok Min, and Hyeonjoon Moon. 2019. "A Survey on Internet of Things and Cloud Computing for Healthcare." *Electronics* 8(7): 768. https://doi.org/10.3390/electronics8070768.

de Morais Barroca Filho, Itamir, and Gibeon Soares de Aquino Junior. 2017. "IoT-Based Healthcare Applications: A Review." In *Computational Science and Its Applications – ICCSA 2017*, edited by Osvaldo Gervasi, Beniamino Murgante, Sanjay Misra, Giuseppe Borruso, Carmelo M. Torre, Ana Maria A.C. Rocha, David Taniar, Bernady O. Apduhan, Elena Stankova, and Alfredo Cuzzocrea, 47–62. Lecture Notes in Computer Science. Cham: Springer International Publishing. https://doi.org/10.1007/978-3-319-62407-5_4.

de Tácio Pereira Gomes, Berto, Luiz Carlos Melo Muniz, Francisco José da Silva e Silva, Davi Viana Dos Santos, Rafael Fernandes Lopes, Luciano Reis Coutinho, Felipe Oliveira Carvalho, and Markus Endler.2017. "A Middleware with Comprehensive Quality of Context Support for the Internet of Things Applications." *Sensors* 17(12): 2853. https://doi.org/10.3390/s17122853.

Dimitrov, Dimiter V. 2016. "Medical Internet of Things and Big Data in Healthcare." *Healthcare Informatics Research* 22 (3): 156–63. https://doi.org/10.4258/hir.2016.22.3.156.

Evalan. 2017. "FDA Approval Sensemedic for US." Evalan. April 15, 2017. https://evalan.com/en/fda-approval-sensemedic/.

Farahani, Bahar, Farshad Firouzi, Victor Chang, Mustafa Badaroglu, Nicholas Constant, and Kunal Mankodiya. 2018. "Towards Fog-Driven IoT EHealth: Promises and Challenges of IoT in Medicine and Healthcare." *Future Generation Computer Systems* 78 (January): 659–76. https://doi.org/10.1016/j.future.2017.04.036.

Gatouillat, Arthur, Youakim Badr, Bertrand Massot, and Ervin Sejdić. 2018. "Internet of Medical Things: A Review of Recent Contributions Dealing With Cyber-Physical Systems in Medicine." *IEEE Internet of Things Journal* 5 (5): 3810–22. https://doi.org/10.1109/JIOT.2018.2849014.

Gopal, Gayatri, Clemens Suter-Crazzolara, Luca Toldo, and Werner Eberhardt. 2019. "Digital Transformation in Healthcare – Architectures of Present and Future Information Technologies." *Clinical Chemistry and Laboratory Medicine (CCLM)* 57 (3): 328–35. https://doi.org/10.1515/cclm-2018-0658.

Goyal, Sukriti, Nikhil Sharma, Bharat Bhushan, Achyut Shankar, and Martin Sagayam. 2021. "IoT Enabled Technology in Secured Healthcare: Applications, Challenges and Future Directions." In *Cognitive Internet of Medical Things for Smart Healthcare: Services and Applications*, edited by Aboul Ella Hassanien, Aditya Khamparia, Deepak Gupta, K. Shankar, and Adam Slowik, 25–48. Studies in Systems, Decision and Control. Cham: Springer International Publishing. https://doi.org/10.1007/978-3-030-55833-8_2.

Guan, Zhitao, Zefang Lv, Xiaojiang Du, Longfei Wu, and Mohsen Guizani. 2019. "Achieving Data Utility-Privacy Tradeoff in Internet of Medical Things: A Machine Learning Approach." *Future Generation Computer Systems* 98 (September): 60–68. https://doi.org/10.1016/j.future.2019.01.058.

Haoyu, Li, Li Jianxing, N. Arunkumar, Ahmed Faeq Hussein, and Mustafa Musa Jaber. 2019. "An IoMT Cloud-Based Real Time Sleep Apnea Detection Scheme by Using the SpO2 Estimation Supported by Heart Rate Variability." *Future Generation Computer Systems* 98 (September): 69–77. https://doi.org/10.1016/j.future.2018.12.001.

Hutter, Elizabeth J. 2012. "Proteus Digital Health Announces FDA Clearance of Ingestible Sensor." July 30, 2012. https://www.businesswire.com/news/home/20120730005440/en/Proteus-Digital-Health-Announces-FDA-Clearance-of-Ingestible-Sensor.

Islam, S. M. Riazul, Daehan Kwak, MD. Humaun Kabir, Mahmud Hossain, and Kyung-Sup Kwak. 2015. "The Internet of Things for Health Care: A Comprehensive Survey." *IEEE Access* 3: 678–708. https://doi.org/10.1109/ACCESS.2015.2437951.

Jain, Shikha, Monika Nehra, Rajesh Kumar, Neeraj Dilbaghi, Tony Y. Hu, Sandeep Kumar, Ajeet Kaushik, and Chen-Zhong Li. 2021. "Internet of Medical Things (IoMT)-Integrated Biosensors for Point-of-Care Testing of Infectious Diseases." *Biosensors & Bioelectronics* 179 (May): 113074. https://doi.org/10.1016/j.bios.2021.113074.

Javaid, Mohd, and Ibrahim Haleem Khan. 2021. "Internet of Things (IoT) Enabled Healthcare Helps to Take the Challenges of COVID-19 Pandemic." *Journal of Oral Biology and Craniofacial Research* 11 (2): 209–14. https://doi.org/10.1016/j.jobcr.2021.01.015.

Joyia, Gulraiz J., Rao M. Liaqat, Aftab Farooq, and Saad Rehman. 2017. "Internet of Medical Things (IOMT): Applications, Benefits and Future Challenges in Healthcare Domain." Journal of Communications. https://doi.org/10.12720/JCM.12.4.240-247.

Kaur, Harleen, Mohd. Atif, and Ritu Chauhan. 2020. "An Internet of Healthcare Things (IoHT)-Based Healthcare Monitoring System." In *Advances in Intelligent Computing and Communication*, edited by Mihir Narayan Mohanty and Swagatam Das, 475–82. Lecture Notes in Networks and Systems. Singapore: Springer. https://doi.org/10.1007/978-981-15-2774-6_56.

Kelly, Jaimon T., Katrina L. Campbell, Enying Gong, and Paul Scuffham. 2020. "The Internet of Things: Impact and Implications for Health Care Delivery." *Journal of Medical Internet Research* 22 (11): e20135. https://doi.org/10.2196/20135.

Khamparia, Aditya, Prakash Kumar Singh, Poonam Rani, Debabrata Samanta, Ashish Khanna, and Bharat Bhushan. n.d. "An Internet of Health Things-Driven Deep Learning Framework for Detection and Classification of Skin Cancer Using Transfer Learning." *Transactions on Emerging Telecommunications Technologies* 32: e3963. https://doi.org/10.1002/ett.3963.

Khanna, Abhishek, and Sanmeet Kaur. 2020. "Internet of Things (IoT), Applications and Challenges: A Comprehensive Review." *Wireless Personal Communications* 114(2): 1687–1762. https://doi.org/10.1007/s11277-020-07446-4.

Lantronix. 2016. "Lantronix Enables IoT Connectivity for Life-Saving Defibrillators." December 13, 2016. https://www.lantronix.com/resources/application-spotlights/medical-device-networking-solutions/.

Madanian, Samaneh, and Dave Parry. 2019. "IoT, Cloud Computing and Big Data: Integrated Framework for Healthcare in Disasters." *Studies in Health Technology and Informatics* 264 (August): 998–1002. https://doi.org/10.3233/SHTI190374.

Maras, Elliot. 2015. "RFID: A Tool For Tracking Products, Assets and More." Food Logistics. December 17, 2015. https://www.foodlogistics.com/software-technology/article/12141721/rfid-a-tool-for-tracking-products-assets-and-more.

Michalakis, Konstantinos, and George Caridakis. 2017. "IoT Contextual Factors on Healthcare." *Advances in Experimental Medicine and Biology* 989: 189–200. https://doi.org/10.1007/978-3-319-57348-9_16.

Miorandi, Daniele, Sabrina Sicari, Francesco De Pellegrini, and Imrich Chlamtac. 2012. "Internet of Things: Vision, Applications and Research Challenges." *Ad Hoc Networks* 10(7): 1497–1516. https://doi.org/10.1016/j.adhoc.2012.02.016.

Mutlag, Ammar Awad, Mohd Khanapi Abd Ghani, N. Arunkumar, Mazin Abed Mohammed, and Othman Mohd. 2019. "Enabling Technologies for Fog Computing in Healthcare IoT Systems." *Future Generation Computer Systems* 90 (January): 62–78. https://doi.org/10.1016/j.future.2018.07.049.

Nasajpour, Mohammad, Seyedamin Pouriyeh, Reza M. Parizi, Mohsen Dorodchi, Maria Valero, and Hamid R. Arabnia. 2020. "Internet of Things for Current COVID-19 and Future Pandemics: An Exploratory Study." *Journal of Healthcare Informatics Research* 4(4): 325–64. https://doi.org/10.1007/s41666-020-00080-6.

Nazir, Shah, Yasir Ali, Naeem Ullah, and Iván García-Magariño. 2019. "Internet of Things for Healthcare Using Effects of Mobile Computing: A Systematic Literature Review." *Wireless Communications and Mobile Computing* 2019 (November): e5931315. https://doi.org/10.1155/2019/5931315.

Qi, Jun, Po Yang, Geyong Min, Oliver Amft, Feng Dong, and Lida Xu. 2017. "Advanced Internet of Things for Personalised Healthcare Systems: A Survey." *Pervasive and Mobile Computing* 41 (October): 132–49. https://doi.org/10.1016/j.pmcj.2017.06.018.

Rajasekaran, Manikandan, Abdulsalam Yassine, M. Shamim Hossain, Mohammed F. Alhamid, and Mohsen Guizani. 2019. "Autonomous Monitoring in Healthcare Environment: Reward-Based Energy Charging Mechanism for IoMT Wireless Sensing Nodes." *Future Generation Computer Systems* 98 (September): 565–76. https://doi.org/10.1016/j.future.2019.01.021.

S. Rubí, Jesús N., and Paulo R. L. Gondim. 2019. "IoMT Platform for Pervasive Healthcare Data Aggregation, Processing, and Sharing Based on OneM2M and OpenEHR." *Sensors* 19(19): 4283. https://doi.org/10.3390/s19194283.

Sadoughi, Farahnaz, Ali Behmanesh, and Nasrin Sayfouri. 2020. "Internet of Things in Medicine: A Systematic Mapping Study." *Journal of Biomedical Informatics* 103 (March): 103383. https://doi.org/10.1016/j.jbi.2020.103383.

Sharma, Nikhil, Ila Kaushik, Bharat Bhushan, Siddharth Gautam, and Aditya Khamparia. 2020. "Applicability of WSN and Biometric Models in the Field of Healthcare." 304–29. https://doi.org/10.4018/978-1-7998-5068-7.ch016.

Siddiqui, Mohd Faizan. 2021. "IoMT Potential Impact in COVID-19: Combating a Pandemic with Innovation." In *Computational Intelligence Methods in COVID-19: Surveillance, Prevention, Prediction and Diagnosis*, edited by Khalid Raza, 349–61. Studies in Computational Intelligence. Singapore: Springer. https://doi.org/10.1007/978-981-15-8534-0_18.

SmartMat. 2014. "SmartMat Info." 2014. http://iotlineup.com/device/smartmat.

Telekom Healthcare Solutions. 2019. "Internet of Medical Things for Hospitals." 2019. https://www.telekom-healthcare.com/en/e-health/internet-of-medical-things.

Toma, Cristian, Andrei Alexandru, Marius Popa, and Alin Zamfiroiu. 2019. "IoT Solution for Smart Cities' Pollution Monitoring and the Security Challenges." *Sensors (Basel, Switzerland)* 19 (15): 3401. https://doi.org/10.3390/s19153401.

TrackR. 2016. "TrackR Bravo Info." 2016. http://iotlineup.com/device/trackr_bravo.

Vishnu, S., S.R. Jino Ramson, and R. Jegan. 2020. "Internet of Medical Things (IoMT) – An Overview." In *2020 5th International Conference on Devices, Circuits and Systems (ICDCS)*, 101–4. https://doi.org/10.1109/ICDCS48716.2020.243558.

12 Secured Multimedia and IoT in Healthcare Computing Paradigms

Hemant Kumar Saini and Himanshu Swarnakar
Government Engineering College, Banswara, India

Dr Kusumlata Jain
Manipal University Jaipur, Jaipur, India

CONTENTS

DOI: 10.1201/9781003219620-12

12.1 INTRODUCTION

In Ice and Sullivan's idea, 20–30 billion IoT and clinical contraptions are depended on to would a piece by human association characteristic framework by 2020. This a lot of related gadgets in the sufficiently vulnerable remedial organizations part is a making pressure for human organizations suppliers, clinical gadget makers, the association, and people generally running free. The climb of internet of medical things (IoMT) as a piece of regular helpful organizations practice is disturbing resulting to neglect to acknowledge central security affirmations [1]. Human organizations will expand its assault surface as opposed to shrink its assault surface. Deliberate endeavors to address such issues have been generally rotated around pre-advertise contraptions, yet the confirmation of the presence and utilization of enormous armadas of heritage clinical gadgets has accomplished new post-highlight rules and rules, including reviews. Neither substitution nor re-confirmation of clinical contraptions before long being used is an enchanting choice taking into account the expense and flightiness included. Inventive ways to deal with oversee structure security and get ready for the confirmation of the IoMT are needed to guarantee quality which contemplates development, predictable results and the cash related ampleness of restorative organizations affiliations [2].

Since the rising of clinical gadgets has changed the substance of human organizations. Perhaps than a standard visit to an emergency community, checking our own thriving is speedily available at this point. Despite the way that this cutoff change is a lot of fortifying we need to return a stage to survey the security of such gadgets. Security and confirmation of these gadgets is in danger. The results are essential concerning security of clinical contraptions comparable number of patient's lives rely on genuine working of these gadgets [3]. Henceforth, security in remedial organizations is of most breaking point significance. Wearable Internet of clinical stuff is sharp gadgets patient's improved by wearing and its personality of success. Such contraptions are follow essentially all–significant turn of events, hotness, sugar, rest, beat with out and out extra. Such contraptions could accessible beginning skull to toe in different plans, for example, sharp handband, wristwatch, shows, support, additional items, piece. Wearable designs join receiving wires, memory, sun filled work area territory and movement. They help in information assortment, show, and far away spread of the in gathering amassed. Such gadgets can screen the flourishing indications of the detestable/clients and fire straightforwardly to the experts to slice down an individual course of action [4].

This innovation, named as internet of things (IoT), "gives an integration approach to this load of actual items that contain implanted advancements to be coherently associated and empowers them to convey and detect or interact with the actual world, and furthermore among themselves". The IoT is an idea that is mirrors an "associated set of anybody, anything, anytime, anyplace, any help, and any organization".

One of the most alluring applications fields for IoT is the Healthcare, giving to us the chance of numerous clinical applications, for example, distant wellbeing monitoring, fitness programs, ongoing sicknesses, and old consideration [5].

The rest of the chapter is organized as follows: in Sections 12.2 and 12.3 a brief description of various IoT devices threat and challenges with the different attacks and their solutions to counter. Section 12.4 includes the various trainings which is mandatory for withstand the security in IoT healthcare. Latest medical arenas that arises IoT and Healthcare requirement discussed in Section 12.5. This innovation brings a bunch of services and applications that we describe in Section 12.6. Section 12.7 described the methodologies to examine their products security which is more necessary than its use in daily life.

12.2 CHALLENGES AND THREATS

The web is a huge in general system that can be utilized to chat with one another. It very well may be utilized to send messages, messages and offer information. The information is sent from customer contraptions like PC, PC or cells and it send the information extra. The web is made three way on-screen characters individuals, customer gadget, and a specialist. Regardless, the IoT is novel added to web. An IoT is any vigilant thing to join actuators. The sensor accumulates the data and sends it to the cloud for extra assessment. The data destroyed can be used to pick. An "amazing contraption" is any mechanical device that has the decision to get its choices. The "radio wire" is a little gear that perceives the message in a "selector" is an extra little hardware which analyzes to the recognized information [6]. Occurrence of that gadgets join advanced PDAs, insightful TV, marvelous obliged air structure, sharp vehicle, and so on Web of things is an arrangement of frameworks. All of the electronic contraptions are connected with one another laying out a framework and further these constructions will be connected with one another embellishment a more important system structure. There are four plan squares of IoT.

End contraptions/centers: this is a fundamental piece of the IoT. Those are dynamic seeing gadgets and selector assembles messages and organizing at the position level. Models sensor the hotness, perception at roadways, etc.

Entryways/neighborhood getting ready center centers: these accessory the end spots to the framework or cloud. A couple of segments basically trade the data to web that is amassed from the sensors, a few doorways correspondingly measure the data acceptably and thusly ahead the pertinent data to specialist for creation checks. It likewise gives the information at the completion place centers by taking from the fogs [7].

Transparency: As Internet Things is an arranged plan, figuring out is a fundamental work. Prodigies choose diverse strategies around IOT to interface the centers to the fragments and areas to the make indistinct. Duplex development suggests the correspondence streams in the midst of occupations [8, 9]. Data or the sign other than streams in bring down. Receptiveness has the choice to a far off or a resentful part. Model Bluetooth, Wi-Fi, ZigBee, etc. cloud-based mechanical assembly and cutoff: the cloud or cloud-based application is used to choose the gathered information;

separate it and a short period of time later make needs. Such mechanical assemblies in addition stock up the gathered data and can be sensibly open from any spot at whatever point [10, 11].

Inconveniences of IoMT: a few issues for web of gear are there. The fundamental specific test in it contraptions is gathering and communicating information. The inconveniences can in like way be equipment and programming issues. As per Internet Security Operations Center, best inconveniences of IoT intertwine care, space to yourself, principles, system, advance [12, 13]. Included opposition are insightfulness an environment, organization, rule, and involvedness.

12.2.1 Challenges of Wearable IoMT

- regardless of whether a device is lost, anyone can move to the information set aside close by
- wireless transmission of data
- data flanked by the social affairs can be caught by developers

There is general be near nothing and sensibly attentive, which works on it to consider fragile information taking everything, extent of Patient-made Health Data (PGHD) aggregated utilizing wearables is ~310 MB per individual reliably. For 100,000 patients, the all-out total will be ~31 TB reliably. Considering overwhelming utilization of wearables in amicable insurance, the extent of patient information will essentially make it [14, 15]. This examines about the security and affirmation shortcomings in wellbeing following wearable gadgets. "The need of wellbeing and security in web related gadgets gathers that unyieldingly sensitive data that happens in IoT correspondences ought to be requested, directed, and ensured with high need." The confirmation data from the client information is being sent by the web. An immense number of the wearable gadgets are with the improvement of mechanization. The course toward computerizing the information develops the arrangement of accommodation yet likewise keeps the information at genuine danger.

12.2.2 Various Short Pits in IoMT Devices

In [16], producers a few shortcomings in FitBit by figuring out the correspondence show, putting away subtleties, activity codes what's more alluded to about the security and affirmation stresses in sharing success information in the social affiliations. "By hoarding FitBite, a set-up of mechanical gatherings that maltreatment these stipulations to dispatch a sweeping degree of assaults close to Fitbit. Other than tuning in, implantation and renouncement of, several assaults can incite prizes and monetary benefits. FitLock was made, a lightweight hindrance structure that guarantees Fitbit while forcing just a little overhead."

The structure of shared receiving wire systems is in Figure 12.1. Flourishing radio wire gadgets like FitBit and the relating information of the client are addressed showed up and took an interest in social affiliations. The client's past person's name is anonym zed. The easygoing organizations can't be utilized unquestionably to share

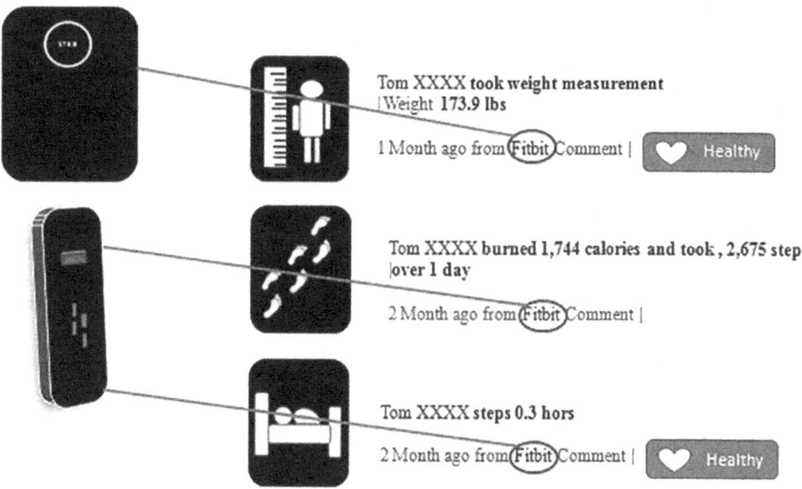

FIGURE 12.1 Shared antenna arrangement [4].

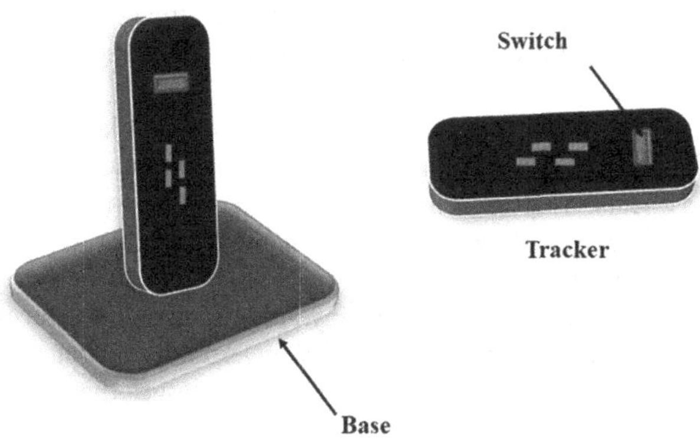

FIGURE 12.2 FitBit systems [4].

solitary subtleties like area, status, and so forth yet can in addition be utilized to share flourishing related information.

Figure 12.2 shows the areas of FitBit tracker. A FitBit tracker is utilized to screen bit by bit genuine action like the segment voyaged, steps got, steps climbed, calories ate up, and so forth It consolidate of trackers, the base, and a customer PC. Base is utilized to extend the tracker. Chaser has a control to is utilized to show the wellbeing information of the client. The client PC is utilized to recover the information as of the ally notwithstanding lay up the information [17].

Figure 12.3 shows the administration firewood of healthy bit. The featured content demonstrates that the apply qualifications are send in simple satisfied that has report id and secret phrase.

```
07/29/ 04:37:38 Reset Channel.
07/29/ 04:37:38 Starting session in pairing mode…
07/29/ 04:37:38 [CTX] CommunicationManager::ResetSession: setting context 00000000
07/29/ 04:37:38 [CTX] CommunicationManager::StartSession: setting context 00EF6D08
07/29/ 04:37:38 Processing request…
07/29/ 04:37:38 Connecting[89]:POST to http://client.fitbit.com:80/device/tracker/uploadData with data:p%5flc
07/29/ 04:37:39 Processing action 'http'..
07/29/ 04:37:39 Sending 8487 bytes of HTML to UI…
07/29/ 04:37:39 Processing request…
07/29/ 04:37:39 Waiting for minimum display time to elase[1000ms]…
07/29/ 04:37:40 Waiting for form input…
07/29/ 04:38:00 UI[\\.\pipe\Fitbitvallagenah]:F
07/29/ 04:38:00 Processing action 'form'…
07/29/ 04:38:00 Received from input:email=networkcrazy13@gmail.com&password=shashi 13&login=%3CSP/
07/29/ 04:38:00 Connecting[90]:POST to http://client.fitbit.com:80/device/tracker/pairing/singupHandler with dr
07/29/ 04:38:01 Processing action 'http'…
07/29/ 04:38:01 Sending 26882byte of HTML to UI…
07/29/ 04:38:01 Processing request…
07/29/ 04:38:01 Waiting for minimum display time to clapse[1000ms]…
```

FIGURE 12.3 FitBit service logs.

12.2.3 Layer-Based Architecture of IoMT

In [18], the paper clarifies about the state of oversight and security of social insurance information. The security and affirmation stresses in the automated success time and in addition breaks down the instruments and strategies that are available to help lessen the danger. It is clarified about the information at different levels across different stages as appeared in Figure 12.4.

It is a three-way tolerant information. The fundamental layer is precious focus where the patient, specialists, human organizations suppliers, prepared experts, escorts are the clients. The resulting covering is the correspondence channel wherever the information was ships off various stages through various shows. These shows can be Bluetooth, Wi-Fi, Broadband, and so on. The third sheet held association focus point that solidifies the cutoff any spot the patient's information is dealt with. Disregarding the way that the creators investigated about the security and affirmation stresses in front line social assurance, the issues that they examined are

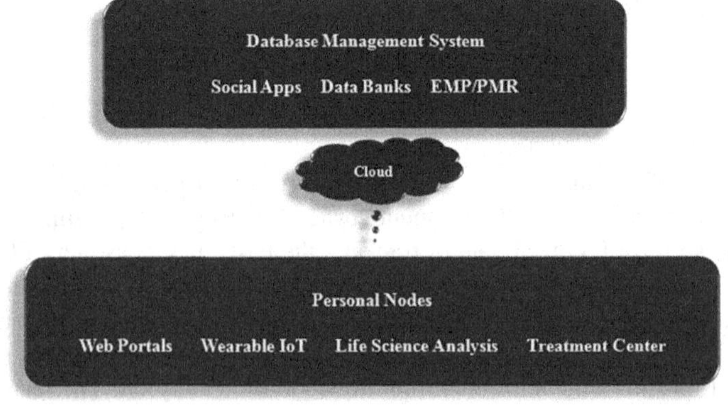

FIGURE 12.4 Translational research in medicine.

exceptionally sweeping to the IOT contraptions, Phishing assaults, correspondence starting with one side then onto the next SSL, and so forth and not communicate to wearable catch of clinical things [19].

12.3 VARIOUS SETTINGS IN IoMT SETUP

Also some setting focused assessments are shown to help in evaluations The FDA is regulating post-show clinical gadgets on the grounds that most gadgets as of now utilized in emergency offices were never proposed to plan outside a solitary clinical focus or even past an individual workstation. As necessities be, these clinical contraptions don't contain inserted security limits. For example, different more settled clinical gadgets run unsupported working frameworks (OSs, for example, Windows 95, 98, 2000—and can never again be fixed nearby. Other fundamental issue are coded or breakdown to apportion code word where not defiled likewise as a shortage of against infection or individual firewall limits. Additionally, some clinical methodology be obliged through intelligent delegates outside of the IT office's space before the veritable clinical instrument creator, while extra require an actual connection with a precise workplace (that doesn't have scrambling limits). Moves to such clinical contraptions that improve security mean an absurd and very time-consuming re-ensure measure. Moreover, the supreme proportion of weak clinical gadgets makes them absurd and restrictive to abrogate [20]. Accordingly, the conceivable reaction to working clinical contraptions has been inaction, paying little mind to the way that the consideration of these gadgets is a responsibility and a danger to clinical, operational, and cash-related segments of each human organizations office.

As exhibited by the AHA's rapid points of interest on US places 2019 story present an aggregate of 931,203 beds. In case, an average emergency community space has amidst 15–20 clinical contraptions which would derive to commonly 18.6 million clinical gadgets in United States clinical offices. Such figure evades any precious gadgets from legitimate and non-consistent staff, for example, tablets, workstations, and cells, which are connected with the emergency community's structure, becoming the full scale number of gadgets inside the clinical office's IT set up. Clinical gadgets join implantation siphons, insulin siphons, tireless screens, sifting frameworks (PET, CT, and so on), breathing apparatus and oxygenation machines, to give a couple of models. Each room participates in any event one imbuement siphon and different screens. Since imbuement siphons and screens are connected with a particular patient and clearly impact a patient's idea, when stood apart from other clinical gadgets, it is fundamental that such clinical gadgets are gotten. Insulin or blend siphons are clinical contraptions that location an upper security risk [21]. Not exclusively are the siphons related with a particular enduring amidst whole emergency community remain, yet insulin and implantation siphons can in like manner be seized to pass on a startling bit in contrast with supported (either lower, higher or no bit using any and all means). This may provoke patient passing or injury.

Indulgent screens might be gotten to through an assertion dodge assault and can be encouraged to give a phony caution, cripple alarms, or show inaccurate patient vitals. Mixed up thriving data could incite a mistaken assessment of the patient's burden, accomplishing anticipated passing or injury. Siphon and patient screen weakness can straightforwardly impact enthusiastic security, a clinical focus' standing and responsibility. While tending to structures denied to a positive anguish, they handle chose message. Imaging structures consolidate any clinical picture get, taking care of, or scattering gear; any instrument relaxing up from X-sections to electronic imaging and correspondence (DICOM) workstations; and picture recording and correspondence workers (PACS). Imaging understanding data is moved to different divisions inside a crisis office and is shipped off recommending specialists now and again engineered outside of the inpatient setting. Clinical contraptions and applications related with outside parts address a more genuine peril than applications only related inside. Imaging data could be sabotaged similarly changed, affecting a patient's prosperity history or recognizing erroneously thriving related decisions. Endeavored by structure, clinical imaging producers have relied on the inside and out slight Windows OS. Most imaging structures use the Microsoft Windows Embedded Standard Edition. Imaging dealers deftly fixes for this specific OS; in any case, the fix ought to be executed at the workplace or flourishing development to be useful. In a progressing KLAS analyze report, CIO and CISO respondents revealed that about 33% of all planned related clinical contraptions at their workplaces are "unpatchable." Unpatchable clinical devices are those that are not, by and by kept up by their creators since they were executed, i.e., not, presently in progress or an intelligently current agreement is open watching out. These unpatchable clinical contraptions address an immense capital endeavor for crisis office working conditions. The partition between the crisis office's typical thing lifecycle and the maker's customary thing life is vexing. Until this partition is settled, crisis workplaces are so far expected to guarantee these unpatchable clinical contraptions. Both patchable and unpatchable clinical contraptions are vulnerable against ambushes.

As shown by the really kept Open Source Cybersecurity Intelligence Network and Resource (OSCINR), there is a customary of 6.2 inadequacies per clinical device, and the FDA has starting late gave surveys for pacemakers and insulin guides with perceived security issues. Created by the FDA, Sensato-ISAO and H-ISAC, the OSCINR showed that overall 60% of clinical contraptions are around the finish of-life stage with no security fixes or empowers open. Whether or not unquestionably the whole of end-of-life things is 33% as basic necessity by the KLAS get some data about or the 60% overviewed by the OSCINR, it is as yet a puzzling complete going between 6.1 million and 11.1 million devices. Substantially more from an overall perspective, clinical contraptions were unequivocally associated by electronic aggressors starting late as April 2018. In the event that previous bleeding edge ambushes have not been based on supportive associations and have been so irritating to government managed retirement working conditions, the impact of a mechanized attack zeroing in on a human associations office could be shocking.

12.4 VULNERABILITY IN IoT

Unmistakable new security will be invaded considering explicit weaknesses in structures. One such was seen in May 12, 2017, where WannaCry ransomware referred to bitcoin as a result from those impacted in kind for the unencryption of their own PC reports. WannaCry sullied in excess of 300,000 PCs across 150 and a few relationship in less than 24 hours, costing incalculable dollars. The WannaCry cryptoworm zeroed in on Windows OS PCs without current security revives (i.e., unpatched) or those running unsupported assortments of Microsoft Windows, for instance, Windows XP (when executed really), Windows 7, and Windows Server 2003. An amount of three WannaCry groupings made. Following five days, new torments were dropped down with different counter attack checks executed by security experts going out of control. Without a doubt, even regardless the catch was not revolved around healing associations workplaces, the business' consistent shortcoming to help its IT structures with the most forward and backward advancement security refreshes and ordinary online assurance best practices made it particularly weak. One structure influenced was the UK National Health Service (NHS). In this, a couple of NHS territories were covered by their shortcoming to see influenced contraptions. Subsequently, these NHS crisis workplaces couldn't offer idea to their patients. Following one month, the Petya wiper malware, which endlessly hurts incidents' IT systems, affected two US flourishing developments among various setbacks.

Because of patient thriving and confirmation concerns accomplished by clinical contraption security shortcomings, government experts have paid respect to pre-advance gadgets. Settlement activity is a reformatory methodology to address postmarket clinical contraption mess up. Notwithstanding, recently, Congress has created oversight sheets of trustees and held hearings to perceive chances and consider extra real activities to address the making issue. Since 2016, the FDA has made approaches and has presented suggestions to help clinical gadget makers (MDMs) in directing on the web insurance rules. This joins the Postmarket Management of Cybersecurity in Medical Devices (Postmarket Cybersecurity Guidance) and the 2018 Medical Device Safety Action Plan. The Action Plan shows the need for required reviews when fitting.

Medical services sensors are anticipated that would oversee crucial private data and individual clinical benefits information. Moreover, such savvy sensors might be related with by and large data applications for their admittance to whenever and anyplace. To enable the full utilization of IoT in the clinical medical care area, it is important to perceive and inspect explicit characteristics of IoT including security necessities, weaknesses, and countermeasures, from the human medical care viewpoint. In various security dangers to medical care, the aggressors attempt to (i) take data of the patient (ii) prevent the administrations from getting the framework, and updates information.

There are two sorts of aggressors in IoT Health frameworks (i) Internal Attackers, and (ii) External Attackers which includes various attacks as seen in Figure 12.5. The discovery of the assailant is very troublesome because of its reality out from the framework. It quietly analyzes the framework tasks and afterward performs malignant exercises.

There are for the most part two sorts of assault in eHealth environment, for example, steering assault and area-based assault. Directing assaults incorporate switch assault, select and sending assault and replay assault. Area-based assault incorporates forswearing of administration assault, finger, and timing-based sneaking around assault and sensor assault.

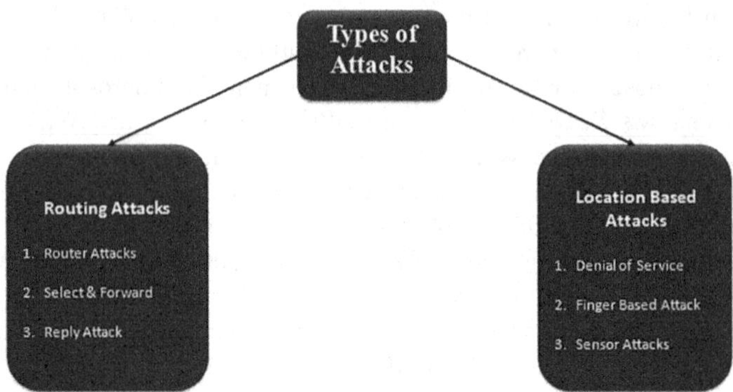

FIGURE 12.5 Types of attacks in smart healthcare.

In the directing assaults, generally the gatecrashers focuses on the course of information to send or drop information bundles [22]. In the area-based assault for the most part the gatecrashers assaults on the objective hub to keep the administrations from getting the framework.

12.4.1 Denial of Service Attack (DoS)

In denial of organization attack, the aggressor over-inconveniences the structure's data transmission with dark traffic, which makes resources far off for others, because various centers won't be talented to send their information in the wake of identifying the clamoring channel. In the denial of organization attack, the patient data can be gotten to without the check and agree of induction to data. The denial of organization attack also makes the system channel of data busy with the objective that no other information can pass to some other sensor in the association. The denial of organization attacks to cause transmission of data between center points partners with be lost or far off. This sort of attack sabotages system or clinical consideration organizations transparency, network handiness, and sensors responsibility. In the denial of organization attack, the assailant can temper the data of the patient, mislead the authorities about the patient's information, can send counterfeit information of a patient, and can add false information of a patient and an adversary replays existing messages to think twice about freshness.

12.4.2 Fingerprint and Timing-Based Snooping (FATS)

The distant association can't recognize the new security risks that get information from investigating the data transmission sensor to sensor, sensor to private region when the transmission is mixed. This genuine layer risk essentially requires broadcast time and special finger impression of each message, where a finger impression is a lot of features of an RF waveform that are exceptional to a specific transmitter. To direct this attack, a gatecrasher listens unobtrusively the whole of sensors' data transmission with timestamps and fingerprints. In the wake of seeing this, the intruder uses the exceptional imprint to affix with each message to a specific transmitter and utilizations different events of deduction for every sensor region. It is right when the gatecrasher can get this information and can upset the clinical issue.

12.4.3 Router Attack

The coordinating of data is huge for clinical consideration based systems in the light of the way that it licenses far off information movement and it engages network versatility in tremendous centers. In any case, it is coordinating two or three issues, essentially taking into account the open thought of far-off systems. In this attack, the attacker attacks on the information that is controlling between the sensors in a distant sensor association. This is relied upon to in a clinical benefits structures subject to far off the most central need is the gotten movement of patient information at the not exactly helpful end that can be trained professional and crisis facility. In this assault, controlling of focal and essential information showing human organizations status of the patients is considered, and there are barely any applications that use multi-trust planning which can be portrayed in Figure 12.6. Multi-trust planning is essential in expanding the combination space of the system thusly offering Àexibility to the disservice of multifaceted design.

12.4.4 Select Forwarding (SF)

In this danger, the interloper gains admittance to single or different sensors to relieve this assault. Thus, this is known as local area arranged explicit sending. In this assault, when the gatecrasher gains admittance to any sensor, it drops the information parcels and furthermore send these bundles to neighbor sensors to make it doubt. This assault influences the framework severely, if the sensor is near base station. Thus, due to parcel drop from the select forwarding (SF) assault, it will in general be difficult to perceive the reason for bundle drop. This assault can be destructive for any persistent or keen clinical wellbeing framework by fragmented information coming to at the beneficiary end. This assault can be more destructive than no information. This is considering the way that in clinical wellbeing terms one may not see the

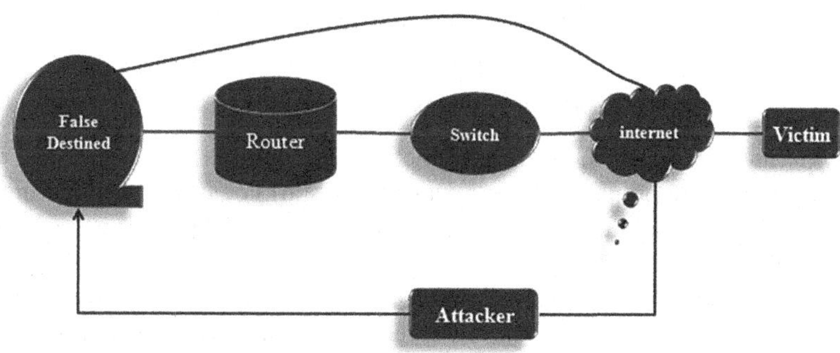

FIGURE 12.6 Routing attacks in e healthcare.

whole picture without complete data. The changed patient information may ship off the collector end. This could bring about some unacceptable treatment of the patient.

12.4.5 Sensor Attack

Because of the inadvertent disappointment of sensors in the remote organization and the pernicious exercises performed by the outside assailants, sensor as often as possible left or join the organization. In the remote organization, the absence of force sensor may pass on. For this situation, the keen aggressor can undoubtedly supplant the sensor with the genuine one and enter in the organization, can perform pernicious exercises without any problem. Accordingly, the patient information on the off chance that not well places at various sensor, the aggressor can change the information to the extent gatecrasher needs. Likewise, bogus information can be embedded or filled in as lawful because of absence of confirmation blueprint.

12.4.6 Replay Attack

A replay assault can alleviate when a gatecrasher gets unapproved admittance to framework. The interloper looks at the exercises on framework and afterward send message to collector when the transmitter stops to send information by then it begin to convey message as first sender. The primary target of the interloper in this assault is to fabricate trust in network. The assailant makes an impression on beneficiary that is generally use in the entrance cycle. A replay attack is depicted as a burst of safety in which some information is put away with no approval and thereafter retransmitted to the beneficiary remembering the ultimate objective to trap the last into unapproved, for instance, bogus acknowledgment or check or a copy exchange.

Based on the above types of internal and external attackers Tables 12.1 and 12.2 mentioned the various attacks based on different principles of security as well as the threats with their counter solutions in brief discussed in Ref. [23].

12.5 VARIOUS IoMT SECURITY MEASURES

Defeating the rising IoMT security issues and difficulties is a difficult undertaking. Nonetheless, relieving them can be accomplished by carrying out various safety efforts, some being specialized and others non-specialized measures.

12.5.1 Non-Technical Security Measures

It incorporates preparing the staff and shielding the patients' private clinical well-being records where preparing the clinical and IT staff could be cultivated in three distinct manners: bringing issues to light, leading specialized preparing, and raising the schooling level.

For that awareness must be raised among the clinical representatives and staff, for the most part the IT division to know and recognize a happening assault from typical organization conduct. Nonetheless, this isn't sufficient, as there is a more serious requirement for characterizing what is a danger, hazard, and a weakness. For such cases,

TABLE 12.1
Various Location and Routing Based Attacks

Attack	Effectiveness	Security Requirements	Approach
Denial of service	Disable system services	Early intrusion detection	Distributed attack uses and deny the system' service
Routing attack	Change route information	Route monitoring	Change the routing table information and drop packet
Sensor attack	Data modification	Node failure and replacement detection	Attacker finds a sensor with low power or failure. Then replace the sensor and enter in network
Replay attack	Unauthorized access, duplicate transmission	Secure authorization	Evaluate the system activities then send message to receiver after completion of transmitter message
Select forwarding attack	Drop data packets	Sensor detection	Intruder drops data packets at desire location by behave as destination end

it is needed to evaluate the dangers in advanced take safety efforts to manage them. But for this also rising the technical training to the clinical staff and workers of the IT division, just after the instructing stage so that they can comply with. Such compliancy are enabled and divided into seven phases which are as follows:

1. Identification Phase where the IT is fit for distinguishing a dubious conduct from an unusual conduct.
2. Confirmation Phase that depends on the capacity to affirm that an assault is happening.
3. Classification Phase that depends on the capacity to recognize the kind of the happening assault.
4. Reaction or Responsive Phase depends on the capacity of the Computer Emergency Response Team (CERT) to rapidly respond to a given assault utilizing the right security safeguarding efforts and keep an assault from heightening.
5. Containment Phase depends on containing the assault occurrence and beating it.
6. Investigation Phase is the execution of criminological confirmations where an examination cycle happens to recognize the reason for the assault, its effect and harm.
7. Enhancement Phase depends on gaining from the exercises of past assaults.

We can't stop here with all the above measures, and the Education Level should also be raised by teaching digital protection and IT staff the important procedures to arrange each assault and what it targets (secrecy, uprightness, accessibility, or

TABLE 12.2
Threats and Countermeasures in IoHT

	Attacks Types	Countermeasure	Reasons	Threats
Data confidentiality attack	Eavesdropping	Encryption	• Broadcast nature of messages via wireless channels • Unencrypted communication channel	Intercept the data driven
	Data interception	Encryption	Nonsecure channels • Open wireless communications	Capture the information and reuse
	Packet capturing	Encryption	• Open wireless communications • Nonsecure channels • Lack of encryption	Use of keys
	Wiretapping	• Secure communications • Closed communications	• Open wireless communication • Nonsecure channels	Wrong connections
	Dumpster diving	• Enhanced employee training • Paperless process	• Lack of employee training • Lack of awareness	Unknowingly sends the data to anyone
	Social engineering	Training staff against baiting/pretexting	Poor training of employees	May affect the confidentiality and privacy
Social engineering attack	Reverse social engineering	Training staff against strangers' questions	No identification and verification processes	Depends on the asked questions, primarily targets confidentiality and privacy. In addition, to affecting authentication and availability
	Error debug	Limit appearing information	Different error questions giving additional information	May affect (data/system's) confidentiality and privacy
	Traffic analysis	VPNs and proxies • Non-linkability • Pseudonyms	• Source and destination information are not encrypted • Lack of secure channels • Weak encryption algorithm	Firewalls break

(Continued)

TABLE 12.2 *(Continued)*
Threats and Countermeasures in IoHT

	Attacks Types	Countermeasure	Reasons	Threats
Privacy attack	Identity/location tracking	• Anonymity • Non-linkability • Pseudonyms	• Lack of secure channels • Location and identification parameters are not encrypted	Locations threaten
Message integrity and authentication attack	• Message tampering alteration • Malicious data injection • Malicious script injection • Cloning and spoofing	• Keyed hash function (HMAC); • Message authentication algorithms	No data integrity and source authentication protection scheme	Prescriptions of hospitals changes
Availability attack	Jamming	Frequency Hooping, direct sequence spread spectrum, beam-forming	Targets Access Points or wireless IoMT devices	Devices will be jammed to control over
	Distributed denial of service (DDOS)	DDOS detection solutions. Increase the security levels of devices to avoid becoming bots	Exploiting devices turning them into bots	Overburden the devices to failures
	De-authentication	Firewalls, Intrusion Detection Systems, Encryption	Captures a handshake to Launch DoS or Password Cracking Attack	Impose itself as legitimate
	Flood	Timestamps, certificate authority, IDS	Overwhelms and Exhausts IoMT's Resources through False Information Injection	Wrong monitoring occurred
	Delay	Firewalls, Timestamps, IDS	Overwhelms & Prevent or Severely Delays any Transceiving of Medical Information	Delay messages which causes the patient may be no more

(Continued)

TABLE 12.2 *(Continued)*
Threats and Countermeasures in IoHT

	Attacks Types	Countermeasure	Reasons	Threats
Authentication attack	Man-in-the-middle	Multi-Factor authentication scheme	Poor authentication scheme (one factor)	Poor authentication scheme (one factor)
	Masquerading	Multi-factor authentication scheme P	Poor authentication scheme (one factor)	May affect data's confidentiality
	Cracking	Multi-factor authentication scheme	Poor authentication scheme (one factor)	May affect the data's confidentiality and integrity
	Replay	• Timestamp or a new random number for each session connection • Multi-factor authentication scheme	Weakness in the authentication protocol	May affect system's availability
	Dictionary	• Strong password • sufficient size of secret key	Weak password and one authentication factor	May affect the data's confidentiality and integrity
	Brute force	• Strong and long password • sufficient size of secret key • Multi-factor authentication scheme	• Weak password • and one authentication factor	May affect data's confidentiality and integrity
	Rainbow table	Long salt passwords	• Weak usernames/password • Short salt passwords	May affect data's confidentiality and integrity
	Birthday	Secure hash algorithm	Weak hashing	May affect data
	Session hijacking	• Encryption • Sniffing filters	• Lack of/poor encryption • Nonsecure channels	May affect data's confidentiality, integrity, and availability

potentially verification). Since aggressors are additionally isolated into insiders or pariahs. To restrict the chance of insider assaults, the right approval and validation methods ought to be applied, alongside the best Intrusion Detection Systems (IDS) to identify any assault dependent on one or the other mark, peculiarity, or conduct.

12.5.2 Specialized Security Measures

There are examination methods that target guaranteeing IoMT information and frameworks security. (1) Multi-factor identification and verification: in request to forestall any conceivable unapproved admittance to IoMT frameworks, guarantee a solid recognizable proof, and check component. The best arrangement is to depend on biometric frameworks. There is additionally the requirement for an information base to store the biometric formats securely and safely for later. However, accomplishing distinguishing proof and check requires a few biometric procedures, which can be isolated into physical and conduct biometric strategies.

12.5.3 Actual Biometric Techniques

Secure physical biometric methods can be embraced and used to shield and keep up with patients' clinical protection without being inclined to any insider danger. This incorporates facial acknowledgment, retina output, or iris examine.

- Facial recognition: Facial acknowledgment figured out how to demonstrate a high check rate. Thus, it was used to perceive an individual's facial design, utilizing a specific advanced camcorder that recognizes and measures the face's construction. This likewise incorporates the distance between the triangle of eyes, nose, and mouth. Consequently, it can confirm real clients from non-real clients by contrasting an examined face and the approved appearances enrolled in the information base.
- Retina scan: A retinal acknowledgment examination depends on dissecting the vein area situated behind the natural eye. It ends up being a precise and secure confirmation strategy.
- Iris scan ends up being fundamental for both recognizable proof and confirmation purposes, because of its capacity to create exact and exact estimations. Iris filter works by dissecting and examining the hued tissue around a particular eye student to check in the event that it coordinates with the put away information to either concede access or not.

12.5.4 Conduct Biometric Technique

A safe social biometric procedure that can be utilized for both distinguishing proof and confirmation stages is the hand calculation. Such biometric frameworks depend close by estimations, including palm size, hand shape, and finger measurements. Then, at that point, it is contrasted with the arrangement of put away information in a data set to check clients. On the off chance that there is match, a given staff will be conceded admittance. If not, access will be denied. In any case, such frameworks are simply restricted to one-to-one frameworks. Truth be told, current frameworks are equipped for separating between a living hand and a dead hand. This keeps enemies from attempting to trick the framework and gain any illicit access [24].

12.5.5 Multifaceted Authentication Techniques

Different studies zeroed in on patients' protection infringement, with the dependence on encryption, verification and access control components as countermeasures. Confirmation is named the principal line of guard that verifies the source and objective the same. Indeed, verification can be a solitary factor confirmation that just depends on a secret phrase as the lone safety effort, which isn't best. It can likewise be a two-factor validation that depends on another safety effort beside the secret key to get to a given framework. At last, it tends to be a multifaceted validation where a third security component is needed to get to a framework. Along these lines, confirmation assumes a vital part in giving security to the available assets on a given organization. Verification can be either incorporated where two hubs confirm themselves through a confided in outsider, or it very well may be circulated where two hubs utilize a pre-characterized secret key to verify one another, without depending on a confided in outsider.

Accessibility Techniques: The significance of keeping up with accessibility against any conceivable disturbance or/and interference of signs is an unquestionable requirement. In any case, keeping up with the worker's accessibility requires the execution of computational gadgets that go about as reinforcement gadgets, along a checked reinforcement and Emergency Response Plans (ERP) if there should arise an occurrence of any abrupt framework disappointment.

12.5.6 Lightweight Intrusion Detection Systems

IoMT gadgets are inclined to various sorts of safety dangers and difficulties. To ensure IoMT frameworks against interlopers, the exercises of IoMT gadgets should be checked and investigated. Commonly, an IDS is the main line of safeguard toward identifying assaults. The various IDS types that can be applied inside IoMT frameworks are Host-based IDS (HIDS), and Network-based IDS (NIDS). While HIDS is connected to a given IoMT gadget to screen any conceivable malevolent action, NIDS screens the organization traffic of a few IoMT gadgets toward recognizing any vindictive movement.

IoMT frameworks and organizations ought to be secured by carrying out IDS to recognize unusual exercises as ahead of schedule as could be expected and to start the right activities to stop any occurrence. An IDS can be either abnormality based, signature-based, or detail based. Mark-based and determination-based discovery techniques require low overhead contrasted with the oddity based one. Sadly, because of the restricted registering power and the high number of interconnected gadgets, a customary inconsistency based IDS isn't proficient in the IoMT case. Irregularity-based identification is the most proficient in recognizing zero-day assaults, which is beyond the realm of imagination through signature based or detail based discovery strategies. Fostering a lightweight inconsistency based IDS is fundamental for the discovery of obscure assaults inside the IoMT setting. Such lightweight methods will be utilized to settle on brief choices in an asset obliged environment, just like the case in IoMT organizations. Without a proficient oddity IDS, IoMT gadgets can be compromised prompting extraordinary impacts particularly for patients [8]. This raised a genuine security worry about current IoMT arrangements by and large, and the requirement for a strong and lightweight IDS. Exploration and modern networks are as yet confronting difficulties in planning a solid and proficient IDS for IoT frameworks since a lot of information should be dealt with in a continuous way. Lightweight and cross breed agreeable IDS with half and half position and crossover discovery procedures are competitor arrangements that can make IoT networks tough against different sorts of assaults including zero-day assaults.

12.6 NEW PERSPECTIVE OF IoT IN MEDICAL USE

Web of medical things (IoMT) pay between unique effects to different people inside a human administrations affiliation or over the social protection environment to acquire aggregate and separate data in order to accumulate IoT critical encounters. The more common use cases in friendly protection incorporate partner people, clients, clinicians, and parental figures. Therapeutic administrations affiliations have guided many related prosperity adventures essentially zeroing in on purchaser responsibility. The ability to relate clients and patients and impact their lead will ask them to make more helpful decisions, which in this manner will incite better outcomes and lower human administrations costs. Seeing to purchasers fundamental signs and development and in this manner thinking of them as liable for therapeutic administrations decisions soul help out additional drive consistence [25]. Preposterous, in participation, is a creating focus on enlightening the prosperity of the general population to control restorative administrations costs. A more grounded revolution around buyer responsibility and imaginative approaches to manage facilitate IoT based health care into new thought movement models is engaging the gathering of related prosperity advancement.

Past attempts to ensure remedial organizations security structures zeroed in on email pollutions and site phishing assaults as in Figure 12.7. Making weight from the current post-highlight administrative condition, reviews, and tremendous settlements

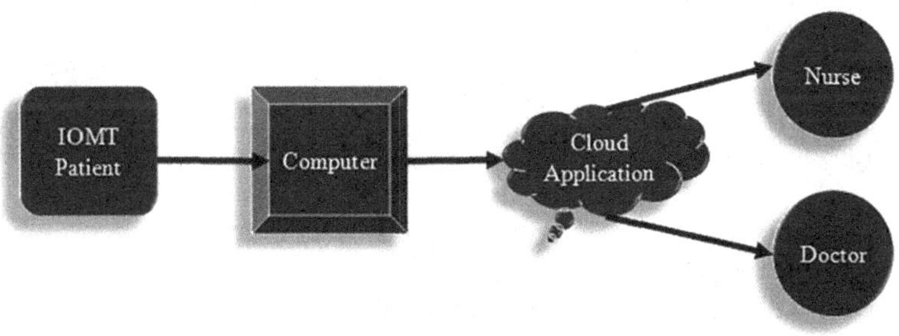

FIGURE 12.7 IoMT architecture.

obliged for HIPAA infringement present more veritable dangers than at later. These dangers need support with a customized security arrange approach. Human organizations structures should pick between restricted choices yet to acknowledge innovative system security to diminish risk from virtual prosperity shortcomings.

12.6.1 Steps to Withstand Security Dangers

To coordinate the current web security dangers, it is proposed to assess the pitfalls with the IoMT-controlled correspondence down to "who" and "how." Also with the scaling of sensible and even minded system security limits, the existing resources utilized to cover such dangers.

IoMT is applied in different fields of medical services, including care for pediatric and elderly patients, automated diagnosis of cancer patients such as lung nodule detection, breast malignancy discovery, the oversight of persistent illnesses, and the administration of private wellbeing and wellness, among others. Various IoMT entangled devices to withstand, including: (a) sensing the glucose level to deal with the diabetes patients. (b) Monitoring the Electrocardiogram (ECG) to quantify the pulse and rhythm. (c) Blood pressure checking for pressure delicate patients. (d) Body temperature observing for viral fever or fever from cold. (e) Oxygen immersion checking all such can be detected and controlled by on board design model as seen in Figure 12.8.

12.6.2 Various Sensors of IoMT

Grouped condition uses of IoMT include: (a) Rehabilitation management, (b) Medication management, (c) Wheelchair.

These incorporate the advancement of a wellbeing observing framework like I-NXGeVita that uses IoT and AI calculations. The ECG sensor recorded the patient's (pulse signals). These signs are put away in a cloud system synchronized with the neighborhood worker. The association between the patient's cloud and the cloud is affirmed by IoMT gadgets like Arduino Uno and Raspberry Pi3.

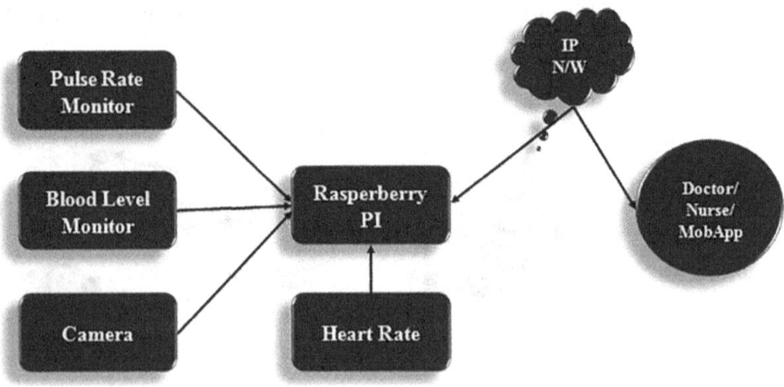

FIGURE 12.8 IoMT design.

12.6.2.1 DBN Sensors

The DBN calculation works with this information to recognize ordinary and abnormal pulses, which arrange and recognize unusual rhythms. In addition, i-NXGeVita illuminates specialists progressively about the most ideal clinical consideration and assistance. The essential objective of the patient framework is to decrease the quantity of delegations, to give constant consideration, to lessen mortality, to keep away from unexpected heart attacks, to diminish transportation costs, and to make clinical judgments. Likewise, the primary goal of doctors "professions is to screen and control patients" wellbeing [26].

12.6.2.2 ECG Sensors

Another ECG Sensor (Module AD8232) is utilized where ECG sensor gathers the patient's heartbeat (heart signals). These codes are stored inside the cloud, which synchronizes with the closest worker. It is the undertaking of the ECG sensors to check the patient's circulatory flows (and developments) by estimating normal and unusual signs. A normal ECG signal is introduced as a moderate period of PQRS-T, so, all things considered the P wave starts: for the QRS complex, it rises 2–3 points and closes at 0.06–0.12 s. The PR stretch reaches from 0.12 to 0.20 s. The whole public relations staff is pivotal. The QRS complex follows a transitory PR of 5 to 30 points, ranging from 0.06 to 0.10 s. The ST fragment reaches out from the S wave to the T wave. A regular T wave has a mass of up to 0.5 MV. The QT stretch ordinarily ranges from 0.36 to 0.44 s. This work depends on information from a few data sets, including the Massachusetts Institute of Technology-Beth Israel Hospital (MIT-BIH) Arrhythmia Database, MIT-BIH Malignant Ventricular Arrhythmia Database, and Arrhythmia. Specifications1.

One of equivalent to for Single Lead Heart Rate Monitor AD8232 is the least expensive card used to measure heartrate. This electric development can be written as straightforward ECG or ECG reading and execution, the ECG can be extremely amazing, and the AD8232 single-step pulse monitories introduced as a functioning speaker to assist you with moving from the intermediate times of PR and QT.

12.6.2.3 AD8232 Sensors

AD8232 is a hindrance to the arrangement of ECG markers and different evaluations of biological potential. It plans to stifle, fortify and target little bioenergy pointers, taking into account the genuine conditions brought about by the advancement of the terminal or a remote area. The AD8232 pulse screen intrudes on the association of nine frameworks, which are dependent on different contacts, wires or connectors. This board incorporates RA (right hand), LA (left hand) and RL (right foot) contacts, which can be utilized to meddle with individual execution. Likewise, there is a LED that reacts to the mood of the heartbeat. To utilize the heart screen, we will require a biomedical touchpad and a sensor connector. Applications such as Monitoring Welfare and Sports Signals, Versatile ECG, Remote wellbeing screen, play gear and bioelectrical signal assortment work S with various actual ancient rarities by means of modern wire-less organization and sensors, the connected IoT gadgets interface.

Various Cardio vascular issues including the heart illnesses, mortal human infection in coronary illness, oxygen level induction, body fatigue, feet get swollen and sluggishness all such measured and getting the concussion of obesity, blood pressure, elevated cholesterol, smoking, eating unfortunate eating routine that all will be handled taxingly with justifiably IoMT which plans an expectation system of coronary illness is finished utilizing arrangement procedures of AI.

At sometimes where the sudden strokes in the heart patients where the clinical team not reached timely and may cause to heart failures in such cases essential ailment requires steady control on the state of the patient's prosperity by directing IoMT system which can be permitted on patient to be distantly followed. The live health information of the patient is caught and verified through the identifiers and sent to the specialist. Likewise bosom malignant growth (BC) is a huge reason for high mortality rate among ladies IoMT and digital mammography find helpful to diagnose breast malignant growth viably first and foremost level itself. Different an intelligent IoMT based bosom malignant growth discovery and determination utilizing profound learning model.

12.7 METHODOLOGIES TO EXAMINE SECURITY IN IoMT

The fundamental objective of the evaluation is to direct reluctant clients while picking an ensured wearable web related clinical contraption (IoMT), connect more great rivalry among makers of IoMT gadgets and thusly liven up the security of IoMT gadgets. MCDM (Multiple Criteria Decision Making) incline toward was utilized in this appraisal. With assessment and confirmation of Internet related wearable clinical gadgets. Each brand name was clearly depicted as "what is it?", "for what reason is it basic?", and "how is it possible that it would be tremendous?". Under "how is it possible that it would be basic?", each property was depicted with a huge load of contemplations. These contemplations are a huge load of Questionnaire (with a Y or N answer) but this approach help the accessories of wearable IoMT gadget. Every assistant's correspondence with the contraption is intriguing. The various assistants for these gadgets meld patient, topic master, clinical focus, support, maker, security scientist, definitive prepared experts, affirmation. This has a 2-advance system.

These clinical gadgets are also calibrated by different observations on the quality and figuring the standardization where the client is also needed to permit the information with some key protection that the gadget catches on encryption procedures to encode their information for safety.

12.8 DEFENDING IoMT

The unnecessary strategy for IoT was orchestrated particularly for feeble IoT gadgets in the helpful organizations advertise. Such is actually that the IoMT condition need. Safeguard utilizes non impeded frameworks to confine who every contraption can talk by and pardon? every contraption can give, and accordingly screens every one trade amidst the gadget to guarantee that the gadget is working as shown by put profile. Furthermore, protector packages IoMT gadgets into safe blended territory which release up as of contraptions to the specialist farm to separate and defend gadgets data on or after the remainder of the business system. Safeguard is relied upon to be not hard to send and direct, and its UI licenses explicit and consistent agents to locally open and go legitimate and extra IoT contraptions effectively notwithstanding safely with no IT laborers intervention. Supported clients can additionally will arranged data on each IoMT gadget, checking region. such keeps clinical agent away from expecting to look at for gadgets. In particular, safeguard have the alternative to be passed on any system foundation with no the need for arrange redesigns or change—a fundamental of additional sellers' responsibilities—drawing in friendly assurance relationship to rapidly improve their security without agonizing over a full construction update. Preposterous structures is a pioneer in the irritated and removed LAN affirmation correspondences business focus notwithstanding dish up different undertakings, counting social security. Each human organizations affiliation ought to consider Extreme Networks a potential accomplice to guarantee about systems and manage IoMT contraptions.

12.8.1 Case Study of WannaCry Cryptoworm

Various new security will be entered considering explicit weaknesses in frameworks. One such was seen in May 12, 2017, where WannaCry ransomware referred to bitcoin as result from those impacted in kind for the unencryption of their own PC records. WannaCry dirtied in excess of 300,000 PCs across 150 and several relationship in less than 24 hours, costing inestimable dollars. The WannaCry cryptoworm zeroed in on Windows OS PCs without current security revives (i.e., unpatched) or those running unsupported assortments of Microsoft Windows, for instance, Windows XP (when executed truly), Windows 7, and Windows Server 2003. A total of three WannaCry arrangements made. Following five days, new afflictions were dropped down with different counter attack checks executed by security pros running free. Undoubtedly, even regardless the catch was not revolved around therapeutic associations working conditions, the business' consistent weakness to help its IT structures with the most back and forth advancement security updates and typical organization wellbeing best practices made it particularly feeble. One system influenced was the UK National Health Service (NHS). A few NHS people group were covered by their shortcoming to see affected contraptions. Thusly, these NHS crisis workplaces couldn't offer idea to their patients. Following one month, the Petya wiper malware, which unendingly harms misfortunes' IT frameworks, impacted two US thriving constructions among different losses. In "The Internet of Things for Health Care, a Comprehensive Survey" [27], study discretionary piece of IoT-base reestablished information where the thriving supplies and phones flourishing applications are mentioned with the flexibility and discussed standard IoT clinical difficulties [27].

It gives a fundamental understanding of prosperity subject in IoT and will not clear to clinical. There critical assaults a bit of the prosperity and guardian question are citated in web with supportive checking gadgets as clear happy entry information and clear satisfied HTTP data status. Such include the goliath principle part of the checking wearable contraptions have section in missing on or after and the keys of these gadgets are confirmation in logfiles as plaintext. The in gathering convey amidst a variety of district is send as direct happy and excusal security tortures are utilize like scrambling all the while as imparted the data. The seclusion and safe house subject in IoT medical applications for the immobilize customer a review gives the assessment of IoT manufacturing and suggest arrange and prosperity supplies of IoT contraptions. It yields IoT, things and address novel. So the essential interest basically on IoT gadgets for crippled customers. Security supplies were future with no evaluation methodology [28]. Also referred to information place of refuge and shield of inventive flourishing applications on IOS and android. "not in any way like sorts of mHealth applications assemble and bid essential, prickly, orchestrated quantifiable data, midpoint for the most part data insurance and prosperity of mHealth work show that 95.63% of utilizations are powerless agreed with some underhandedness from first to last data security and sanctuary perspective. The better bits of the information in applications are set missing in dark or the applications store record. The security and gatekeeper of the enduring all together is significantly in hazard dealt with in the legitimate flourishing assurance [29].

12.8.2 Wireless Sensor Networks for Healthcare Applications

In "Security and Privacy Issues in Wireless Sensor Networks for Healthcare Applications" far off sensor systems (WSN) in clinical military sales take a gander at how Wireless sensors meet in gathering from the patients. Government work environments will examine alliance and creator need to beat these issue to have a level usage. Figure 12.9 shows the common plans of eliminated sensor coordinate in man military applications. It show unavailable machine express through the web. Time taken by the accepting wires is sent removed to the doorway. The ways drive the data to the dull circumstance or really to the wine worker. Such grouping is placed in the document and can be taken to be everyone when by the web [30].

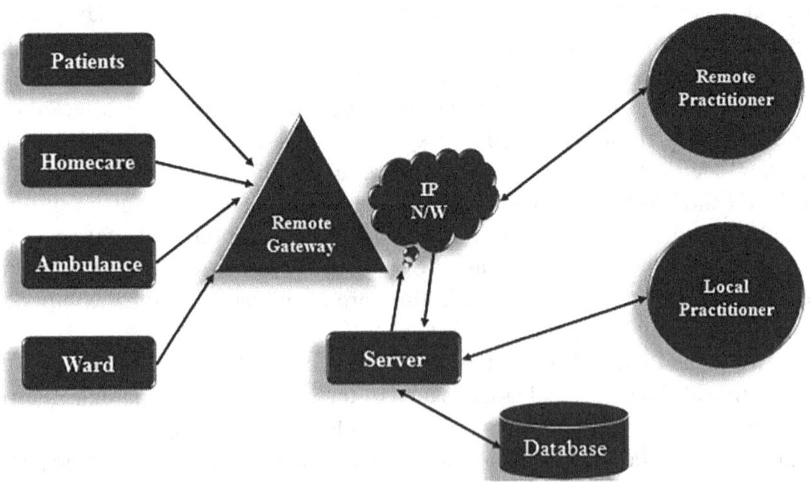

FIGURE 12.9 Typical architecture of WSN in healthcare application.

Source: Wearable internet of medical devices.

Here a portion of the instances of the wearable clinical gadgets is being explained which shows the monitoring framework devices of the body

- Heartbeat scrutinize, e.g., Tinke
- pressure height scrutinize, e.g., Pip device
- Gluco impel, e.g., Animas Vibe
- siesta scrutinize, e.g., Pebble time
- Blood pressure scrutinize, e.g., Withings
- neural scrutinize systems, e.g., Muse
- Hotness scrutinize, e.g., Temptraq
- pulsation scrutinize e.g. Wahoo device
- corporeal behavior scrutinize, e.g., FitBit
- sensitivity electrical scrutinize, e.g., EKG an electrocardiogram

12.9 CONCLUSION

The conversation about utilizing IoT wellbeing based framework makes the medical care framework more interesting by offering portable based types of assistance, figuring applications, and easy to understand interfaces by which a client can without much of a stretch record information about their wellbeing and can dissect them for various purposes.

The inclusion of m-wellbeing in IoT makes the medical care framework smaller by bringing different administrations, applications, outsider APIs, and versatile sensors. The IoHT can give more protection and security to the clinical contraptions that are helpless against any hacking assault. The portable processing gives start to finish security to get the huge measure of information that are sent to the IoT gadget to the worker or from the worker to IoT. The gigantic applications, for example, glucose level detecting, oxygen immersion checking framework, and internal heat level observing of IoT with the help of portable registering contribute toward acquiring upgrades the medical services framework. The accessible examination work in Internet of Healthcare Things is additionally getting well known with the idea of portable figuring technology. As with the upsides of IoT, a few difficulties likewise exist which should be figure out. A portion of the difficulties for IoT gadgets alongside the proposed arrangement are portrayed The explores are as yet going on to concentrate how IoT applications in medical services offers greater security in the medical services framework, acquires protection and security wellbeing IoT devices. The goals of the part are to contemplate the effects IoT in medical services environment or shrewd clinics considering our orderly writing survey convention.

REFERENCES

1. Frost & Sullivan. (2017), Internet of Medical Things, Forecast to 2021.
2. Healthcare IT News (2018), https://www.healthcareitnews.com/news/wannacry-petya-1-year-later-good-bad-and-ugly.
3. AHA (2019), Fast Facts on US Hospitals, www.aha.org/statistics/fast-facts-us-hospitals

4. Elizabeth O'Dowd (2017), HIT Infrastructure News, IoT Sensors Critical to Successful Health IT Infrastructure, https://hitinfrastructure.com/news/iot-sensors-critical-to-successful-health-it-infrastructure
5. Miliard, M. (2016), Healthcare IT News Security, Cybersecurity for Networked Medical Devices Pose Huge Risks to Patient Safety, https://www.healthcareitnews.com/news/cybersecurity-pro-networked-medical-devices-pose-huge-risks-patient-safety
6. Frost & Sullivan (2016), Cybersecurity Threats and Medical Device Connectivity, www.frost.com
7. Red Bank, N. J. (2018), AEHIS, News, FDA Partners with Sensato-ISAO and H-ISAC to create Open Source Cybersecurity Intelligent Network and Resource, https://aehis.org/fda-partners-with-sensato-isao-and-h-isac-to-create-open-source-cybersecurity-intelligence-network-and-resource/
8. Malwareless (2017), https://malwareless.com/wannacry-ransomware-massively-attacks-computer-systems-world/
9. Davis, J. (2018), Healthcare IT News, https://www.healthcareitnews.com/news/md-anderson-pay-43-million-settlement-ocr-hipaa-violations
10. Medical Device Cybersecurity Report (2018), Advancing Coordinated Vulnerability Disclosure. Medical Device Innovation Consortium. http://mdic.org/wpcontent/uploads/2018/10/MDIC-CybersecurityReport.pdf.
11. FDA (2016), Postmarket Management of Cybersecurity in Medical Devices, https://www.fda.gov/downloads/MedicalDevices/DeviceRegulationandGuidance/GuidanceDocuments/UCM482022.pdf
12. FDA (2018), Medical Device Safety Action Plan, https://www.fda.gov/downloads/AboutFDA/CentersOffices/OfficeofMedicalProductsandTobacco/CDRH/CDRHReports/UCM604690.pdf
13. TI.com. (2020), Challenges in the internet of things, http://www.ti.com/ww/en/internet_of_things/iot-challenges.html
14. Zhou, W., & Piramuthu, S. (2014), Security/privacy of wearable fitness tracking IoT devices. In *2014 9th Iberian Conference on Information Systems and Technologies (CISTI)* (pp. 1–5). doi: 10.1109/CISTI.2014.6877073
15. Altoros Offices. (2020), https://www.google.com/maps/d/viewer?mid=1pBwJc_lTuAZKIuEmmnCK57PYH18
16. Rahman, M., Carbunar, B., & Banik, M. (2016). Fit and vulnerable: Attacks and defenses for a health monitoring device. IEEE Transactions on Mobile Computing, 15(2), 447–459. doi: 10.1109/TMC.2015.2418774
17. Filkins, B. L., Kim, J. Y., Roberts, B., Armstrong, W., Miller, M. A., Hultner, M. L., Steinhubl, S. R. (2016). Privacy and security in the era of digital health: What should translational researchers know and do about it? American Journal of Translational Research, 8(3), 1560–1580.
18. Khamparia, A., Singh, P. K., Rani, P., Samanta, D., Khanna, A., & Bhushan, B. (2020). An internet of health things driven deep learning framework for detection and classification of skin cancer using transfer learning. Transactions on Emerging Telecommunications Technologies. doi: 10.1002/ett.3963
19. Soni, S., & Bhushan, B. (2019). A Comprehensive survey on Blockchain: Working, security analysis, privacy threats and potential applications, *2nd International Conference on Intelligent Computing, Instrumentation and Control Technologies (ICICICT)*, pp. 922–926, doi: 10.1109/ICICICT46008.2019.8993210
20. Bhushan, B., Sahoo, C., Sinha, P., & Khamparia, A. (2021). Unification of Blockchain and Internet of Things (BIoT): requirements, working model, challenges and future directions. Wireless Networks 27, 55–90. doi: 10.1007/s11276-020-02445-6

21. Butt, S. A., Diaz-Martinez, J. L., Jamal, T., Ali, A., De-La-Hoz-Franco, E., & Shoaib, M. (2019). IoT Smart Health Security Threats. *19th International Conference on Computational Science and Its Applications (ICCSA)*. doi: 10.1109/iccsa.2019.000-8
22. Islam, S. M. R., Kwak, D., Kabir, M. H., Hossain, M., & Kwak, K. S. (2015). The internet of things for health care: A comprehensive survey. IEEE Access, 3, 678–708. doi: 10.1109/ACCESS.2015.2437951
23. Ameen, M. A., Liu, J., & Kwak, K. (2012), Security and privacy issues in wireless sensor networks for healthcare applications. Journal of Medical Systems, 36(1), 93–101. doi: 10.1007/s10916-010-9449-4
24. AL-Mawee, W. (2015). Privacy and security issues in IoT healthcare applications for the disabled users a survey, p. 57. Western Michigan University.
25. Saxena S., Bhushan, B., & Ahad, Mohd A. (2021). Blockchain based solutions to secure IoT: Background, integration trends and a way forward, Journal of Network and Computer Applications, Volume 181, 103050, ISSN 1084–8045. doi: 10.1016/j.jnca.2021.103050
26. Khamparia, A., Singh, P. K., Rani, P., Samanta, D., Khanna, A., & Bhushan, B. (2020). An internet of health things driven deep learning framework for detection and classification of skin cancer using transfer learning. Transactions on Emerging Telecommunications Technologies. doi: 10.1002/ett.3963
27. Lake, D., Milito, R. M. R., Morrow, M., & Vargheese, R. (2014). Internet of things: Architectural framework for ehealth security. Journal of ICT Standardization, 1(3), 301–328. doi: 10.13052/jicts2245-800X.133
28. Borgohain, T., Kumar, U., & Sanyal, S. (2015). Survey of security and privacy issues of internet of things. International Journal of Advanced Networking Applications, 6, 2372–2378.
29. Khandelwal, S. (2017). Over 8,600 vulnerabilities found in pacemakers. Retrieved April 13, 2018, from https://thehackernews.com/2017/06/pacemaker-vulnerability.html.
30. Peck, M. E. (2011). Medical devices are vulnerable to hacks, but risk is low overall, https://spectrum.ieee.org/biomedical/devices/medicaldevices-are-vulnerable-to-hacks-but-risk-is-low-overall

13 Designing Contactless Automated Systems Using IoT, Sensors, and Artificial Intelligence to Mitigate COVID-19

Sourabh Shastri, Sachin Kumar, Kuljeet Singh, and Vibhakar Mansotra
University of Jammu, Jammu, India

CONTENTS

13.1 INTRODUCTION

Coronaviruses (CoVs) are a large family of viruses that include the Middle East respiratory syndrome (MERS)-CoV, severe acute respiratory syndrome (SARS)-CoV (Fan et al. 2019; Al-Turaiki, Alshahrani, and Almutairi 2016; Albahri et al. 2020), and the current virus, i.e., SARS coronavirus 2 (SARS-CoV-2) ("Coronavirus Disease 2019 (COVID-19) | CDC" 2020) which was announced as a global pandemic on 11 March 2020 ("WHO | World Health Organization," n.d.) (Ucar and Korkmaz

DOI: 10.1201/9781003219620-13

2020). People have not overlooked a pandemic like this in history that spreads so fast all over the world. The related disease was called COVID-19 by the World Health Organization (WHO).

Before looking for the technological solutions for the COVID-19 pandemic, we came up with a systematic review on the COVID-19, including its origin, symptoms, diagnosis, treatment, and applications of Internet of Things (IoT), sensors, and artificial intelligence (AI)-based automation for combating with this pandemic.

The COVID-19 eruption began on 31 December 2019 when 27 cases of pneumonia with unknown etiology were noted at the WHO's country office in China (Chamola et al. 2020). The origin of the COVID-19 outbreak is Huanan Seafood wholesale market in Wuhan city, China (Charmain Butt 2020). Subsequently, the spreading rate of coronavirus cases has been increased like a wildfire. At present, COVID-19 is the prime cause of millions of deaths globally. The large-scale deaths were found in countries including the United States, Brazil, United Kingdom, India, Italy, Spain, Iran, Russia, France, and Mexico (Toğaçar, Ergen, and Cömert 2020; Shastri et al. 2020). There are over and above 157 million cases of COVID-19 affecting 222 countries and territories and more than 3 million deaths as on 7 May 2021, according to the last update of the WHO and world meters report (Teo 2020). Therefore, there is a vital necessity to provide monitoring, treatment, diagnosis, and follow-ups during the pandemic to manage and contain COVID-19 (Bokolo Anthony Jnr. 2020).

The statistics for the utmost affected countries and regions of the world are illustrated in Figure 13.1 (Source: Worldometers ("Coronavirus Graphs: Worldwide Cases and Deaths – Worldometer" 2020)).

As on 20 February 2020, and based on 55,924 laboratory-confirmed cases, the most appeared symptoms of COVID-19 infected persons found were fever, fatigue, dry cough, and sputum production. Moreover, shortness of breathing, arthralgia, sore throat, headache, chills, etc. are also noted in some cases (Source: World Health Organization (WHO)). The symptoms are categorized into three categories, viz., most-, less-, and rare-appearing symptoms. The detailed symptoms that appeared in COVID-19 infected persons are shown in Table 13.1.

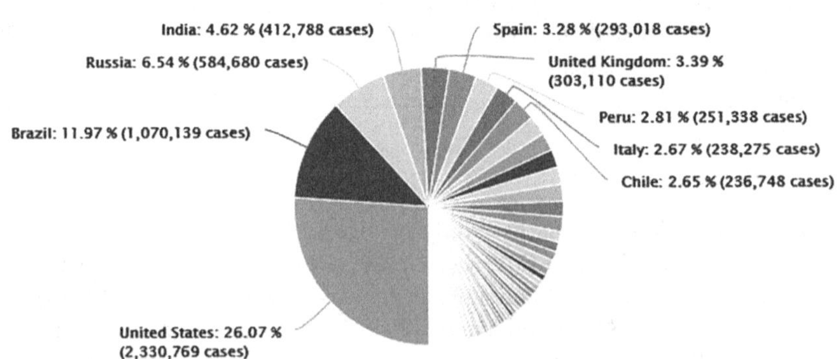

FIGURE 13.1 Country-wise case distributions.

TABLE 13.1
COVID-19 Symptoms

Most Appearing Symptoms	Fever	87.9%
	Dry cough	67.7%
	Fatigue	38.1%
	Sputum production	33.4%
Less Appearing Symptoms	Shortness of breath	18.6%
	Arthralgia	14.8%
	Sore throat	13.9%
	Headache	13.6%
	Chills	11.4%
Rare Appearing Symptoms	Nausea	5.0%
	Nasal congestion	4.8%
	Diarrhea	3.7%
	Hemoptysis	0.9%
	Conjunctival congestion	0.8%

Source: World Health Organization (WHO).

13.1.1 Diagnosis and Treatment

SARS-CoV 2 is an emerging virus. At the time of writing this review, there is no vaccine, drug, or effective antiviral treatment for this infection (Singhal 2020). The primary treatment of COVID-19 is symptomatic (He, Deng, and Li 2020; Tang, Tambyah, and Hui 2020), and for diagnosis purposes, there exist only two main types of COVID-19 tests: viral tests and serology tests. A viral test is a saliva or oral test, and the two main types of viral tests are the antigen test and the polymerase chain reaction (PCR) test. The current standard method that is mostly applied for COVID-19 diagnosis is RT-PCR (Ozturk et al. 2020; Ardakani et al. 2020). Still, they seem not to be much effective for diagnosing and treating the growing number of cases. A critical problem faced by the countries with high coronavirus disease infection rates is the lack of proper diagnosis facilities. In particular, it complicates the social distancing measures and the hospitals' capacity management (Latif et al. 2020). In this context, IoT, sensors, and AI-based solutions can provide an accurate diagnosis and treatment for coronavirus disease and other types of infections such as pneumonia (Pereira et al. 2020).

13.1.2 Prevention

COVID-19 is considered an infectious disease (Li, Hu, and Gu 2020) that transmits from one person to another through different modes of transmission such as airborne zoonotic droplets (Kumar 2020) disseminated by cough, sneezing, physical contact, etc. (Cuevas 2020). As discussed, there is no proper treatment existing for this pandemic till now. The only ways of prevention are staying safe at home, always wearing

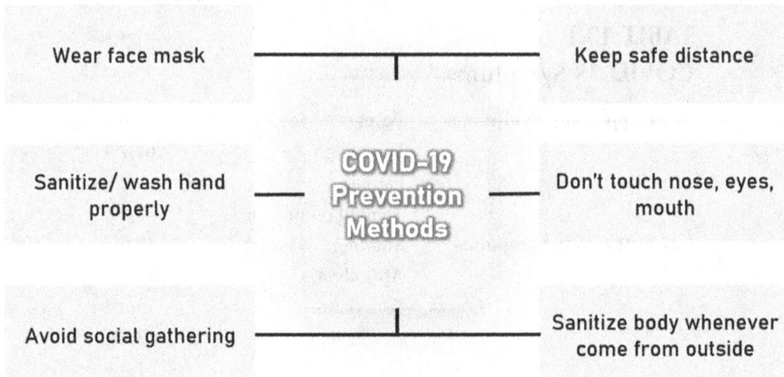

FIGURE 13.2 Preventive methods for COVID-19.

a face mask to cover the nose and mouth in public, practicing social distancing, sanitizing or washing hands properly, etc., as depicted in Figure 13.2.

The rest of the chapter is organized as follows: Section 13.2 presents literature survey. Overview and discussion on IoT, sensors, and AI-based automation are illustrated in Section 13.3. COVID-19 preventive automation system based on IoT, sensors, and AI is discussed in Section 13.4. The conclusion of the chapter is discussed in Section 13.5.

13.2 LITERATURE REVIEW

The IoT is a system of interrelated, wireless, and connected devices that can collect, transfer, analyze, and store data over the network without requiring human-to-human or human-to-computer interaction with the help of sensors integrated with smartphones, drones, robots, etc. (Javaid et al. 2020; Kelly et al. 2020). It is a well-proven and authoritative technology that acts as a junction to the numerous tactics, philosophy of AI, instantaneous analytics, sensory products, etc. (Singh et al. 2020). IoT's primary functional feature is its ever-present connectivity, while its three basic processes are reliable transmission, comprehensive perception, and intelligent processing (Bai et al. 2020). Besides, IoT and healthcare are now coherently working together, named as Internet of Medical Things (IoMT) (Zaidi and Kumar Prasad 2020). As AI is visible in every sector of present-day society (Dey, Cheng, and Tan 2020), it is a forthcoming and practical tool or device to classify early infections due to COVID-19. It can enhance treatment uniformly and assists the decision-making by developing convenient algorithms. AI is not only accommodative in the treatment of coronavirus-infected patients but also for their proper health monitoring with the help of IoT sensors, cloud computing, and blockchain. AI can assist to develop appropriate treatment, drug, vaccine, and prevention strategies (Vaishya et al. 2020). Medical image analysis, disease surveillance, curative research, and risk prediction are some of the AI applications. Recently, fascinating innovations have emerged by

TABLE 13.2
Related Automation Systems for the Prevention of COVID-19

S. No.	Related Work	Technologies Used	References
1	Disinfection tunnel	IoT and sensor-based	(Verma 2020; Biswal et al. 2020; Tran 2020)
2	Automatic hand sanitizer dispenser	IoT and sensor–based	(Sarkar 2020) ("DIY Easy Non-Contact Automatic Hand Sanitizer Dispenser or Automatic Soap Dispenser With Arduino : 6 Steps – Instructables" 2020)[a]
3	Face touch avoidance	AI and others	("Do Not Touch Your Face" 2020; "How to Stop Touching Your Face in the Times of COVID-19 – Firstpost" 2020; "Don't Touch Your Face – APIC" 2020; "Coronavirus – This Is How Much You Touch Your Face I World Economic Forum" 2020; "You Probably Touch Your Face 16 Times an Hour: Here's How to Stop" 2020; "How to Avoid Touching Your Face so Much – BBC Future" 2020)[a] (Mueller, Martin, and Grunwald 2019)
4	Social distancing	Deep learning, IoT, sensors, and others	("RAKSH – Social Distancing Device" 2020; "COVID-19: AI-Enabled Social Distancing Detector Using OpenCV" 2020; "Home I Ministry of Health and Family Welfare I GOI" 2020)[a] (Qian and Jiang 2020; Saunders 2020; Adolphs 2001)
5	Face mask detection	Deep learning	(Wang et al. 2020; Jiang and Fan 2020; Lin et al. 2020)

Note:
[a] Media Reports and Web Reference.

a concerted effort among AI, IoT, big data, blockchain, cloud computing, and many other technologies.

In this chapter, we discuss a total of four automation systems based on IoT, sensors, and AI technologies, which can be helpful in fighting coronavirus. The literature that is somewhat related to the current study and the prementioned technologies is reviewed in Table 13.2.

13.3 IOT, SENSORS, AND AI-BASED AUTOMATION

In this busy world, everyone has a shortage of time. Owing to their frantic work schedules, people forget to take proper care of their health. Everyone wants the family to be safe, secure, and home to be a smart home. With the advancement of new technologies, life is getting simpler and easier in all aspects. The main thing that is preferred in all situations is the time and accuracy that can be possible through automation. That's why, in today's world, automation systems are being preferred over manual systems.

The present time is of IoT and AI-based technologies. A massive number of automated tools and software are currently running in every field, such as education, industry, agriculture, research, and development. The primary use of these technologies is in robotics, drones, smartphones, wearable devices, online learning, and many more.

13.3.1 Present Automation in Healthcare

Successful implementation of the technologies mentioned above can improve the medical staff's efficiency while reducing their workload. The doctors can take care of their patients from a remote location. E-mobile computing enhances the functionalities of IoT in healthcare by bringing immediate support in the form of mobile health (m-health) (Nazir et al. 2019). Some other automations for hospitals and healthcare are IoT-based electrocardiogram (ECG) monitoring, diabetes prevention, healthcare registers (i.e., database of patients), and IoT in orthopedics.

13.3.2 Automation Systems for COVID-19

Automation can play an essential role in tackling this major outbreak of COVID-19. The apex example of IoT and AI-based automation is internet hospitals. It offers key medical support to the public during the COVID-19 outbreak, promotes social distancing, reduces social panic, prevents improper medical-seeking behaviors, enhances the public's ability to self-protect, and facilitates epidemiological screening. Therefore, it can play an indispensable role in preventing and controlling coronavirus. The immense literature has emerged on early prevention, diagnosis, treatment, forecasting, prediction, and analysis of COVID-19 using IoT and AI-based tools and automation (Shastri, Singh, Kumar, et al. 2021).

Aarogya Setu's application developed by the National Informatics Centre (NIC), Government of India (GoI), is an example of AI that helps in this challenging situation (Jhunjhunwala 2020). This application was tested and implemented as a solution over a week using progressively custom development approaches as the requirements, and its use case became more solidified. Faezipour and Abuzneid (2020) provides a self-testing COVID-19 application that records breathing sounds to check coronavirus symptoms. This deep-learning-based application recognizes breathing patterns. The real-time early detection and monitoring systems of coronavirus disease using IoT-based smart glasses were presented in Mohammed, Hazairin, et al. (2020). This smart glass can identify high body temperature even in the group and send the data to be displayed on a mobile application. Hegde et al. (2020) proposed a system that can find cyanosis and fever while it estimates respiratory effort and heart rate using a merger of visible light and thermal cameras operating on an edge computation platform. The name of their proposed system is Auto Triage. In Sun et al. (2020), the authors explored the utility of a radar-based platform as a toolbox to test the effect and impact of non-pharmaceutical interventions (NPIs) to limit the spread of infectious diseases such as COVID-19. They looked over the parameters derived from mobile phones, including phone usage and Global Positioning System (GPS). Moreover, they investigate wearable devices' parameters, including heart rate, step counts, and sleep patterns. They found that most participants spent more time at home, travelled less, and were more active on their phones, particularly interacting with others using social networking applications.

The application of cognitive radio-based IoT specific for the medical domain is known as Cognitive Internet of Medical Things (CIoMT) and first presented in Swayamsiddha and Mohanty (2020). It is an auspicious technology for speedy diagnosis, better treatment, dynamic tracking, and monitoring without spreading the virus to others. Tuli et al. (2020) discussed improved mathematical modeling, cloud computing, and machine learning to forecast the growth and trend of coronavirus disease. They showed that their model could make statistically better predictions than the baseline. Real-time early detection of coronavirus and monitoring systems using an IoT-based smart helmet integrated with a thermal imaging system was developed by Mohammed, Syamsudin, et al. (2020). It can also classify high body temperature in the crowd and send the data to be displayed on a smart mobile application. Celesti et al. (2020) established how a telemedical service could be developed through a healthcare workflow running in Federated Hospital IoT Clouds (FHCs) environment leveraging blockchain.

Moreover, the different concepts of smart cities based on machine learning to contain coronavirus are given by James et al. (2020) and Abaker et al. (2020). Mobile cabin hospitals also play a key role in China's battle against coronavirus (Zhou et al. 2020). They can control the limitedness of medical resources and the high risk of cross-infection.

13.4 COVID-19 PREVENTIVE AUTOMATION SYSTEM BASED ON IoT, SENSORS, AND AI

The numerous automation systems and applications of IoT, sensors, and AI have been discussed in the above sections. Figure 13.3 exhibits the various applications of the automation system for COVID-19. Few are already implemented, while many are still not present and can be designed using the prementioned technologies.

Hereafter, the four applications of leading technologies to design the COVID-19 preventive automation system are explained in Sections 13.4.1–13.4.4. All automation systems follow the same proposed general framework, displayed in Figure 13.4, and are based on the technologies discussed above. The proposed framework is based on IoT, sensors, and AI-based techniques. The input taken by the camera and sensors is sent to the microcontroller/Raspberry Pi for the processing, which is coded with Arduino sketch and Python script. Based on the logic used in the detection system, the alert devices notify with the help of a beep sound or buzzer and LED.

13.4.1 Auto-Sanitization Tunnel with Face Mask Detection

Auto-sanitization or disinfection tunnel can be employed at the entry point of home, office, school, etc. The purpose of this tunnel is to sanitize people coming from outside.

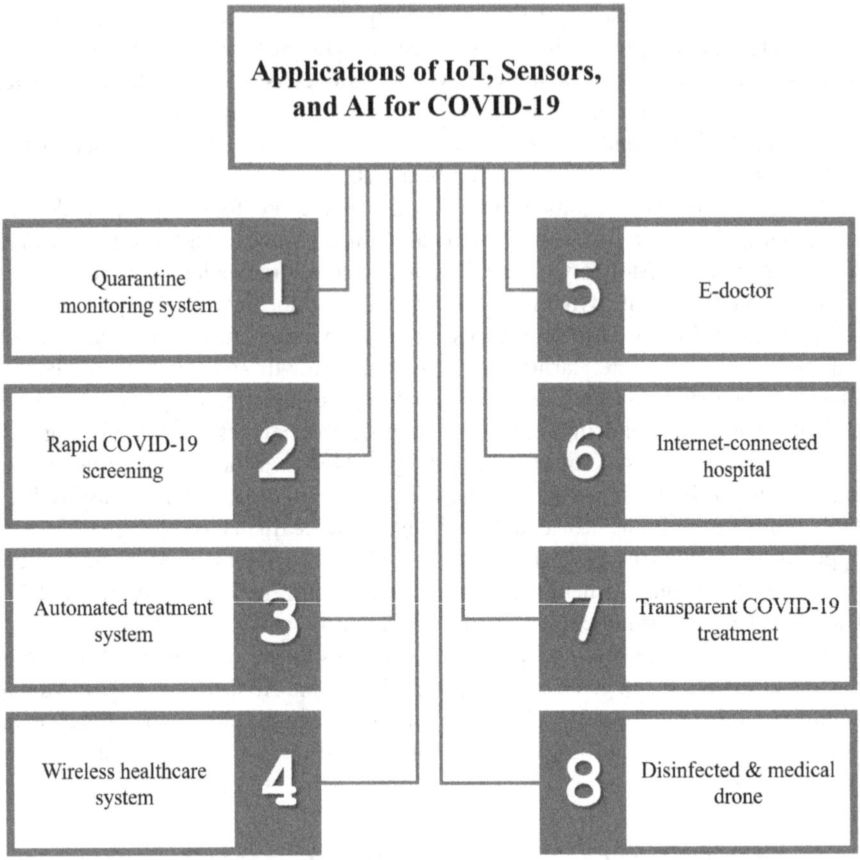

Applications of IoT, Sensors, and AI for COVID-19

1 Quarantine monitoring system
2 Rapid COVID-19 screening
3 Automated treatment system
4 Wireless healthcare system
5 E-doctor
6 Internet-connected hospital
7 Transparent COVID-19 treatment
8 Disinfected & medical drone

FIGURE 13.3 Applications of IoT, sensors, and AI for COVID-19.

This system employs IoT, computer vision, and deep learning technologies in development. On the entry point, a passive infrared (PIR) sensor can be mounted, which detects if a human has moved in or out of the sensor range ("How PIRs Work | PIR Motion Sensor | Adafruit Learning System" 2020; "Arduino Lesson – PIR Motion Sensor « osoyoo.Com" 2020). A sensitivity adjustment potentiometer is used to set

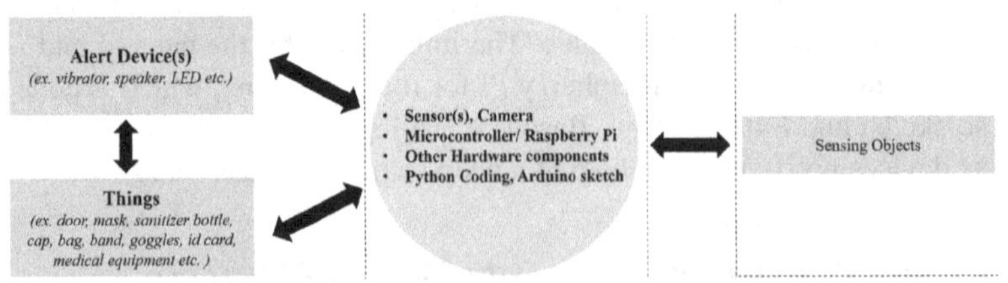

FIGURE 13.4 The proposed general framework for the COVID-19 preventive automation system.

at what distance the sensor should detect the human body. This distance is known as threshold distance, and the area under threshold distance is termed as the detecting area or range of the sensor. Moreover, the sensor's threshold time can also be adjusted, determining how much time the sensor takes to detect a body. Whenever a body is detected, a trigger is generated by the sensor, which turns on the sprinkler nozzles. Sprinkler nozzles produce a sanitizing mist of sanitizing liquid coming from the sanitizer tank. They sanitize the human body coming in contact by spraying the sanitized mist. The sanitization time can be set through the PIR sensor or Arduino sketch uploaded in the Raspberry Pi. On the other end of the tunnel, a camera is mounted above the door connected with the same Raspberry Pi performing face mask detection. A Python script is applied to train and test face mask detectors using OpenCV, deep learning, and TensorFlow ("COVID-19: Face Mask Detector with OpenCV, Keras/TensorFlow, and Deep Learning – PyImageSearch" 2020). Suppose the person inside the tunnel wears the face mask, then the main door will automatically open. Otherwise, a speaking alert will be generated, "Please wear the mask." The flow diagram of its methodology is depicted in Figure 13.5.

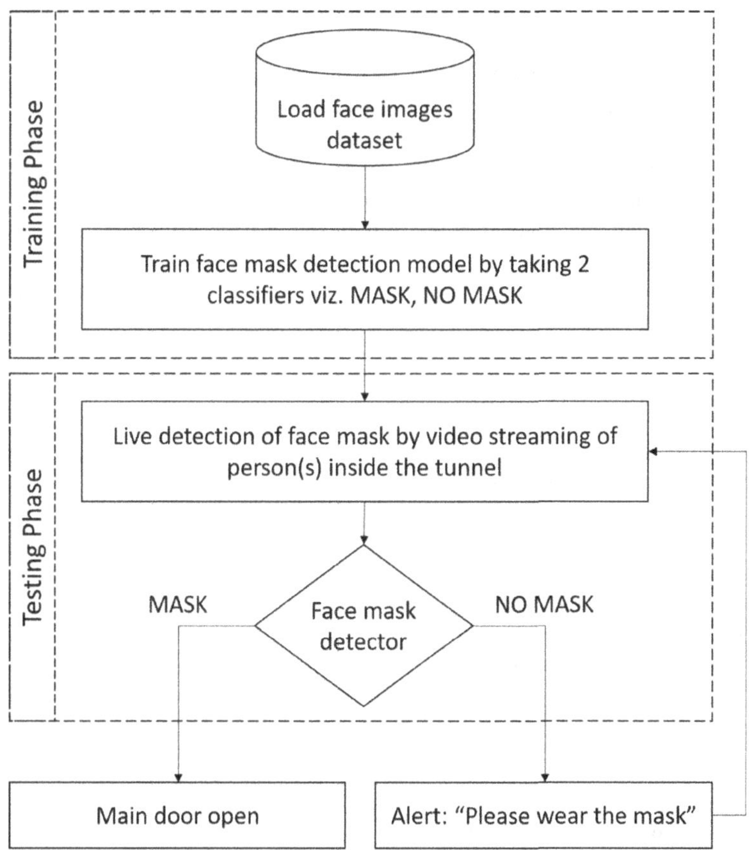

FIGURE 13.5 Flow diagram of face mask detection system.

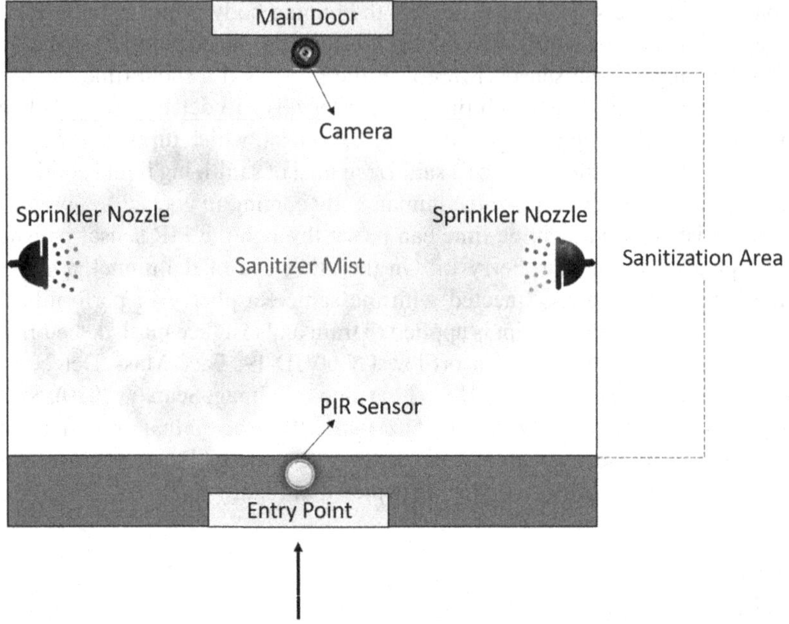

FIGURE 13.6 Auto-sanitization tunnel with face mask detection.

The 2D top view of the auto-sanitization tunnel with face mask detection is shown in Figure 13.6, while its algorithm is summarized in Algorithm 13.1.

Algorithm 13.1

Initially,
Main door = CLOSED;
Camera = OFF;
Sprinkler nozzles = CLOSED;
PIR sensor detecting body:

1. Begin
2. if (body comes inside the detecting area of PIR sensor) then
3. Body detected
4. Sprinkler nozzle = OPEN
5. Start sanitization
6. Wait (x sec.)
7. Camera = ON
8. if (camera is activated)
9. Start detecting mask on face
10. if (mask detected) then
11. Main door = OPEN

12. else
13. Alert: "Please wear the mask"
14. else
15. goto step 6
16. else
17. Continue detecting body
18. End

13.4.2 Auto Contactless Hand Sanitizer Dispenser

The other helpful IoT application to prevent the COVID-19 pandemic is an automatic and contactless hand sanitizer/soap dispenser. The massive number of sources related to this application is mentioned in Table 13.2. This auto-sanitization dispenser uses some hardware components like proximity/ultrasonic sensor, servo motor, microcontroller (Arduino Uno or Arduino Nano), wires, power source, and nozzle. The diagram of this automated device is shown in Figure 13.7.

Here, a proximity or ultrasonic sensor is placed close to the nozzle of the sanitizer bottle. The sensor's job is to detect a human hand and trigger a signal, which starts the servo motor to automatically flush out the liquid sanitizer with the help of a pipe from the bottle through the nozzle. When the hand is taken away from the sensor, the sensor will stop sensing, which automatically turns OFF the servo motor, and hence, the flow of liquid from the sanitizer is stopped. A microcontroller controls the whole system. NodeMCU is also used instead of a microcontroller. The short algorithm of this device is shown in Algorithm 13.2.

Algorithm 13.2

Initially,
Nozzle = CLOSED
Servo motor = OFF

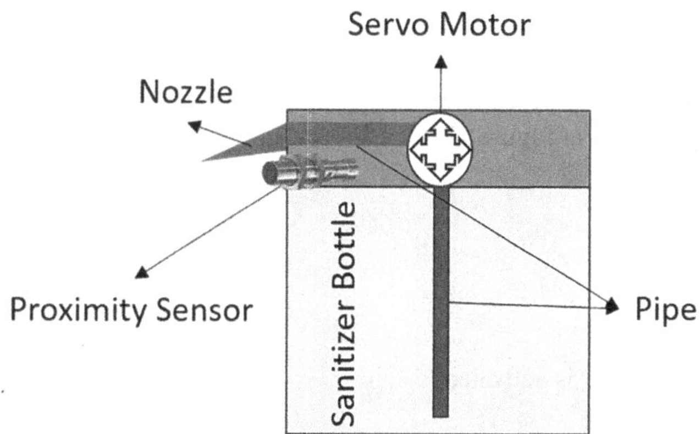

FIGURE 13.7 Contactless automatic hand sanitizer dispenser.

Proximity sensor detecting hand:

1. Begin
2. if (hand detected by proximity sensor) then
3. Servo motor = ON
4. Nozzle = OPEN
5. Pour sanitizer
6. else
7. Continue detecting hand
8. End

13.4.3 Face Touch Avoidance System

One of the utmost precautions that can prevent coronavirus infection is to avoid touching the face because the nose and mouth are the only areas from where the viruses can quickly enter the body. Indeed, touching mucous membranes on the face with dirty or unwashed hands allows the germs that cause respiratory infections to enter the body ("How to Stop Touching Your Face in the Times of COVID-19 – Firstpost" 2020). A recent study has revealed that ordinary people touch their faces approximately 23 times/hour (Kwok, Gralton, and McLaws 2015). Ironically, another research concludes that people touch their face 15.7 times/hour (Nicas and Best 2008). Whatever the exact number is, touching the face several times can lead the coronavirus or other infectious germs to enter the body. A simple sensor attached to a microcontroller can be used on a face mask to make a sensor-based mask device that restricts people from touching their faces. Several sensors can perform this task, but the most effective one is the ultrasonic sensor. Here, the target object is hand, but the problem is that it can detect any object. The range of the sensor can be reduced by a few centimeters (e.g., 5 cm) to resolve this issue. This can be done by making a change in the Arduino sketch uploaded into the microcontroller. The other components required for the same are buzzer/beeper, wires, and battery. The ultrasonic sensor can be attached to the face mask and would be controlled by microcontroller. Whenever the sensor detects the hand/object closer to 5 cm, a beeper starts beeping to make an alert. Note that instead of ultrasonic sensor, another sensor can also be used. The step-by-step procedure of working is shown in Algorithm 13.3, and the sketch is represented at Figure 13.8.

Algorithm 13.3

Initially,
Beep sound = OFF
The ultrasonic sensor is activated:

1. Begin
2. if (hand(s) distance (from ultrasonic sensor) <= sensing range) then

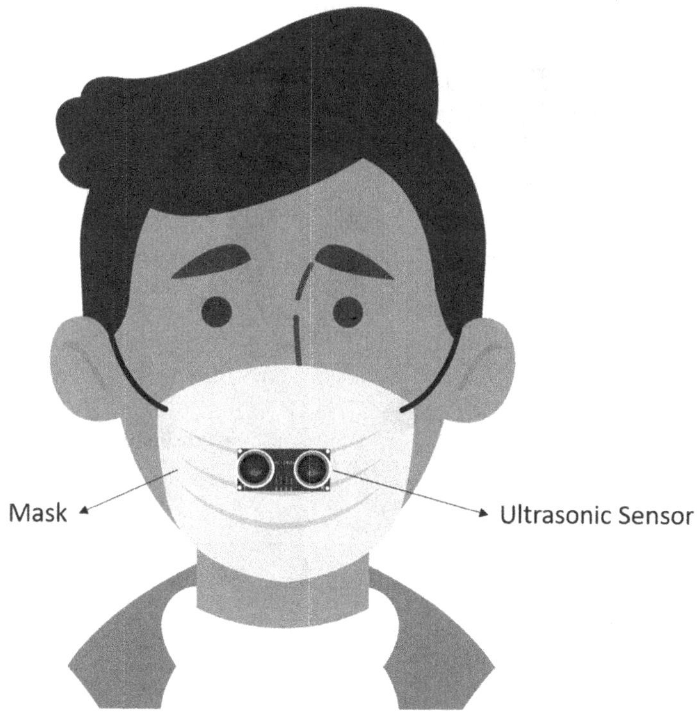

FIGURE 13.8 Face touch avoidance system.

3. Hand detected
4. Alert: "Beeping sound = ON"
5. else
6. goto step 1
7. End

13.4.4 Smart Social Distancing Device

As the spread of COVID-19 is still unstoppable, the government, doctors, cops, and communities are being asked to reduce close contact between people by respecting social distancing. Social distancing means "physical distancing," which is a NPIs for the prevention of infection ("Home | Ministry of Health and Family Welfare | GOI" 2020), and this is also an important and efficacious way to slow down the spread of coronavirus.

A minimum distance of 6 ft (1.8 m) was recommended to maintain social distancing. But people sometimes forget to maintain social distancing. To warn and alert people to maintain a specific distance between them, IoT and sensor-based technology play an important role, limiting the people not to come close (Shastri, Singh, Deswal, et al. 2021). The sensor, mostly used for smart social distancing, is a proximity sensor, which enables detecting similar devices in its proximity. As soon

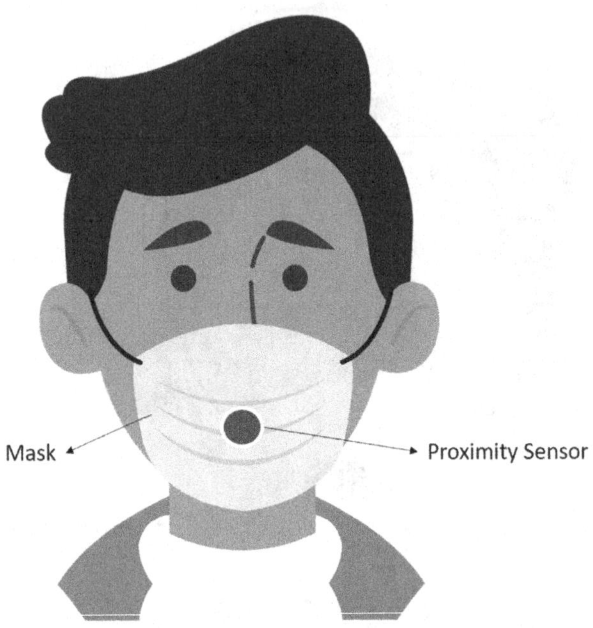

FIGURE 13.9 Social distancing mask.

as two or more devices come within the vicinity, it alerts the users through vibration, LED, and sound alarms to maintain social distancing. Several social distancing devices are already designed, but the most used gadgets are the smart wrist bands. The other ways of making intelligent social distancing devices are on objects like face masks, caps, bags, ID cards, and goggles. One can also embed a proximity sensor on these things. Figures 13.9–13.14 show the daily use of things working as

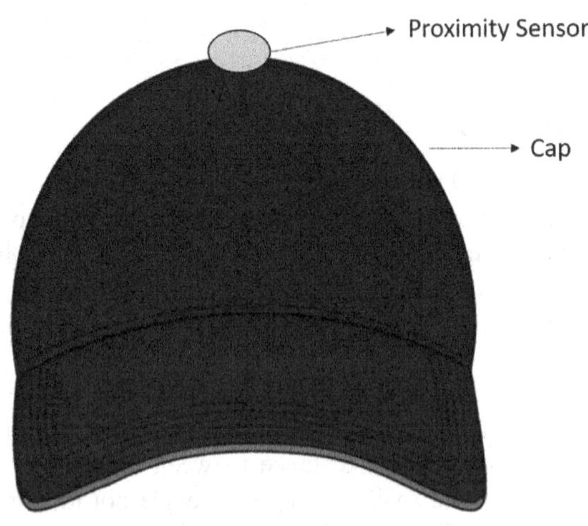

FIGURE 13.10 Social distancing cap.

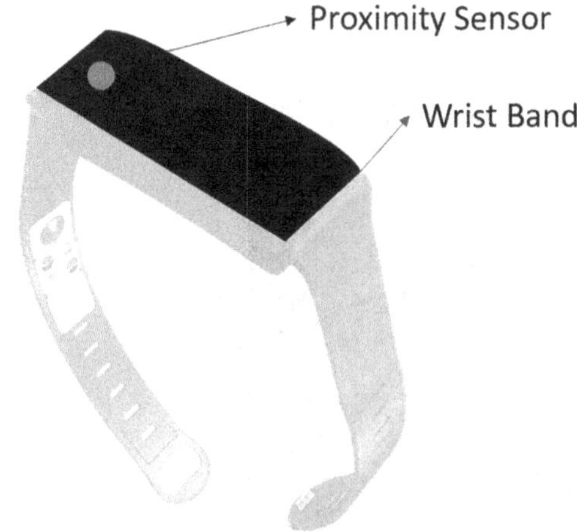

FIGURE 13.11 Social distancing wrist band.

FIGURE 13.12 Social distancing goggles.

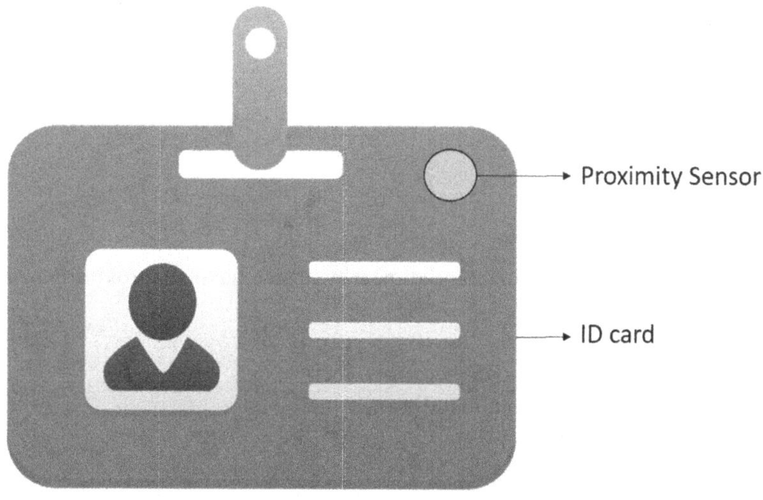

FIGURE 13.13 Social distancing ID card.

FIGURE 13.14 Social distancing bag.

a smart social distancing device. All follow the same working principle, as depicted in Algorithm 13.4.

Algorithm 13.4

Initially,
Vibration = OFF
Sound = OFF
A proximity sensor is detecting another nearby device having a proximity sensor:

1. Begin
2. If (distance between two devices <= 6 feet)
3. Alert: "Please maintain 6 feet distance" and
4. Vibration = ON
5. else
6. Continue detecting nearby device
7. End

13.5 CONCLUSION

This chapter comes up with some of the latest insights on infectious COVID-19. Beginning with a review of COVID-19, we discussed its origin, current situation, symptoms, diagnosis, treatment, prevention, and available technology devices. Primarily, our discussion focuses on the use of emerging technologies like IoT, sensors, and AI in boosting the prevention of coronavirus by designing and extending automated systems. A general framework is proposed, which can help design other IoT, sensors, and AI-based devices. Moreover, we give some ideas about automated devices like social distancing masks, goggles, caps, bags, and identity cards, which can be conceptualized for future use. These types of devices are hugely beneficial to students, workers, doctors, etc. to prevent the cause of coronavirus. There is no need to wear or take extra devices; one can make the available daily-used things or objects automated with the help of the prementioned technologies and by applying the proposed framework. The reviewed methods can significantly help to decrease the spreading rate of COVID-19.

These technologies are beneficial to build healthcare and hospital devices and various contactless tools for schools, colleges, industries, airports, railways, and shops. We primarily focus on prevention, but these technologies can also help the diagnosis and treatment of COVID-19. With the help of contactless treatment and monitoring, we can save frontline workers and patients as well. Moreover, automation tools can also be developed to facilitate screening in large gatherings, such as bus stands, railway stations, airports, markets, and all public spots with AI sub-technologies, viz., deep learning and computer vision collaboration with IoT and sensors.

Authors' Statements:
Conflict of interest - None declared.
Funding - No funding was received.
Ethical Approval - Not required.

REFERENCES

Abaker, Ibrahim, Targio Hashem, Absalom E. Ezugwu, Mohammed A. Al-Garadi, N. Idris, Ibrahim A. T. Hashem, Absalom E. Ezugwu, et al. 2020. *A Machine Learning Solution Framework for Combatting COVID-19 in Smart Cities from Multiple Dimensions.* Medrxiv, 2020.05.18.20105577. doi:10.1101/2020.05.18.20105577.

Adolphs, R. 2001. "The Neurobiology of Social Cognition." *Current Opinion in Neurobiology* 11(2): 231–239. doi:10.1016/S0959-4388(00)00202-6.

Al-Turaiki, Isra, Mona Alshahrani, and Tahani Almutairi. 2016. "Building Predictive Models for MERS-CoV Infections Using Data Mining Techniques." *Journal of Infection and Public Health* 9(6). King Saud Bin Abdulaziz University for Health Sciences: 744–748. doi:10.1016/j.jiph.2016.09.007.

Albahri, A. S., Rula A. Hamid, Jwan K. Alwan, Z. T. Al-qays, A. A. Zaidan, B. B. Zaidan, A. O.S. Albahri, et al. 2020. "Role of Biological Data Mining and Machine Learning Techniques in Detecting and Diagnosing the Novel Coronavirus (COVID-19): A Systematic Review." *Journal of Medical Systems* 44(7). doi:10.1007/s10916-020-01582-x.

Ardakani, Ali A., Alireza R. Kanafi, U. R. Acharya, Nazanin Khadem, and Afshin Mohammadi. 2020. "Application of Deep Learning Technique to Manage COVID-19 in Routine Clinical Practice Using CT Images: Results of 10 Convolutional Neural Networks." *Computers in Biology and Medicine* 121(April). Elsevier Ltd: 103795. doi: 10.1016/j.compbiomed.2020.103795.

"Arduino Lesson – PIR Motion Sensor osoyoo.com." 2020. Accessed June 18. https://osoyoo.com/2017/07/27/arduino-lesson-pir-motion-sensor/.

Bai, Li, Dawei Yang, Xun Wang, Lin Tong, Xiaodan Zhu, Nanshan Zhong, Chunxue Bai, et al. 2020. "Chinese Experts' Consensus on the Internet of Things-Aided Diagnosis and Treatment of Coronavirus Disease 2019 (COVID-19)." Clinical EHealth 3. KeAi Communications Co., Ltd: 7–15. doi:10.1016/j.ceh.2020.03.001.

Biswal, M., R. Kanaujia, A. Angrup, P. Ray, and S. M. Singh. 2020. "Disinfection Tunnels: Potentially Counterproductive in the Context of a Prolonged Pandemic of COVID-19." *Public Health* 183: 48–49. doi:10.1016/j.puhe.2020.04.045.

Bokolo Anthony Jnr. 2020. "Use of Telemedicine and Virtual Care for Remote Treatment in Response to COVID-19 Pandemic." *Journal of Medical Systems* 44(7): 132. doi:10.1007/s10916-020-01596-5.

Butt, Charmain. 2020. "Covid19: Deep Learning System to Screen Coronavirus Disease 2019 Pneumonia." Applied Intelligence. http://arxiv.org/abs/2002.09334.

Celesti, Antonio, Armando Ruggeri, Maria Fazio, Antonino Galletta, Massimo Villari, and Agata Romano. 2020. "Blockchain-Based Healthcare Workflow for Tele-Medical Laboratory in Federated Hospital IoT Clouds." *Sensors (Switzerland)* 20(9). doi:10.3390/s20092590.

Chamola, Vinay, Vikas Hassija, Vatsal Gupta, and Mohsen Guizani. 2020. "A Comprehensive Review of the COVID-19 Pandemic and the Role of IoT, Drones, AI, Blockchain, and 5G in Managing Its Impact." *IEEE Access* 8(April): 90225–90265. doi:10.1109/access.2020.2992341.

"Coronavirus – This Is How Much You Touch Your Face | World Economic Forum." 2020. Accessed June 12. https://www.weforum.org/agenda/2020/03/covid-19-prevention-touching-face/.

"Coronavirus Disease 2019 (COVID-19) | CDC." 2020. Accessed June 1. https://www.cdc.gov/coronavirus/2019-ncov/.

"Coronavirus Graphs: Worldwide Cases and Deaths – Worldometer." 2020. Accessed June 4. https://www.worldometers.info/coronavirus/worldwide-graphs/#case-distribution.

"COVID-19: AI-Enabled Social Distancing Detector Using OpenCV." 2020. Accessed June 15. https://towardsdatascience.com/covid-19-ai-enabled-social-distancing-detector-using-opencv-ea2abd827d34.

"COVID-19: Face Mask Detector with OpenCV, Keras/TensorFlow, and Deep Learning – PyImageSearch." 2020. Accessed June 20. https://www.pyimagesearch.com/2020/05/04/covid-19-face-mask-detector-with-opencv-keras-tensorflow-and-deep-learning/.

Cuevas, Erik. 2020. "An Agent-Based Model to Evaluate the COVID-19 Transmission Risks in Facilities." *Computers in Biology and Medicine* 121(April). Elsevier Ltd: 103827. doi:10.1016/j.compbiomed.2020.103827.

Dey, Sukhen, Qiang Cheng, and Joseph Tan. 2020. "All for One and One for All: Why a Pandemic Preparedness League of Nations?" *Health Policy and Technology* 9(2). Elsevier Ltd: 179–184. doi:10.1016/j.hlpt.2020.04.009.

"DIY Easy Non-Contact Automatic Hand Sanitizer Dispenser or Automatic Soap Dispenser With Arduino : 6 Steps – Instructables." 2020. Accessed June 11. https://www.instructables.com/id/DIY-Easy-Non-Contact-Automatic-Hand-Sanitizer-Disp/.

"Do Not Touch Your Face." 2020. Accessed June 11. https://donottouchyourface.com/.

"Don't Touch Your Face – APIC." 2020. Accessed June 11. https://apic.org/monthly_alerts/dont-touch-your-face/.

Faezipour, Miad, and Abdelshakour Abuzneid. 2020. "Smartphone-Based Self-Testing of COVID-19 Using Breathing Sounds." *Telemedicine and E-Health* 26(10): 1202–1205. doi:10.1089/tmj.2020.0114.

Fan, Yi, Kai Zhao, Zheng Li Shi, and Peng Zhou. 2019. "Bat Coronaviruses in China." *Viruses* 11(3): 27–32. doi:10.3390/v11030210.

He, Feng, Yu Deng, and Weina Li. 2020. "Coronavirus Disease 2019: What We Know?" *Journal of Medical Virology*, no. March: 719–725. doi:10.1002/jmv.25766.

Hegde, Chaitra, Zifan Jiang, Pradyumna B. Suresha, Jacob Zelko, Salman Seyedi, Monique A. Smith, David W. Wright, Rishikesan Kamaleswaran, Matt A. Reyna, and Gari D. Clifford. 2020. *AutoTriage – An Open Source Edge Computing Raspberry Pi-Based Clinical Screening System*. MedRxiv, 2020.04.09.20059840. doi:10.1101/2020.04.09.20059840.

"Home | Ministry of Health and Family Welfare | GOI." 2020. Accessed June 15. https://www.mohfw.gov.in/pdf/.

"How PIRs Work | PIR Motion Sensor | Adafruit Learning System." 2020. Accessed June 18. https://learn.adafruit.com/pir-passive-infrared-proximity-motion-sensor/how-pirs-work.

"How to Avoid Touching Your Face so Much – BBC Future." 2020. Accessed June 13. https://www.bbc.com/future/article/20200317-how-to-stop-touching-your-face.

"How to Stop Touching Your Face in the Times of COVID-19 – Firstpost." 2020. Accessed June 11. https://www.firstpost.com/health/how-to-stop-touching-your-face-in-the-times-of-covid-19-8247921.html.

James, Philip, Ronnie Das, Agata Jalosinska, and Luke Smith. 2020. "Smart Cities and a Data-Driven Response to COVID-19." *Dialogues in Human Geography*, 2043820620934211. doi:10.1177/2043820620934211.

Javaid, Mohd, Abid Haleem, Raju Vaishya, Shashi Bahl, Rajiv Suman, and Abhishek Vaish. 2020. "Industry 4.0 Technologies and Their Applications in Fighting COVID-19 Pandemic." Diabetes and Metabolic Syndrome: Clinical Research and Reviews 14(4). Elsevier Ltd: 419–422. doi:10.1016/j.dsx.2020.04.032.

Jhunjhunwala, Ashok. 2020. "Role of Telecom Network to Manage COVID-19 in India: Aarogya Setu." Transactions of the Indian National Academy of Engineering, no. 0123456789. Springer, Singapore. doi:10.1007/s41403-020-00109-7.

Jiang, Mingjie, and Xinqi Fan. 2020. "RetinaMask: A Face Mask Detector." http://arxiv.org/abs/2005.03950.

Kelly, Jaimon T., Katrina L. Campbell, Enying Gong, and Paul Scuffham. 2020. "The Internet of Things : Impact and Implications for Healthcare Delivery Table of Contents."

Kumar, Dharmendra. 2020. "Corona Virus: A Review of COVID-19." *Eurasian Journal of Medicine and Oncology* 4(2): 8–25. doi:10.14744/ejmo.2020.51418.

Kwok, Yen L. A., Jan Gralton, and Mary L. McLaws. 2015. "Face Touching: A Frequent Habit That Has Implications for Hand Hygiene." *American Journal of Infection Control* 43(2). Elsevier Inc: 112–114. doi:10.1016/j.ajic.2014.10.015.

Latif, Siddique, Muhammad Usman, Waleed Iqbal, Junaid Qadir, Sanaullah Manzoor, Gareth Tyson, Ignacio Castro, et al. 2020. "Leveraging Data Science To Combat COVID-19: A Comprehensive Review TESCON (Tools for Enforcement of Smart Contracts) View Project CONTRIVE View Project Leveraging Data Science To Combat COVID-19: A Comprehensive Review," no. April. doi:10.13140/RG.2.2.12685.28644/4.

Li, Gang, Rui Hu, and Xuefang Gu. 2020. "A Close-up on COVID-19 and Cardiovascular Diseases." *Nutrition, Metabolism & Cardiovascular Diseases* 30(7): 1057–1060. doi: 10.1016/j.numecd.2020.04.001.

Lin, Kaihan, Huimin Zhao, Jujian Lv, Canyao Li, Xiaoyong Liu, Rongjun Chen, Ruoyan Zhao, and Zheng Wang. 2020. "Face Detection and Segmentation Based on Improved Mask R-CNN." *Discrete Dynamics in Nature and Society* 2020. doi:10.1155/2020/9242917.

Mohammed, M. N., Nurul A. Hazairin, Halim Syamsudin, S. Al-Zubaidi, A. K. Sairah, Safinaz Mustapha, and Eddy Yusuf. 2020. "2019 Novel Coronavirus Disease (Covid-19): Detection and Diagnosis System Using Iot Based Smart Glasses." *International Journal of Advanced Science and Technology* 29(7 Special Issue): 954–960.

Mohammed, M. N., Halim Syamsudin, S. Al-Zubaidi, A. K. Sairah, Rusyaizila Ramli, and Eddy Yusuf. 2020. "Novel Covid-19 Detection and Diagnosis System Using IoT Based Smart Helmet." *International Journal of Psychosocial Rehabilitation* 24(7): 2296–2303. doi:10.37200/IJPR/V24I7/PR270221.

Mueller, Stephanie M., Sven Martin, and Martin Grunwald. 2019. "Self-Touch: Contact Durations and Point of Touch of Spontaneous Facial Self-Touches Differ Depending on Cognitive and Emotional Load." *PLoS ONE* 14(3). Public Library of Science. doi:10.1371/journal.pone.0213677.

Nazir, Shah, Yasir Ali, Naeem Ullah, and Iván García-Magariño. 2019. "Internet of Things for Healthcare Using Effects of Mobile Computing: A Systematic Literature Review." *Wireless Communications and Mobile Computing* 2019. doi:10.1155/2019/5931315.

Nicas, Mark, and Daniel Best. 2008. "A Study Quantifying the Hand-to-Face Contact Rate and Its Potential Application to Predicting Respiratory Tract Infection." *Journal of Occupational and Environmental Hygiene* 5(6): 347–352. doi:10.1080/15459620802003896.

Ozturk, Tulin, Muhammed Talo, Eylul A. Yildirim, Ulas B. Baloglu, Ozal Yildirim, and U. R. Acharya. 2020. "Automated Detection of COVID-19 Cases Using Deep Neural Networks with X-Ray Images." *Computers in Biology and Medicine* 121 (April). Elsevier Ltd: 103792. doi:10.1016/j.compbiomed.2020.103792.

Pereira, Rodolfo M., Diego Bertolini, Lucas O. Teixeira, Carlos N. Silla, and Yandre M.G. Costa. 2020. "COVID-19 Identification in Chest X-Ray Images on Flat and Hierarchical Classification Scenarios." *Computer Methods and Programs in Biomedicine* 194. Elsevier B.V.: 105532. doi:10.1016/j.cmpb.2020.105532.

Qian, Meirui, and Jianli Jiang. 2020. "COVID-19 and Social Distancing." *Journal of Public Health (Germany)*, no. Mikulska 2019. doi:10.1007/s10389-020-01321-z.

"RAKSH – Social Distancing Device." 2020. Accessed June 14. https://www.businesswireindia.com/raksh-social-distancing-device-67921.html.

Sarkar, Abhinandan. 2020. "Design of Automatic Hand Sanitizer with Temperature Sensing." *International Journal of Innovative Science and Research Technology* 5(5): 1269–1275.

Saunders, Fenella. 2020. "Social Distancing and Connection." *American Scientist* 108(3): 130. doi:10.1511/2020.108.3.130.

Shastri, Sourabh, Kuljeet Singh, Monu Deswal, Sachin Kumar, and Vibhakar Mansotra. 2021. "CoBiD-Net: A Tailored Deep Learning Ensemble Model for Time Series Forecasting of Covid-19." *Spatial Information Research*, June. Springer, 1–14. doi:10.1007/s41324-021-00408-3.

Shastri, Sourabh, Kuljeet Singh, Sachin Kumar, Paramjit Kour, and Vibhakar Mansotra. 2020. "Time Series Forecasting of Covid-19 Using Deep Learning Models: India-USA Comparative Case Study." *Chaos, Solitons and Fractals* 140 (November). Elsevier Ltd: 110227. doi:10.1016/j.chaos.2020.110227.

Shastri, Sourabh, Kuljeet Singh, Sachin Kumar, Paramjit Kour, and Vibhakar Mansotra. 2021. "Deep-LSTM Ensemble Framework to Forecast Covid-19: An Insight to the Global Pandemic." *International Journal of Information Technology (Singapore)* 13(4): 1291–1301. doi:10.1007/s41870-020-00571-0.

Singh, Ravi P., Mohd Javaid, Abid Haleem, and Rajiv Suman. 2020. "Internet of Things (IoT) Applications to Fight against COVID-19 Pandemic." *Diabetes and Metabolic Syndrome: Clinical Research and Reviews* 14(4). Elsevier Ltd: 521–524. doi:10.1016/j.dsx.2020.04.041.

Singhal, Tanu. 2020. "A Review of Coronavirus Disease-2019 (COVID-19)." Indian Journal of Pediatrics 87(4): 281–286. doi:10.1007/s12098-020-03263-6.

Sun, Shaoxiong, Amos Folarin, Yatharth Ranjan, Zulqarnain Rashid, Pauline Conde, Callum Stewart, Nicholas Cummins, et al. 2020. "Using Smartphones and Wearable Devices to Monitor Behavioural Changes during COVID-19," no. April. http://arxiv.org/abs/2004.14331.

Swayamsiddha, Swati, and Chandana Mohanty. 2020. "Application of Cognitive Internet of Medical Things for COVID-19 Pandemic." *Diabetes & Metabolic Syndrome: Clinical Research & Reviews*. Diabetes India. doi:10.1016/j.dsx.2020.06.014.

Tang, Julian W., Paul A. Tambyah, and David S.C. Hui. 2020. "Emergence of a Novel Coronavirus Causing Respiratory Illness from Wuhan, China." Journal of Infection 80(3): 350–371. doi:10.1016/j.jinf.2020.01.014.

Teo, Jason. 2020. "Early Detection of Silent Hypoxia in Covid-19 Pneumonia Using Smartphone Pulse Oximetry." *Journal of Medical Systems* 44(8): 134. doi:10.1007/s10916-020-01587-6.

Toğaçar, Mesut, Burhan Ergen, and Zafer Cömert. 2020. "COVID-19 Detection Using Deep Learning Models to Exploit Social Mimic Optimization and Structured Chest X-Ray Images Using Fuzzy Color and Stacking Approaches." *Computers in Biology and Medicine* 121 (May). doi:10.1016/j.compbiomed.2020.103805.

Tran, Trung Pham, Nguyen Manh Ha, Nguyen Ngoc Sang. 2020. "The Disinfectant Solution System Preventing SARSCOV-2 Epidemic." International Journal of Scientific Engineering and Science 4(6): 11–14.

Tuli, Shreshth, Shikhar Tuli, Rakesh Tuli, and Sukhpal S. Gill. 2020. "Predicting the Growth and Trend of COVID-19 Pandemic Using Machine Learning and Cloud Computing." *Internet of Things* 11. Elsevier B.V.: 100222. doi:10.1016/j.iot.2020.100222.

Ucar, Ferhat, and Deniz Korkmaz. 2020. "COVIDiagnosis-Net: Deep Bayes-SqueezeNet Based Diagnosis of the Coronavirus Disease 2019 (COVID-19) from X-Ray Images." *Medical Hypotheses* 140 (April). Elsevier: 109761. doi:10.1016/j.mehy.2020.109761.

Vaishya, Raju, Mohd Javaid, Ibrahim H. Khan, and Abid Haleem. 2020. "Artificial Intelligence (AI) Applications for COVID-19 Pandemic." Diabetes and Metabolic Syndrome: Clinical Research and Reviews 14(4): 337–339. doi:10.1016/j.dsx.2020.04.012.

Verma, Dr Archana. 2020. "COVID-19 Disinfectant Tunnels May Harm Humans More than Virus." *Journal of Medical Science and Clinical Research* 08(04): 313–315. doi:10.18535/jmscr/v8i4.58.

Wang, Zhongyuan, Guangcheng Wang, Baojin Huang, Zhangyang Xiong, Qi Hong, Hao Wu, Peng Yi, et al. 2020. "Masked Face Recognition Dataset and Application," 1–3. http://arxiv.org/abs/2003.09093.

"WHO | World Health Organization." n.d. https://www.who.int/.

"You Probably Touch Your Face 16 Times an Hour: Here's How to Stop." 2020. Accessed June 12. https://www.healthline.com/health-news/how-to-not-touch-your-face.

Zaidi, Kashif, and Sanjeev K. Prasad. 2020. "Impact of IoT Adoption in Healthcare: COVID-19 and Online Medical Learning Environments." *SSRN Electronic Journal*, 1–5. doi:10.2139/ssrn.3605015.

Zhou, Ying, Lingling Wang, Lieyun Ding, and Zhouping Tang. 2020. "Intelligent Technologies Help Operating Mobile Cabin Hospitals Effectively Cope with COVID-19." *Frontiers of Engineering Management*. doi:10.1007/s42524-020-0113-5.

14 Analysis of the Framework for the Development, Security and Efficacy of IoT-Based Mobile Health-Care Solutions for Antenatal Care

J. O. Nehinbe
ICT Security Solutions, West Africa, (W/A)

J. A. Benson
NHS England, Keynes, England

L. Chibuzor
NHS Milton Keynes CCG, Keynes, England

CONTENTS

DOI: 10.1201/9781003219620-14

14.1 INTRODUCTION

The rates of miscarriage, maternal mortality, birth defects, neonatal infections and other preventable health problems are worrisome according to the global review of recent years [1–6]. These health issues are frequently associated with failure of the huge numbers of pregnant women to complete or surpass the minimum antenatal visits recommended by the World Health Organization (WHO) across the globe [6, 7]. With the advent of Internet and various developments in electronic industry, the health-care delivery systems should gradually shift from treatment delays, limited access of poor pregnant women to health-care services and prolong period on queues (without receiving sound medical services from specialists) to the practice of excellent service delivery under the platform of the IoT-based health-care devices [1, 2, 8].

Fundamentally, the perceived needs, experience and physiological system of every pregnant woman dynamically change throughout various stages of pregnancy. The experience may possibly be painful for psychotic pregnant women with critical mental disorders. Some pregnant women unconsciously exhibit behaviors like swollen legs, spit out of saliva, over eating, dizziness, anxiety and loss of appetite [2, 9]. Consequently, clinicians and pregnant women require a kind of rapid monitoring system and platform that can swiftly facilitate forum for health workers and pregnant women to remotely contact each other. The premise is that pregnant woman should not require to book for appointment that would eventually necessitate them to wait sometimes for weeks before they can visit their clinicians.

Interviews with pregnant women suggest that pregnancy can be traumatizing to some expectant mothers and health workers irrespective of their historical experience with pregnancy. As a result, modern antenatal cares that are built on the platforms of Internet of Things (IoT), wearable devices, computer systems and Global System of Mobile (GSM) communication to assist pregnant women to actively participate in routine preventive checkup from their health workers deserve broad review [8, 10]. The IoT-based mobile health-care solutions for antenatal cares are multipurpose health-care solutions that are embedded software components so that they would be able to work cooperatively with smart devices to lessen perennial challenges with parent-hood. These kinds of health-care solutions can serve the ultimate goals of promoting the well-being and social welfare of expectant mothers and their babies. In Mwaura (2018), the logic is that IoT cannot exist without Internet facilities. Hence, the front end of the IoT-based mobile health solutions is made up of the interconnectivity of the graphical user interface (GUI) with portable devices such as smart phones, smart watches and smart pendants while the back end of the technology is built on a complex integration of many devices such as the Internet, computer and wireless networks; cryptographic algorithms, chips and sensor technologies so that pregnant women and health workers can use IoT-based mobile health solutions to interactively and swiftly exchange health-care information without stress [8, 10, 11, 12].

Adequate data security and better quality control of the above technology is inevitable because they encompass the use some inbuilt sensors to systematically track, gather and analyze physiological changes, symptomatic expressions, body temperature; blood pressure, heart beat and fetus movement in the womb of the wearers [12, 13]. The sensors can promptly transmit the above indicators or statistical variables (in the form of alerts) to the mobile devices of midwives, general practitioner (GP) or other designated health workers. Another fact is that health workers can instantly spot and communicate with the pregnant women who require urgent medical attention without the need for them to assign colleagues that must travel to examine the patients face to face [14]. Interventions that may make some pregnant women contented, happy, healthy and thriving may just require medical recommendations or tips on how they can maintain healthy lifestyles with their babies during and after pregnancy rather than traveling to meet midwives in the various hospitals [8, 10]. It is also important to note that the above technology will facilitate the development of smart counseling and smart therapeutic relationships between the pregnant woman and their health workers. This is really important because some pregnant women still rely heavily on ‘old wives tales and remedies’ that can be harmful to the unborn children in some circumstances. Similarly, the review of some symptoms and physiological changes in the bodies of some pregnant women may imply they merely require supplementing their diets with a kind of food that is readily available to them [6, 7]. Some interventions may require the pregnant women to do more exercises to get on well. In terms of health-care promotions, midwives and gynecologists can employ mobile health-care devices that operate on the platform of IoT to receive and relay these vital clues to educate and increase the awareness of the required pregnant women that possess the enabling technology [2]. IoT-based mobile health-care solutions can support epidemiological inquiries that will ultimately augment the well-being and upbeat of pregnant women [4, 6, 15]. Receiving text messages and vital tips from midwives or GP in pregnancy can help some pregnant women to be less anxious and enable them to overcome prenatal deficiencies, survive domestic abuses and potential health problems that may originate from pregnancy.

In a developed country, pregnant women can book appointments to attend antenatal care at home, hospital, Children's and maternity centers [6]. The settings are made safe for pregnant women to discuss sensitive issues and this is particularly helpful in the situations of teenage pregnancies [1, 7]. So, IoT-based health-care solutions for antenatal cares can enable some pregnant women to privately discuss the results of their pregnancy scans and other confidential issues on addiction, domestic abuse, mental health and economic problems with health professionals and receive professional comments at their leisure time and during holidays once the pregnant women are within the coverage of their mobile networks. Essentially, IoT-based mobile health-care solutions for antenatal cares are promising areas that can tremendously be helpful to pregnant women in the developing and developed countries in countless ways. We believe that financial investment on these health-care solutions can help health management board to tackle unemployment and boost internally generated revenue.

Conversely, intrusions are serious threats that are inevitable in the usage of IoT-based health-care solutions across the globe. With the pervasiveness of the Internet, it is plausible that analysts may need to design models that would require several iterations to classify unknown intrusions and intruders against the above health-care solutions [11, 12]. Some classifiers may need to be augmented to simultaneously predict complex intrusions. Thus, numerous data mining approaches and contrasting techniques have been devised for grouping intrusions on cyber resources in recent years [16]. Meanwhile, some of the existing methods are only suitable for grouping intrusions that will form convex clusters and clusters that will not converge rapidly. Besides, there are several other technical issues and social problems that may associate with the above health-care solutions [15]. Unfortunately, most of them may elude the imaginations of researchers and manufacturers during their designs and implementations. Also, factors like costs, management of enabling applications and availability of the products are key indicators that can determine the rate of patronizing the IoT-based mobile health-care solutions that are proposed in this chapter in a global market. In addition, discrepancies in the quality of service delivery and health-care policies may determine how primary health-care centers, policy makers and hospitals in the developed and developing countries would adopt them for improving patients' experiences. Studies have shown that high rates of miscarriages, maternal mortality, birth defects, neonatal infections, prevalence of low birth weight and other preventable health problems are worrisome in global assessment of the development in health sector in recent years [2, 6]. Thus, feelers begin to contemplate about the fundamental ways of designing these devices together with the selection, management and security of IoT mobile health-care solutions for antenatal care and how they correlate with the above challenges [15, 17]. Accordingly, this chapter uses mixed methods to qualitatively and quantitatively review the above issues on the IoT-based mobile health-care solutions for antenatal cares. One of the substantial contributions of this chapter is its ability to propose feasible measures on how to design IoT-based mobile health-care solutions for antenatal cares in other to reduce the above challenges. The chapter also explicates how analysts can as well determine the market trend and usefulness of attributes for describing inbound and outbound packets that migrate across IoT-based networks.

14.2 DEFINITIONS OF TERMS

Antenatal care is used to represent midwifery service that a professional like midwife or obstetrician offered on appointment to a pregnant lady to ensure she has a healthy pregnancy, safe delivery and a healthy baby. Antenatal care is also known as pregnancy or maternity care.

Mobile healthcare (m-healthcare) is used to describe the use of mobile devices like cellular phones, personal digital assistants (PDAs) and tablet in health-care systems to enable health professionals in remote locations to keep patients with health conditions under constant watch and examination without the needs for the patients to meet the health professionals face to face in the a clinical setting [18]. It is a technology whereby patients can upload, update, receive, access and verify historical health data and personal information about their health on their individual mobile phones.

IoT-based mobile healthcare is the integration of mobile healthcare with IoT. It is a type of mobile health-care services that health professionals deliver to patients under the platform of the IoT [19].

Health-care solutions are the application of computer systems to solve some challenges in health-care delivery system. Health-care delivery systems are usually large and their departments are interrelated [20]. Hence, manual operations are ineffective in a densely populated health sector. Computer service providers may decide to computerize the departments, management, process flow and medical activities in the form of health-care solutions to improve the efficiency of health delivery services. Health-care solutions come with different brand and names depending on their functionalities and capabilities. Some health-care solutions focus on electronic health record, treatment tools, health-care supply chain management, health-care therapeutic, monitoring and diagnostic toolkits, health-care networking and clinical workflow.

IoT-based mobile health-care solutions are computerized mobile health-care delivery products that clinicians use to deliver medical services to patients and patients receive through the platform of the IoT [13, 21]. They are well-groomed portable devices that are also connected to the mobile phones of doctors and patients that wear or use them.

Wearable devices are electronic devices that patients can wear or use to detect, analyze and convey the body signals of the wearers to dedicated health workers so that the health workers can remotely respond to the wearers [16, 10]. There are many types and categories of wearable devices. Some of them are activity trackers, seizures and intelligent wristwatches.

Category utility is a statistical concept for estimating the degree of predictability of attributes of categorical datasets [16].

Categorical dataset is a kind of dataset that its attributes consist of a combination of numeric and alphanumeric entries.

Information gain is a statistical concept for calculating the quantity of information that certain attributes of categorical datasets can furnish analysts [16].

Efficacy of IoT-based mobile health-care solutions is a concept that represents the degree of the efficiency (ability) and effectiveness (helpfulness and usefulness) of IoT-based mobile health-care solutions.

Smart Intrusion Detection Systems (IDSs) are mechanisms that can detect, log and send the reports about intrusions to the emails or mobile phones of human operators so that the operators can remotely analyze and thwart the intrusions in progress.

Smart Mobile Intrusion Prevention Systems (MIPSs) are wireless mechanisms that can detect, log and prevent intrusions and subsequently send the reports about them to the emails or mobile devices of human operators so that the operators can remotely analyze and verify the veracity of the intrusions that the mechanisms have stopped from migrating into the networks.

14.3 THEORETICAL MODEL FOR IoT-BASED MOBILE HEALTH-CARE SOLUTIONS FOR ANTENATAL CARE

Wearable devices, tablets and mobile phones are usually the front-end components of IoT-based mobile health-care solutions for antenatal care [8].

Figure 14.1 demonstrates that the above components are dynamically interrelated in a complex manner. Nowadays, hospital management is going through gradual revolution that is driven by Information Communication Technology (ICT) and electronic industries. Historically, hospitals delegated mobile health workers to attend to the pregnant women in remote places [10]. Mobility of health workers was a major impediment to community health-care services for pregnant women in this era. Remote working experience can be stressful for case managers with limited human and nonhuman resources. So, it was getting difficult and unattainable for most case managers to urgently complement their colleagues and deliver health-care services to pregnant women in offshore locations.

Shortly, there was a paradigm shift in health-care service delivery. Health workers do not necessarily need to migrate to deliver health-care services. Some hospitals still embrace community health-care services for pregnant women while significant of them indirectly emphasize group and individual care to support traditional type

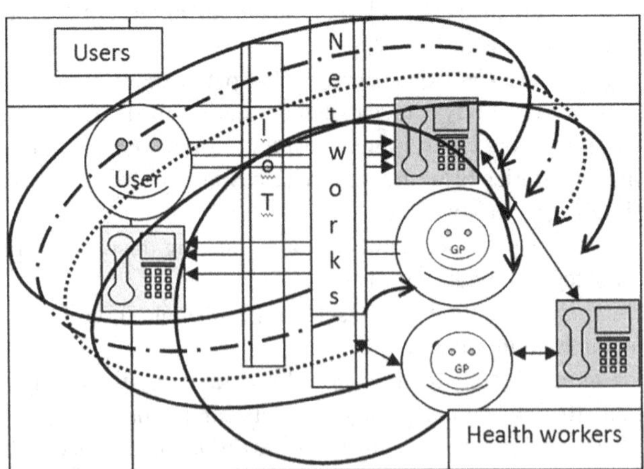

FIGURE 14.1 IoT-based mobile healthcare solutions for antenatal care.

of antenatal care. Computerization of basic operations of hospital management that started with payroll is gradually penetrating all aspects of health sector [8].

Today, most activities like supply chain management, inventory management, personnel department and public health that are central points of control in hospitals are computerized to ensure that they are running smoothly [12, 22]. The computerization of hospital management has also extended to specific jobs of health workers to enhance the quality of health services deliver to patients. For instance, there are vendor-managed computer toolkits to manage the logistics process of diagnosing and dispersing pharmaceutical products to patients. Mobile apps have also been developed to enable health workers relate with their patients through mobile systems.

Studies have stated several benefits of mobile health-care solutions and how they can reduce the tasks and the plight of community health-care workers that must travel to examine pregnant women in need of vital health-care services [8, 10, 12]. Despite poor awareness of some of the existing mobile apps, their efficacies for helping and advising pregnant women during the lockdown due to Covid-19 pandemic are frequently criticized in recent time. Networking analytics suggest that generations of scientific innovations in health sector always possess inherent vulnerabilities and shortcomings that are not often discovered during design and implementation [8]. Though health sector is still witnessing stunted development in the areas of mobile health-care solutions, several studies have indicated that the detection of symptoms associated with pregnancy like swollen legs, blood pressure and conditions of the baby in the womb is not within the functional capability of mobile phones [1, 4, 6, 7, 9]. Thus, mobile health-care services begin to experience dramatic developments. The argument is that with the inclusion of ICT and wearable devices from the electronic industry and integration of many things together, health service delivery will considerably improve. However, competitive advantage and qualitative health service delivery are determined based on timeliness to access and receive support from health-care workers. However, antenatal healthcare is still suffering from acute shortage of IoT-based mobile health-care solutions that can identify and proffer medical advice on common symptoms of pregnant women across 9 months of pregnancy.

14.4 METHODOLOGY FOR DESIGNING IoT-BASED HEALTH-CARE SOLUTIONS FOR ANTENATAL CARES

We support the use of standard software development methodologies like agile and waterfall to develop the health-care solutions proposed in this chapter [15, 17]. We further adopted mixed methods to gather the datasets that we used for supporting the proposed IoT-based mobile health-care solutions for antenatal cares. We recruited and qualitatively interviewed 20 anonymous pregnant women, 10 anonymous health workers and 6 software developers in Nigeria for the survey. Three of these pregnant women confirmed that they were expecting multiple births. We also interviewed 9 anonymous health specialists and 4 anonymous software developers in the United States, Canada and United Kingdom. We transcribed their responses and statistically analyzed them. Thereafter, we conducted thematic analysis of the data and produced graphical illustrations of the results.

Intrusions such as Distribution-Denial-of-Service (DDoS) attacks and password crackers have been identified as major threats to IoT-based health-care solutions in contemporary bulletin in recent time. Besides, the shipment and distribution of IoT-based mobile health-care solutions for antenatal cares in a global market can be statistically studied to improve sales and profitability of the above health-care solutions. Another potential issue that may arise in future is how operators of IoT-based antenatal health-care solutions will design classification or learning programs that will construct representations and enable them to quickly determine the classes of new intrusions without incurring huge overheads [16, 23]. The premise is that suitable statistical measures such as category utility and information gain can be adopted to quantitatively predict the above issues. Thus, this chapter proposes the use of category utility and information gain to lessen the above challenges [16]. We envisage that lack of suitable datasets could be a major challenge in this domain. Hence, we demonstrate how operators can use attributes of alerts from Smart IDSs and Smart IPSs such as Source IP address (SI), Destination IP address (DI), Time to live (TTL) and IP Length (IPL) to analyze and verify potential intrusions against the proposed health-care solutions in future [11, 12].

Consequently, we adopt Snort in Intrusion Detection System's (IDS) mode to generate alerts of standard trace files. We then design C++ programming codes to implement category utility and information gain in other to demonstrate the how to manage the above challenges. Category utility is a measure that can be used to determine the likely numbers of attribute or value that a predictor can accurately predict [16]. In this case, we believe that category utility can also be adapted to predict market dominance of various kinds of wearable devices for antenatal cares. If C_1, C_2 … C_k … C_N; should represent the clusters of attributes that are expressed as: A_1, A_2, … A_j—suppose the values of the ith attribute of these clusters are V_{i1}, V_{i2}, … to V_{ij}, then the expected value of correct attribute a predictor can guess if the predictor has the knowledge of the cluster (C_k) that an attribute belongs to is [16]:

$$\sum_i \sum_j (P(A_i = V_{ij} \mid C_k))^2 \tag{14.1}$$

Likewise, the expected value of correct attribute the predictor can guess without the knowledge of the cluster that the attribute belongs to is expressed as follows:

$$\sum_i \sum_j (P(A_i = V_{ij}))^2 \tag{14.2}$$

Then, the category utility for the clustering scheme that is formed by an attributes A_i is expressed as:

$$\frac{\sum_k P(C_k)\left\{\sum_i \sum_j (P(A_i = V_{ij} \mid C_k))^2 - \sum_i \sum_j (P(A_i = V_{ij}))^2\right\}}{Tn} \tag{14.3}$$

where Tn represents the total number of the groups or clusters that attribute A_i has formed. By using Network Intrusion Detection Systems (NIDSs), hospital management board can safeguard the peripherals of IoT-based antenatal health-care solutions proposed in this chapter.

Similarly, information's theory is an alternative approach we propose to evaluate the attributes of packets that signify intrusions against IoT-based antenatal care solutions [16]. The above packets logically constitute categorical dataset that can into various classes and in accordance to the values held in the attributes of the alerts that originate from them. It is possible to compute the information gain for the attribute of each cluster by using information's theory. This can be made by calculating the uncertainty or information content of the whole dataset using particular attributes that are used for clustering the dataset.

Mathematically, the information content for each attribute of the dataset is:

$$I_{attribute} = -\sum_{i=1}^{n} p(v_i)\ln_2 p(v_i) \tag{14.4}$$

where v_i represents the ith value of the attribute and its sum is taken over all the values of the attribute. The total information content, I_{Total} for the entire dataset is derived by adding the uncertainty of all the chosen attributes of the dataset together.

$$I_{Total} = -\sum_{j=1}^{m}\sum_{i=1}^{n} p(v_{ij})\ln_2 p(v_{ij}) \tag{14.5}$$

In this case, v_{ij} is the ith value of the jth attribute of the dataset and their probabilities of occurrences are estimated for the entire dataset. Essentially, we adopt DARPA 2000 datasets and labeled them as DARPA-1 and DARPA-2 to demonstrate how analysts can adopt the above statistical concepts to investigate and predict market trend or intrusions like DDoS attacks against the proposed health-care solutions in future time. Further still, the design clusters each dataset on the basis of the values held in the SI, DI, TTL and IPL. The classifier then uses the above two concepts to quantify the usefulness of the four attributes mentioned above. The results obtained are discussed below.

14.5 RESULTS AND ANALYSIS

This section provides the results of the qualitative and qualitative analyses of the responses of the participants and experiments mentioned above. Some of the results obtained are graphically illustrated in Figures 14.2–14.6.

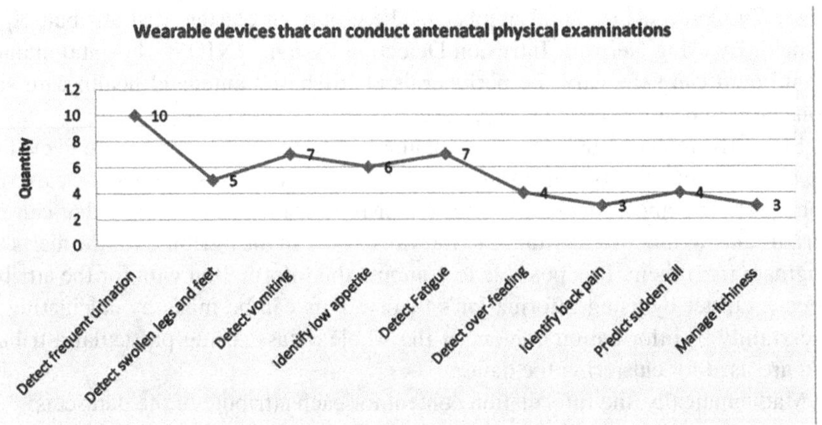

FIGURE 14.2 Wearable devices that can conduct antenatal physical examinations.

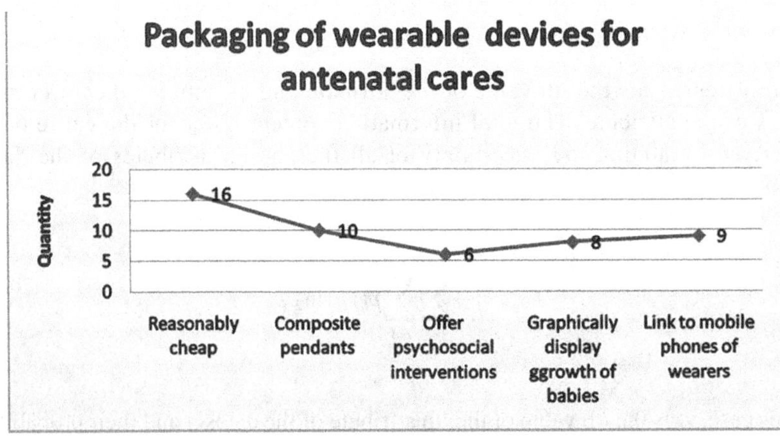

FIGURE 14.3 Packaging of wearable devices for antenatal cares.

14.5.1 Results

Figure 14.2 suggests that most of the respondents in the above survey would prefer and recommend wearable devices that possess certain functional and users' requirements. They would expect the devices to be able to conduct series of antenatal physical examinations in the same manner like doctors and nurses.

In other words, the proposed IoT-based solutions for antenatal cares should be able to detect symptoms of frequent urination, swollen legs, feet; vomiting, fatigue, discomfort, sudden fall and identify low appetite. Similarly, the participants suggest that most pregnant women may prefer wearable devices that can equally detect fatigue, overfeeding; back pain, predict sudden fall and help pregnant women to manage body itchiness and stretch marks. Specifically, the participants suggest

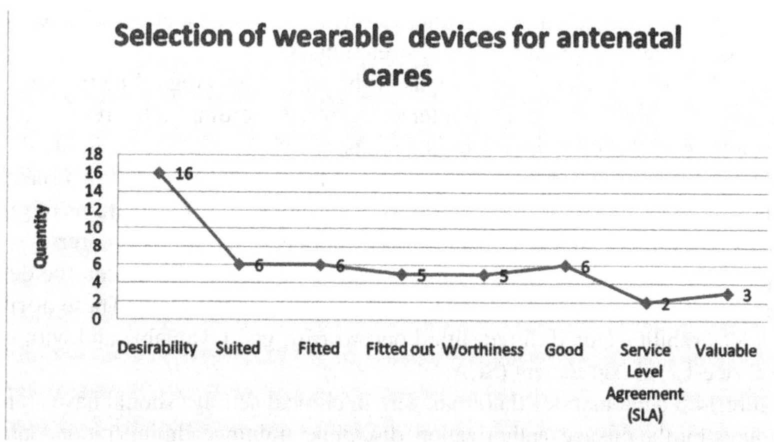

FIGURE 14.4 Selection of wearable devices for antenatal cares.

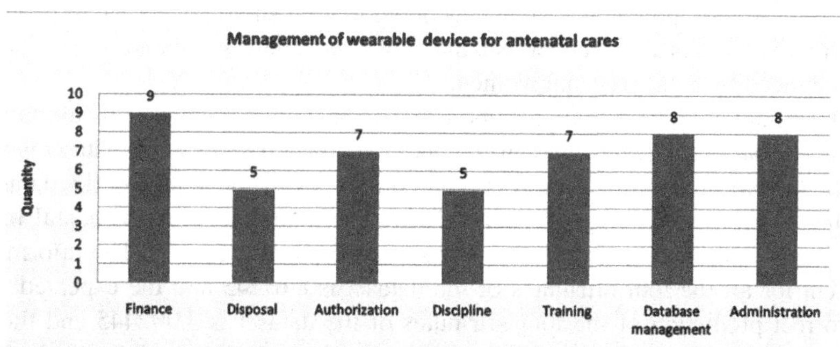

FIGURE 14.5 Management of wearable devices for antenatal cares.

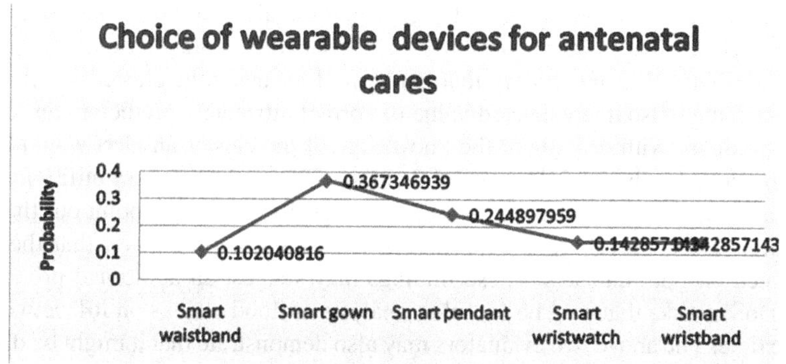

FIGURE 14.6 Choice of wearable devices for antenatal cares.

wearable devices that can detect frequent urination, swollen feet, swollen legs, sensations, back pains and predict fall and stretch marks.

Figure 14.3 illustrates the perceptions of the participants regarding the industrial packaging of wearable devices for antenatal cares. According to the results, significant numbers of the participants prefer industry to manufacture wearable devices for antenatal cares that are reasonably cheap and possess composite functions. Such devices should be able to offer psychosocial counseling, illustrate patterns of growth of babies in the womb and send results to the mobile phones of the wearers.

Figure 14.4 itemizes specific attributes that respondents would want the devices to possess for them to select, use or to possibly recommend them. These attributes include desirability, suited; fitted, fitted out, worthy, good, valuable and with excellent Service-Level Agreement (SLAs).

Figure 14.5 demonstrates that managers in clinical settings should have standard procedures for the finance, authorization, discipline, training, administration, database management and disposal of IoT-based mobile health-care solutions to improve security and efficacy.

Figure 14.6 recommends the potential preference of the pregnant women that would wear the proposed wearable devices for antenatal health-care services. Accordingly, most pregnant women may prefer to wear smart gowns, smart pendants, smart wristwatches and smart wristband. However, smart waistband is less likely to be embraced by most pregnant women.

Quantitative experimental results suggest that Snort triggers 834 alerts from DARPA-1 dataset. The information content for all the attributes in the dataset is 8.023 while the expected value of correct prediction of all the attributes in the dataset is 3.003905 and the category utility is 0.7509. The quantitative experimental results also imply that Snort generates 816 alerts from DARPA-2 dataset. The information content for all the four attributes of the dataset is 8.67243 and the expected value of correct prediction of the four attributes of the dataset is 3.002445 and the category utility is approximately 0.7506. Further analyses of the two datasets show that they are coordinated DDoS attacks of two categories of intruders that may typically intrude into digital networks.

The results further indicate that three of the four attributes of the above datasets (i.e. DI, TTL and IPL) formed similar numbers of clusters but the SI slightly varies in both datasets. The average gain for grouping the alerts of both datasets with the above three attributes is zero. Similarly, the information content for each of the datasets is similar to the uncertainty that is required for using the clustering scheme of each attribute. Also, the expected value of correct attribute a predictor can forecast if the predictor with or without the knowledge of the cluster an alert would belong to is the same for the DI, TTL and IPL belongs. Hence, the category utility for these three attributes is zero for both datasets. The results essentially point out that both evaluators show that there is no best attributes for grouping alerts that the Snort generated on both datasets. These findings may suggest an additional property of the DDoS attacks that may be useful to analysts of flood attacks on IoT networks in future time. The above two evaluators may also demonstrate that it might be difficult to establish the best evaluators for clustering intrusion logs that will indicate DDoS attacks against IoT infrastructure in the future time.

14.5.2 Suggestions for Ensuring Market Dominance

The above results have many technical and social implications for the proposed solutions to dominate markets in developed and developing countries. In terms of functional requirements, the above findings imply that wearable devices that can conduct antenatal physical examinations, diagnosis and prenatal screening tests would be helpful to pregnant women. Pregnant women need IoT-based mobile health-care solutions that can identify symptoms of frequent urination, especially in the last trimester of the pregnancy. The technology should be able to assist them to deal with swollen legs and feet, offer medical advice on how they can get relief, avoid extra fluid retention and hormonal imbalance in their bodies [6, 7, 24]. The sensors should possess capability to complement midwives and obstetricians to conduct IoT-based mobile tests, seeking for diseases and conditions of the fetus in the womb and offer advice to the wearers and time to go for further checkups [1, 2, 12].

Furthermore, IoT-based mobile health-care solutions for pregnant women should also focus on the detection of sensations like tingling or numbness in fingers and hands of pregnant women. Some participants interviewed said that some pregnant women may perceive numbness sensation or discomfort like tingling in their wrists or hands [21]. The respondents also hinted that some pregnant women may experience pressure in their pelvic region as the baby is getting close to the end of gestation period in the womb. Consequently, the pressure may disturb the normal breathing of some pregnant women.

Another veritable area where pregnant women would need IoT-based mobile health-care solutions is in the identification of back pain, dizziness and sudden fall. The rate of food intake and stability of some pregnant women may reduce as belly grows in size. Falls during pregnancy can cause severe injuries that may harm or deform the baby in the womb. Pregnant women often fault the validity of most scanning test. A device that can send alerts or massages to them to remind them of the sex of their babies, kinds of wear, posture and best practices for pregnant women is needed at all times. In addition, the above technologies are needed to identify and advice how to handle itchiness and stretch marks around the belly button area or other parts of bodies of some pregnant women [8].

Moreover, studies have shown that some pregnant women immensely benefit from group antenatal cares for the fact that it is inexpensive compared to one-to-one visits and the fact that it permits them to choose care group of their own [1, 2]. A review looking into perception and self-critical evaluation made by pregnant women suggests that some pregnant women may foresee high risk of complicated pregnancy and would want to make special clinical referral to enable them meet with specialist doctors face to face. In many countries, pregnant women keep track of their visits during antenatal cares. Such background information is helpful to store medical history and monitor stages of immunizations the pregnant women have received and pending or subsequent immunization they must receive. Pregnant women who attend different hospitals also need this information to guide clinicians on past and future interventions for the pregnant women. Therefore, IoT-based mobile health-care solutions are required to automate and incorporate these functional requirements into modern apps and wearable devices for pregnant women to deliver midwife-led group care services to them.

Some experienced and well-educated pregnant women may want to wear affordable and composite pendants that can accurately diagnose pregnancy and compute growth charts. For instance, a pendant should possess many functions such that pregnant women can personally configure and customize to taste. For experienced mothers, some pregnant women prefer to wear devices that should be able to graphically display the height and weight of their babies and correlate these attributes to the weights of previous babies. This functionality can suggest retarded growth and overweight in the baby in the womb. Wearable sensors can assist pregnant mothers with psychosocial interventions necessary to minimize addictions. They can render counseling to the wearers on how to stop smoking and drinking alcohol. These devices could focus on functions of rehabilitation centers and will be able to give incentives in the form of feedback to encourage new life. Essentially, IoT-based mobile health-care solutions for pregnant women are expected to guarantee the security and protect the privacy of users in conjunction to the above functionalities in a modern health-care delivery.

14.5.3 Selection of IoT-Based Mobile Health-Care Solutions for Antenatal Cares

This section offers further suggestions on key security, development, and operational areas that are fundamental to manufacturers and service providers in achieving effective designs of IoT-based mobile health-care solutions for pregnant women [13, 25]. Suitable IoT-based mobile antenatal solutions concern all pregnant women of diverse demographic status, social and economic backgrounds [13]. The acceptance and availability of the above health-care services would have extensive benefits to individual health workers and collaborative health services. Nonetheless, their developments and getting a suitable market for them may suffer setbacks due to socioeconomic problem, lack of awareness and rigid attitudes of patients and health service workers.

The selection, widespread and adoption of the above technologies may be a serious issue across the developing and developed nations due to cultural, religious and demographic variations [13, 22]. Therefore, we proffer eight useful guidelines that hospital management board should examine before selecting any IoT-based mobile health-care solutions for pregnant women.

14.5.3.1 Desirability

The hospital management board should check if they are worthy of chosen the IoT-based mobile health-care solution for antennal care in their hospital. With client-centered approach, hospital management board needs to seek and involve the participation of their pregnant women to ascertain whether they will embrace or discard the proposed solution.

14.5.3.2 Suited

The hospital management board and pregnant women must evaluate the suitability of the IoT-based mobile health-care solutions before accepting or rejecting them. Wearable devices must be built with strong cryptographic algorithms, portable, reliable and attractive.

14.5.3.3 Fitted

IoT-based mobile health-care solutions must be fitted for the desired purpose. Thus, it is imperative for the hospital management board to ensure that wearable devices on the above solutions satisfy the preference of pregnant women. Hospital management board should endeavor to certify the efficiency and effectiveness of the above health-care solution to establish their limitations before they are approved for deployment.

14.5.3.4 Fitted Out

The hospital management board should verify each product to ascertain they are prepared with proper equipment. Wearable devices should be furnished with essential equipment with high durability to enable pregnant women use them throughout their pregnancy period. These devices should have protective covers like casing and clothes.

14.5.3.5 Worthy

The hospital management board should verify if the IoT-based mobile health-care solutions they intend to procure or deploy have merit or value. IoT-based mobile health-care solutions should be worthy of being chosen to support antenatal care. They should have qualities and abilities that merit recognition as health-care solutions for pregnancy women.

14.5.3.6 Valuable

Maintenance of health-care solutions is necessary to cope with recently discovered requirements, bugs and errors. There are license fees and maintenance due to associated with proprietary software. It is important to check if the IoT-based mobile health-care solutions for antenatal care have great material and monetary value for the investment that would be committed to them before taking final decision about them.

14.5.3.7 Good

IoT-based mobile health-care solutions for pregnant women should have desirable and positive qualities. They should be suitable for promoting and enhancing well-being of pregnant women and their babies. Fundamentally, IoT-based mobile health-care solutions are expert systems. They should be verified to ensure they possess equal knowledge, skill and aptitude required by midwives, doctors and other health workers to tackle tasks associated with antenatal care.

14.5.3.8 Service-Level Agreement (SLA)

IoT-based mobile health-care solutions for pregnant women are proprietary software solutions. There are diverse issues with hospitals that are being serviced by local and offshore service providers. Some newly discovered requirements and bugs may consume substantial time and resources of the developers before they can fix them or redesign the software. The selected software vendors must be capable of conducting maintenance of the above health-care solutions on a regular basis [17]. Therefore, hospital management board should carefully verify level of services the service providers would offer them. They must thoroughly review the license and maintenance agreements, reference sites, third party, training facilities and how the service providers intend to manage system downtime before approving their products.

14.5.4 Management of IoT-Based Mobile Health-Care Solutions for Antenatal Cares

The management aspect of the IoT-based mobile health-care solutions for prenatal cares deals with core issues that people in charge of running the proposed health-care solutions should focus on [23, 24].

14.5.4.1 Finance

IoT-based mobile health-care solutions for prenatal cares must be well financed. The choice of wearable devices on the above platform may vary from a pregnant woman to another. Some pregnant women may prefer; smart eyeglasses, smart pendants, smart slippers, smart waistband, smart miniskirt and smart wristband. Therefore, hospital management board must sample the opinions of patients in their antenatal cares before procuring the above solutions. The board must obtain and provide money for regular payment of service providers, annual license and maintenance due to avoid interruptive service supply.

14.5.4.2 Disposal

Wearable devices on the platform of IoT-based mobile health-care solutions are fixed assets of hospital management board. Inventory of the above solutions is highly essential to keep their records. Usually, fixed asset depreciates on monthly basis. Some organizations may decide to dispose their fixed assets after the assets have fully depreciated and keep the records of all the disposed assets in accumulated depreciation accounts. Some fixed assets may be disposed after they have reached their net-booking values.

Therefore, it is imperative for the health management board to decide beforehand on how they intend to get rid of fully or partially depreciated hardware components in an environment-friendly way (like wearable devices, tablets and PDAs) and software component (like mobile apps) of IoT-based mobile health-care solutions for prenatal cares prior to the commitment of financial resources to procure them.

14.5.4.3 Authorization

Specific health workers must be given official permission and approval to manage IoT-based mobile health-care solutions for prenatal cares to ensure proper accountability and to ensure best clinical practices. Hackers often seek for poor clinical practices to exploit. Authorized and well-trained IT professionals are required to operate the above health-care solutions. Health management board must approve and mandate routine audit of IoT-based mobile health-care solutions for prenatal cares. This kind of audit is a panacea to establish and fix vulnerabilities in the deployed solutions and to equally certify the level of compliance of users and health workers with best global security standards. The audit could be quarterly intervals to assess how various products and segments of IoT-based mobile health-care solutions for prenatal cares in the hospitals are achieving the desire objectives stated by the hospital management.

14.5.4.4 Discipline

The health management board should have standard procedures for punishing wrongdoing among health workers and IT staffs that oversee IoT-based mobile health-care solutions for prenatal cares. IoT-based mobile health-care systems are extensive domains. Due to shortage of qualified health professionals, some hospitals may simultaneously engage health workers that are also working in other intensive care units. Consequently, it is plausible that some health workers may mishandle or wrongly disclose information about wearable devices for antenatal cares to unapproved sources. In addition, some IT workers and helpdesk may delay or interrupt the procedures for escalating problems patients have reported to vendors of the above health-care solutions. Thus, hospitals should have a policy in place to punish such misbehavior and wrongful conduct in order to gain control and enforce obedience at workplace. This will tremendously checkmate lapses in the security in IoT-based health-care systems and issues on the retrieval and protection of digital patient information in the databases of the above solutions.

14.5.4.5 Training

Pregnancy is sensitive stages of human life whereby any wrong diagnosis, wrong therapy or delay by health workers to respond to the remote signals they receive from wearable devices of the pregnant women can have protracted impacts on the victims and their babies [12]. Service users and health service providers always require routine training. The health management board can improve the strength of the midwives and specialist doctors on the usage of wearable devices and other toolkits supported by the IoT-based health-care solutions for antenatal care in their hospitals to enable the above health workers gain self-control on their jobs. These health workers can in turn train the pregnant women on the usage, functionalities and maintenance of these solutions.

14.5.4.6 Administration

The health management board should delegate specific workers, committees and departments to take control of the administration of the IoT-based health-care solutions for antenatal care. The collaborating departments and committees may be drawn from the Information Technology (IT), nursing and control. They should be legally bonded by policy documents to guide them in exercising authority and managing the above health-care solutions on behalf of the hospital.

14.5.4.7 Database Management

Conceptually, the IoT-based health-care solutions for antenatal care comprise hardware, software, middleware and digital networking systems. Some vendors may host them locally while some vendors may prefer to host them in the cloud. The insurance and risk classification of the above strategies varies in each case. Data owners have little control over data that is warehoused in the cloud. Contrarily, data owners have full control over the data that is stored locally. Thus, creation and maintenance of the back-end databases for the above solutions should be properly supervised and reviewed by competent Information System (IS) auditors to ensure they are properly implemented with high security in focus.

Usually, two-level passwords are recommended to secure the root directory of the operating systems for the mobile services. The hardware and software must be properly hardened to ensure compliance to the best security standards for implementing IoT-based solutions. Access to the back-end databases and tables for storing the identity of pregnant women, antenatal examinations, number of antenatal visits, midwife-led care, group activity, individual activity, intervention received; medical history, discussions, text messages with health-care workers, care information and pending schedules should be rigorously protected from the back-end of the apps.

14.5.5 Security and Social Challenges with Unsuitable IoT-Based Mobile Health-Care Solutions for Antenatal Care

We provide a quite number of indicators that may correlate to the security and effectiveness of IoT-based mobile health-care solutions for antenatal cares. Some of these indicators are probably due to the pervasive nature of the Internet, or the above health-care solutions are applied wrongly; lack of adherence to security precautions or the devices are not adapted for antenatal care purposes [11, 12].

14.5.5.1 Security Issues with IoT-Based Mobile Health-Care Solutions for Antenatal Care

Users and vendors should take necessary steps so that IoT-based mobile health-care solutions for antenatal cares would be used without threats or dangers. The networks of above solutions should be designed to provide maximum security for the devices and users. Sensors that radiate harmful photons or radiations are not suitable for the design of the devices proposed in this chapter [11].

Intrusions can pose dangers to the effectiveness of the above health-care solutions. Intruders can target the database, middleware, networks and patient information. Studies have identified Distributed-Denial-of-Service (DDoS) attacks, leakages of digital data and thefts of hardware (such as phones and wearable devices) are increasing threats to the usage of IoT-based health-care solutions [12]. Some intruders may focus on how they can compromise or take control of the mobile systems that provide communication and connectivity between health workers and pregnant women. The impacts of sneaky IoT attacks on IoT's infrastructure can be devastated if they cause prolong downtime. Preventive measures for high-level IoT attacks should include the deployment of strong smart Mobile Intrusion Prevention Systems (MIPSs) in the perimeters of hospitals' networks [11, 12].

Installation of strong smart Intrusion Detection Systems (IDSs) at the database and networks levels can complement the capability of IoT-based health-care solutions for antenatal cares to demonstrate resilience and hardiness against attacks. Such toolkits can help IT to initiate phantom processes that will quickly enable the devices to recover and operate in normal ways.

Rambling is a major security frontage and an issue that may be difficult to locate and fix in software that works with complex components. Proper preview of the underlying algorithms, back-end, middleware, front end, hardware visage and various interfaces of IoT-based mobile care solutions for antenatal care can help developers and Information System (IS) auditors to eliminate rambling. Rambling occurs whenever the solutions continue to extract, interchange and send bio-physiological data of pregnant women. An example of rambling in this context is a pregnant woman (that has never given birth) to receive a text message comparing a baby in her womb with the grown-up twins she had 3 years ago. So, the recipients would obviously notice information that such information is fallacy, or it spreads out in a different direction.

14.5.5.2 Social Issues with IoT-Based Mobile Health-Care Solutions for Antenatal Care

There are numerous social and ethical issues that software developers must strongly consider while designing and testing the above solutions [15, 17, 24].

14.5.5.2.1 Stigmatization

It may be possible that some pregnant women or health workers may be victims of stigmatization or open condemnation for using or recommending faulty solutions. So, wearable solutions for antenatal cares should not be marked with a stigma. They should not be stigmatized by society because of their size, functionalities and accuracies. Stigmatization may also arise if users should discover inappropriateness in their functionalities. Therefore, IoT-based mobile health-care solutions for antenatal cares should be attractive and reflect the qualities advertised by their vendors and marketers.

14.5.5.2.2 Addiction

It is possible to notice pregnant women who may be abnormally dependent on certain wearable devices for antenatal cares. A composite of wearable devises that is made up of portable and related parts could address the psychologically and habit that some wearers may form for wearing them.

14.5.5.2.3 Overzealous

Wearable devices may turn out to make some pregnant women to express excessive enthusiasm for them. Health workers may retrieve and replace them at least twice before delivery. This will curb intense devotion and strong emotions of pregnant women to specific types, colors or brands.

14.5.5.2.4 Passionless

Some pregnant women may not be passionate to use IoT-based mobile health-care solutions for antenatal cares for a number of reasons. The above health-care solutions should not be unfitness. Users may feel passionless once they discover that the devices lack the power to perform antenatal cares. Hence, it is imperative that midwives should conduct routine assessment of the usage and feelings of wearers about the device.

Furthermore, wearable devices that are excessively big or heavy may be inconvenient to wear freely. They should be manufactured in different sizes, colors and shapes to eliminate inconvenient discomfort in wearing or using them. They should not cause anxiety and they should possess the quality of being convenient and useful for antenatal cares. Wearable pendants may cause distress or extreme psychological or physical pains if they are not portable. This may disoblige or ignore the wishes of the wearers due to discomfort. Manufacturers and health workers should be aware that wearers of the above health-care solutions would be able to conduct self-critical evaluations of the devices. Once users discover that such devices are irrelevant and have no relevant bearing or connection with antenatal cares may jeopardize continue usage and market penetration of the products. Therefore, software testers must adequately conduct series of tests before the deployment of the above solutions into the market. We recommend users acceptance test to eliminate extraneous issues that are not essential or pertinent to the antenatal cares. IoT-based mobile health-care solutions for antenatal cares should not be digressive. They should not be of superficial relevance and they should not tend to depart or discursive from the main points of ensuring well-being of mothers and babies in the wombs of pregnant women. They can cover a wide range of symptoms of pregnancies, but they must be competent to carry out the tasks of midwives and specialist doctors by reason and high level of accuracy rather than intuition.

The above solutions should not indicate moot even at the countenance of research development. The reports generated from their databases should not open to debate or argument. In other words, synthetic and hypothetical datasets are not suitable for testing the legal significance and veracity of the above solutions. Involvement of many categories of pregnant women across nine months during the testing phase of the above health-care solutions would tremendously help hospital management board to eliminate disputable claims and highly controversial findings at the initial stage that users would have spotted out after deployment. The above software solutions must be accompanied with well-informative documentations. Proper documentations of various kinds of testing that are conducted by qualified software testers, bugs and how they will be fixed must be properly carried out.

14.6 CONCLUSION

Pregnancy may be associated with psychosocial feelings and emotional states such as discomfort, severe stomach pains, headache, swollen legs, loss of appetite, fatigue, nervousness and occasional distress that can endanger babies in the wombs. Inaccessibility of the pregnant women to effective and timely antenatal health-care services can pose a significant threat to the affected mother and parenthood. Accordingly, this chapter mostly informs that antenatal cares involve routine medical checkups for pregnant women to enable them receive diagnosis and medical recommendations from health workers on how they can enjoy healthy lifestyles with their babies during and after pregnancy. Hospital management board faces series of challenges that include acute shortages of health professionals in most countries and critical challenges such as inability of many pregnant women to attend antenatal care programs and high rate of incomplete visits that is recommended by the World Health Organization (WHO) [6, 7].

Unfortunately, whilst IoT-based health-care solutions are gaining wider recognition and acceptability in other aspects of hospital management, we have observed that service providers have mistakenly disregarded investments on the IoT-based health-care solutions that can electronically assist antenatal cares over the years [8, 12, 25]. We further discovered that considerable numbers of pregnant women are unable to compensate for the above lapses while most midwives and other health workers greatly relied on conventional health-care solutions to support and diagnose pregnant women. Meanwhile, significant numbers of pregnant women are yearning for smart gowns that contain intelligent materials and capable of conducting antenatal tests and producing equivalent results that medical experts would obtain by alternatively diagnosing the wearers using face-to-face approach. In other to increase the participation in antenatal cares, this chapter pragmatically gathers and analyzes some of the requirements of potential users and then proposes how software engineers can design, build and test the IoT-software applications that will satisfy the requirements of pregnant women in general. The premise is that smart gowns should be made with elastic clothes and they should contain IoT-based sensors inserted into the gowns. Also, smart pendants, smart wristwatches and smart wristbands should contain intelligent sensors that are capable of accurately diagnosing and analyzing physiological changes in the body of the wearers. Above all, we submit that these devices should be consolidated into fashionable, affordably cheap formats and they should be available to the reach of pregnant women across the globe.

Nevertheless, poor attendance in the conventional antenatal healthcare may possibly correlate to high rates of miscarriage, maternal mortality, birth defects, neonatal infections, prevalence of low birth weight and other preventable health problems that have been identified to surge according to global assessment of the development in health sector in recent years. The chapter believes that goodness of the few IoT-based mobile health-care solutions that health workers can adopt to conduct medical diagnosis on antenatal examinations with the view to identify health needs of pregnant women begin to face strong criticisms.

Thus, the physical layers of the IoT-based automations are fundamental areas that demand due consideration during software development in the above domains. So, we argue that cost, privacy issues and lack of trust on the above health-care solutions have the likelihood to increase the ratio of pregnant women that are unwilling to embrace them. Thus, this chapter broadly reviews relevant social and ethical issues and goes further to propose nine fundamental areas of pregnancy that software developers and service providers can exploit to gain market dominance in the areas of antenatal healthcare. The chapter also proposes theoretical models for the selection, management and security of IoT-based mobile health-care solutions.

Many security and social challenges that may confront pregnant women and health workers for using or recommending unsuitable IoT-based mobile health-care solutions and how they can be controlled are broadly discussed. We believe that the above analyses would tremendously assist hospital management board in the selection and management of the IoT-based mobile health-care solutions in general. Nonetheless, other issues may arise in future regarding the design, upgrade and selection of the above solutions. Presently, lack of suitable datasets is an impediment to research designs in this domain. Significant numbers of attributes can be used to describe network packets that migrate across the IoT-based networks. It is possible that the degree of informative and predictability of these attributes are different. Variability of their usefulness may be articulated from intrusion to intrusion. We therefore suggest that data donors should simulate different kinds of datasets on the above health-care solutions so that analysts and health-care operators would be able to extend the above illustrations and perfectly evaluate the efficacies of their models with category utility, information gain and other metrics for quantifying the classifications of learning programs on the above issues.

Finally, this chapter submits that rather than routine visit of the pregnant women to the clinics or hospitals, they can wear smart devices that would enable their midwifery to relate to them electronically throughout their pregnancy period. We further advocate that global health systems must dynamically change with ICT revolutions. Hence, the IoT-based mobile health-care solutions for pregnant women must be accompanied with the new health and social policies. Routine audit and training for health workers as well as pregnant women are required to ensure compliance of the users of the above antenatal care solutions to the best security and operational standards. Future research should focus on how data donors can generate suitable datasets that researchers can adopt to explore the other areas of the IoT-based mobile health-care solutions for the antenatal and post-antenatal health-care services.

REFERENCES

1. American Pregnancy Association (2015). Mood swings during pregnancy. Available at: https://americanpregnancy.prg/pregnancy-health/mood-swings-during-pregnancy/; Accessed on 16/04/2021
2. American College of Obstetricians and Gynecologists (2019). Bleeding during pregnancy. Available at: https://www.acog.org/women-health/fegs/bleeding-during-pregnancy/; Accessed on 07/05/2021
3. Bagag, M; Taleeb, T., Bernabe, J.B. & Shameta, A. (2020). A machine learning system framework for IoT systems; IEEE Access; DOI:10.1109/ACCESS.2020.2996214
4. Deutchman, M; Tabay, A. T. & Turok, O. (2019). First timester bleeding, American Family Physicians, Vol. 79, pp. 985–994; Available at: https://pubmed. ncbi.nlm.nih.gov/19514696/; Accessed on 07/05/2021
5. Khamparia, A., Singh, P. S., Rani, P., Samanta, D., Khanna, A. & Bhushan, B. (2020). An internet of health things-driven deep learning framework for detection and classification of skin cancer using transfer learning. In Special Issues in Transactions on Emerging Telecommunications Technologies. https://doi.org/10.1002/ett.3963
6. NIH (2021). What Are Some Common Signs of Pregnancy? NICHD Publications; Rockville, USA. Available at: https://www.nichd.nih.gov/ health/topics/pregnancy/conditioninfo/signs; Accessed on 10/05/2021
7. NHS (2020). Signs and Symptoms of Pregnancy; NHS England, United Kingdom. Available at: https://www.nhs.uk/pregnancy/trying-for-a-baby/signs-and-symptoms-of-pregnancy/; Accessed on 08/05/2021
8. Mutlags, A. A., Abd.Ghani, M. K., Arunkumar, N. & Mohammed, M. A. (2019). Enabling techniques for fog computing in healthcare IoT systems; Future Generation Computing Systems, Vol. 90, pp. 62–78
9 Davis, K. (2020). Week 1 of the Pregnancy Symptoms and Testing; Healthline media UK Limited; Gloucestershire, England, UK. Available at: https://www.medicalnewstoday.com/articles/pregnancy-symptoms-week-1; Accessed on 07/05/2021

10. Mwaura, W. (2018). Why IoT is the real; Department of Software development, Andela, USA. Available at: https://internetofthingsagenda.techtarget.com/definition/Internet-of-Things-IoT; Accessed on 07/05/2021
11. McCabe, J. D. (2003). Network Analysis, Architecture, and Design, (2nd Edition), Morgan Kaufmann, USA
12. Tanenbaum, A. S. & Wetherall, D. J. (2021). Computer Networks (5th Edition). Prentice Hall, Indian. ISBN-13: 970-0132553179
13. Robbert, K. W. (2006). Effective Software Project Management (Wiley Desktop Editions) (1st Edition). Wiley. ISBN: 0764596365
14. Kumar S., Arora A. K., Gupta P., & Saini B. S. (2021). A Review of Applications, Security and Challenges of Internet of Medical Things. In: Hassanien A. E., Khamparia A., Gupta D., Shankar K., Slowik A. (eds) Cognitive Internet of Medical Things for Smart Healthcare. Studies in Systems, Decision and Control, Vol. 311. Springer, Cham. https://doi.org/10.1007/978-3-030-55833-8_1
15. Edward, J. B., Feeney, A.B., Denno, P., Flater, D., Libes, D. E., Steves, M. P. & Wallace, E. K. (2008). Concepts for Automating Systems Integration, National Institute of Standards and Technology, US Department of Commerce Publications (NISTIR 6928), USA
16. Han, J. & Kamber, M. (2006). Data Mining: Concepts and Techniques (2nd Edition). Morgan Kaufmann, USA.
17. Kay, R. M. (2021). How to learn the fundamentals of software engineering in a more interesting and less painful way; freeCodeCamp. Available at: https://www.freecodecamp.org/news/learn-the-fundamentals-of-software-engineering/; Accessed on 22/04/2021
18. Sharma, N., Kaushik, I., Bhushan, B. & Gautam, S. (2020). Applicability of WSN and Biometric Models in the Field of Healthcare In book: Deep Learning Strategies for Security Enhancement in Wireless Sensor Networks; pp. 304–329. DOI: 10.4018/978-1-7998-5068-7
19. Swayamsiddha, S. & Mohanty, C. (2020).Application of cognitive Internet of Medical Things for COVID-19 pandemic. Diabetes and Metabolic Syndrome Clinical Research and Reviews 14(5); DOI:10.1016/j.dsx.2020.06.014
20. Goyal, K., Sharma, K., Bhushan, B. & Shankar, A. (2021). IoT Enabled Technology in Secured Healthcare: Applications, Challenges and Future Directions. In Cognitive Internet of Medical Things for Smart Healthcare, pp. 25–48; DOI:10.1007/978-3-030-55833-8_2
21. Robison, K. (1992). Putting the Software Engineering into CASE. New York City, John Wiley & Sons incorporation, USA.
22. Odhlando, D. (2018). Systems design in software development; Department of Software development, Andela, USA. Available at: https://medium.com/the-andela-way/system-design-in-software-development-f360ce6fcbb9; Accessed on 08/05/2021
23. Loucopolous, P. & Kaurakostals, V. (1996). Systems Requirements Engineering. McGraw-Hill International Series in Software Engineering. London.
24. Lumen (2021). Introduction to Computer Applications and Concepts: Module 6: Ethics and software development, US Navy, USA. Available at: https://courses.lumenlearning.com/zeliitel15/chapter/reading-software-development/; Accessed on 08/05/2021
25. Tommy, Early signs and symptoms of pregnancy; Registered charity organization in England, Wales and Scotland, UK, 2020. Available at: https://www.tommys.org/pregnancy-information/im-pregnant/early-pregnancy; Accessed on 05/05/2021

Index

T

V

W

Printed in the United States
by Baker & Taylor Publisher Services